LOGICISM RENEWED

LECTURE NOTES IN LOGIC

A Publication of

THE ASSOCIATION FOR SYMBOLIC LOGIC

LECTURE NOTES IN LOGIC 23

LOGICISM RENEWED

Logical Foundations for Mathematics and Computer Science

Paul C. Gilmore
Department of Computer Science
University of British Columbia, Vancouver

ASSOCIATION FOR SYMBOLIC LOGIC

A K Peters, Ltd. • Wellesley, Massachusetts

Addresses of the Editors of Lecture Notes in Logic and a Statement of Editorial Policy may be found at the back of this book.

Sales and Customer Service:
A K Peters, Ltd.
888 Worcester Street, Suite 230
Wellesley, Massachusetts 02482, USA
http://www.akpeters.com/

Association for Symbolic Logic:
Sam Buss, Publisher
Department of Mathematics
University of California, San Diego
La Jolla, California 92093-0112, USA
http://www.aslonline.org/

Library of Congress Cataloging-in-Publication Data

Gilmore, Paul C. (Paul Carl), 1925-
Logicism renewed: logical foundations for mathematics and computer science / Paul C. Gilmore
p. cm. – (Lecture notes in logic ; 23)
Includes bibliographical references and index.
ISBN-13: 978-1-56881-275-5 (alk. paper)
ISBN-10: 1-56881-275-2 (alk. paper)
ISBN-13: 978-1-56881-276-2 (pbk : alk. paper)
ISBN-10: 1-56881-276-0 (pbk : alk. paper)
1. Logic, Symbolic and mathematical. I. Title. II. Series.

QA9.2.G55 2005
511.3–dc22

2005052008

Publisher's note: This book was typeset in LaTeX, by the ASL Typesetting Office, from electronic files produced by the authors, using the ASL document class `asl.cls`. The fonts are Monotype Times Roman. This book was printed in the USA by Boyd Printing Co., on acid-free paper. The cover design is by Richard Hannus, Hannus Design Associates, Boston, Massachusetts.

14 13 12 11 10 09 08 07 06 05 5 4 3 2 1

CONTENTS

PREFACE

Logicism, as described by Russell in the preface to the first edition of his Principles of Mathematics, is the thesis

> ... that all pure mathematics deals exclusively with concepts definable in terms of a very small number of fundamental logical concepts, and that all its propositions are deducible from a very small number of fundamental logical principles ...

In the introduction to the second edition of Principles of Mathematics [126], Russell reviewed the material presented in the first edition and described how some of the views presented there had been modified by his own and others' work. He concluded the introduction with the following

> The changes in philosophy which seem to me to be called for are partly due to the technical advances of mathematical logic in the intervening thirty-four years, which have simplified the apparatus of primitive ideas and propositions, and have swept away many apparent entities, such as classes, points, and instants. Broadly, the result is an outlook which is less Platonic, or less realist in the medieval sense of the word. How far it is possible to go in the direction of nominalism remains, to my mind, an unsolved question, but one which, whether completely solvable or not, can only be adequately investigated by means of mathematical logic.

A nominalist related view of logic and mathematics was expressed by Ayer in Chapter IV of [11]:

> ... the truths of logic and mathematics are analytic propositions ... we say that a proposition is analytic when its validity depends solely on the definitions of the symbols it contains.

To pursue this thesis more carefully it is necessary to distinguish between what Frege has called the *sense* of a symbol and its *nominatum* or what Carnap [24] has called the *intension* of a symbol and its *extension*. For "the definitions of the symbols" must be understood to mean the definitions of their intensions.

The intension of a predicate name is what a person must know to determine whether the predicate can be said to hold of a particular object; that is whether the object is a member of the extension of the predicate. Thus predicates with the same intensions necessarily have the same extensions. But the converse does not generally hold. The traditional example of Frege concerns the two predicates of being the morning star and being the evening star: The two predicates have distinct intensions while their extensions in each case have a single member, namely the planet Venus. Nevertheless the converse has been accepted in most set theories and logics intended as a foundation for mathematics. For example it is the first axiom of Zermelo's set theory [151], an axiom of Whitehead and Russell's Principia Mathematica [149] and Andrews' version [9] of Church's simple theory of types [28] that evolved from it. An axiom of extensionality is a convenient simplifying assumption but it is far from clear how essential it is for the development of mathematics within a type theory and it is a definite hindrance to the applications of logic in computer science; for example, in the paper [8] extending Robinson's resolution theorem proving method [123] to higher order logic, it and other axioms of the simple theory of types [28] are dropped. Further, as Fitting remarks in his preface to [48] extensionality can get in the way of other applications of type theory as well.

The need for languages that can be understood by both humans and computers requires distinguishing between the intension and extension of predicates. Human interaction with a computer maintained database, for example, is initially by means of predicates whose intensions are only understood by humans, but whose extensions are conveyed to the computer by data entry. Once the extensions of these predicates are available as data within the machine, a query in the form of a name of a defined predicate can be conveyed to the machine and the machine can respond with a listing of the extension of the predicate [60]. Note that the machine responds to the *name* of a predicate and it is this name that conveys all that the machine needs to know of the *intension* of the defined predicate.

Computers are consummate nominalists. Thus an investigation of nominalist motivated logics can be of importance not only for the foundations of mathematics, but also for computer science. A goal of this book is to demonstrate that the logic ITT, *Intensional Type Theory* [63], motivated by a nominalist interpretation of logic can provide a logical foundation for both mathematics and computer science - hence the title "Logicism Renewed". The nominalism of ITT is that suggested by the quote of Ayer, although he acknowledges the prior work of Carnap [23]. It is not the nominalism of Goodman and Quine [70] but is close to Henkin's response in [75] to the earlier paper.

The possibility of a nominalist interpretation of language arises with higher order predication when a predicate is said to apply to a predicate. Consider for example the English sentence 'Yellow is a colour'. In a traditional interpretation the sentence is understood to assert that 'The colour predicate applies to the predicate of being yellow'. In a nominalist interpretation in the style of Ayer, on the other hand, it would be understood to have as its meaning "Yellow' is used in English as a colour word'. In the traditional interpretation, 'Yellow' is being used as the name of a predicate, while in the nominalist interpretation it is being mentioned as the name of the predicate. Carnap in [23] distinguishes between these two interpretations using the example 'Five is a number' and "Five' is a number-word'. Similarly Montague remarks in [105]

> On several occasions it has been proposed that modal terms ('necessity', 'possible' and the like) be treated as predicates of expressions rather than as sentential operators. According to this proposal, we should abandon such sentences as 'Necessarily man is rational' in favor of "Man is rational' is necessary' (or "Man is rational' is a necessary truth'). The proposal thus amounts to the following: to generate a meaningful context a modal term should be prefixed not to a sentence or formula but to a *name* of a sentence or formula ....

The single quotes used in these examples suggests that a nominalist motivated logic will require a quote notation to maintain the distinction between use and mention, but in practice that is not the case; for example, [54] maintains the distinction in a Lisp-like language without the use of a quote operator introduced by McCarthy in [98], and the types of a typed logic can be used as will be described next.

The types of Principia Mathematica and its descendents are in place to prevent the definition of contradictory predicates, the *paradoxes*. Although they eliminate the paradoxes, they are not universally regarded as providing an *explanation* of the source of the paradoxes as suggested by the following quote of Church from [27]:

> ... Frege's criticism seems to me still to retain much of its force, and to deserve serious consideration by those who hold that the simple theory of types is the final answer to the riddle of the paradoxes.

Another explanation is offered by Sellars in [130, 131], namely that the paradoxes arise from a confusion of the use and mention of predicate names. An example of such a confusion would be to not distinguish between the distinct roles of the two occurrences of 'yellow' in the sentence 'This pencil is yellow and yellow is a colour'; in the first occurrence it is being *used* as a predicate name while in the second it is being *mentioned* as a predicate name. Another example is the sentence 'Or and and are conjunctions in English'. At its first occurrence 'and' is being used as a conjunction while at its second occurrence it is being mentioned. A novel use of typing can prevent a confusion of the use and mention of predicate names.

The types of ITT are recursively defined as follows: 1 is the type of names of individuals and $[\tau_1, \dots, \tau_n]$ is the type of predicate names with arguments of the types $\tau_1, \dots, \tau_n$, $n \geq 0$. Thus the type $[\,]$ of the truth values is the type of a predicate with no arguments. Apart from notation, these are the types of the theory $\mathcal{F}^\omega$ of [9] and Schütte's simple type theory [128], without types for functions; they are a proper *subset* of the types of Church's simple theory of types since ITT has no function types. The absence of functions is however only temporary; they are introduced by the traditional method of definite descriptions into a conservative extension ITTε of ITT in Chapter 5.

In ITT it is necessary to distinguish between the name of a predicate in a predicate position, where it is used, and the same name in a subject position, where it is mentioned. The distinction is maintained by allowing certain terms that have the type of a predicate name to also have a *second* type, namely the type 1 of an individual name, referred to as its *secondary* type. A term when typed as a predicate is being *used*, and when typed as an individual is being *mentioned*. This is reminiscent of the *type-shifting* allowed in Montague grammars [112].

A thesis of this monograph is that the types of ITT are implicit in the distinction made between subject and predicate in a natural language; in this sense ITT is a formalized *dialect* of natural language similar to a view held by Montague [104].

ITT is called an *intensional* logic because a name of a predicate is identified with its intension so that intensions in the form of predicate names are treated as individual names. This identification of the intension of a predicate with its name is justified by the assumption that a user of a language discovers the intension of a predicate from its name, and that the intensions of two predicates with the same name in the same context cannot be distinguished. Thus in place of an axiom of *extensionality* in ITT appears a rule of *intensionality* that concludes that the intensions of two predicates are identical when they have identical names. A sufficient condition for two names to be identical is that they have the same normal form under lambda reduction. The justification for this identification is that removable abstractions are irrelevant to the process of verifying the truth or falsehood of a sentence in natural languages; sentences with the same normal form necessarily have the same truth value; see §3.1.1.

The advantages of admitting the secondary typing of predicate names become most apparent in the definition of recursively defined predicates. A predicate 0, *zero*, of any type $[\tau]$ and a binary predicate S, *successor*, of any type $[\tau, \tau]$ can be defined. But without secondary typing the sequence of expressions 0, $S(0)$, $S(S(0))$, $\cdots$, are only terms if the type of the term S occurring first in the expressions changes with each expression: In $S(0)$ the type of S is $[\tau, \tau]$, in $S(S(0))$ it is $[[\tau], [\tau]]$, and so on. With secondary typing, however, each of the terms in the sequence is of type [1] and also of secondary type 1 when S is of type [1, 1]. Further a predicate N, *number*, of type [1] can be defined and all of Peano's axioms [115] derived for it.

The method used to define N in ITT is a departure from the fixed point method that originated with Tarski's paper [141] and is used to define recursive predicates in [111] and [117]. A predicate form of a map of [111] or operator of [103], called a *recursion generator*, is used together with defined predicates Lt, *least predicate operator* and Gt, *greatest predicate operator*. An advantage of the method over the fixed point method is that the definition of a recursive predicate, well-founded or not, appears in the same form as the definition of any other predicate. The fixed point property used to define a predicate in the fixed point method becomes a derivable property when it is defined using the recursion generator method.

Chapter 1 introduces basic concepts needed for the later chapters. It concludes with a presentation of a proof theory and a completeness and Cut-elimination theorem for a first order logic with λ-abstraction called *Elementary Logic* or EL for short. This provides an introduction to the proof theory and completeness and Cut-elimination theorem provided for ITT in Chapter 3. The semantics is described in the style used by Henkin in [74], and now called in computer science *denotational semantics*, [106]. Although Chapter 1 can serve as an introduction to first order logic, it contains much that is not needed for the usual applications of first order logic, and is missing some topics. Readers familiar with first order logic and the lambda calculus can refer to it only as needed.

Chapter 2 describes an elementary type theory TT with the same types as ITT. The properties of the λ-abstraction operator needed for the type theories are confirmed. Intensional and extensional identities are defined along with extensional and intensional models of TT. Polymorphic typing, or *polytyping* as it is called in [100], is a typing in which the type assigned to a symbol can depend upon the context; for TT it is a restatement of what Whitehead and Russell called *typical ambiguity* [149]. A polytyping algorithm is described here for TT.

Chapter 3 provides the basic justification for the logic ITT including an extension of the polymorphic typing introduced for TT. Semantic proofs of the consistency and of completeness of ITT are provided; from the latter the redundancy of the Cut rule follows. These proofs are improved versions of the proofs of [63].

Chapter 3 concludes with a sketch of a consistent extension SITT, *Stongly Impredicative Type Theory*, of ITT. The logic presented in Russell and Whitehead's Principia Mathematica is known as the *ramified* theory of types. It had not only the type restrictions of TT but also level restrictions. The types and levels remove from the logic the ability to form impredicative definitions in the sense of Poincaré [45]. In this strict sense the logic TT, and therefore also ITT, is impredicative. SITT permits the definition of predicates that cannot be defined in ITT, and that are related to the paradoxical predicates that gave rise to the ramified theory of types in the first place; that is why SITT is called a *strongly* impredicative type theory. It is perhaps at best of historical interest because of its treatment of Russell's set; no serious attempt has been made to find a useful application for it.

Chapter 4 provides the foundation for recursion theory based on the recursion generator method of defining recursive predicates. The method is then illustrated with a study of Horn sequent definitions in ITT. Horn sequents are a higher order form of the Horn clauses [94] that are used in logic programming languages such as Prolog [138] and Gödel [80], and generalized for the simple theory of types in [108]. A Horn sequent can be regarded as a "sugared" definition of a recursion generator in terms of which a well-founded and a non-well-founded predicate is defined. Many examples are offered.

In Chapter 5 the conservative extension ITTε of ITT is defined. ITTε provides a formalization of Hilbert's choice operator ε [78] in terms of which the definite description operator ι is defined. Partial functions are defined using ι. The result is a treatment of partial functions that has similarities with the method of dependent types used for example in PVS [110].

A Gentzen sequent calculus formulation of intuitionist first order logic is obtained from the classical by limiting succedents to at most one formula; see for example [140] or [88]. Further, classical first order logic can be developed fully as a sublogic of the intuitionist [88]. Analogous relationships are developed in Chapter 6 between a Gentzen sequent calculus formulation ITTG [ITT*Gentzen*] of ITT and its intuitionist formulation HITTG [*Heyting*ITTG] with restricted sequents. Further HITTG is shown to have the hallmark property of an intuitionist/constructive logic, namely that succedents that are disjunctions or existential formulas have *witnesses* in the following sense: If a sequent $\Gamma \vdash [F \vee G]$ is derivable, then at least one of $\Gamma \vdash F$ and $\Gamma \vdash G$ is also derivable; and if $\Gamma \vdash \exists P$ is derivable then for some term t, $\Gamma \vdash P(t)$ is also derivable.

In order to adapt more easily some of the results described in Chapters 3 to 5 to HITTG, a semantic tree formulation HITT of HITTG is provided. A perhaps surprising conclusion is that the recursion theory developed in Chapter 4 using ITT can also be fully developed within HITT; thus, for example, an intuitionist higher order natural number theory can be developed without the need for any axioms.

This is demonstrated in [64] where an intuitionistic real number theory is developed in HITT. Further, Kripke's possible world semantics for intuitionistic logic [90] is extended for HITT, and the completeness theorem and the redundancy of Cut are proved for this logic.

The last two chapters attempt to justify the subtitle "Logical Foundations for Mathematics and Computer Science". No discussion of the logical foundations of mathematics would be complete without consideration of the implications of Gödel's incompleteness theorems [69]. In Chapter 7 mathematics is described as the study of the properties of predicates defined in an intensional logic such as ITT, thus reflecting the view of Ayer expressed in the quote above; but also reflecting the view expressed by Steen in the introduction to [137]:

> Mathematics is the art of making vague intuitive ideas precise and then studying the result.

For "vague intuitive ideas" are made precise by defining the intensions of predicates and "the result" studied by studying the extensions of the predicates. From this point of view the theorems of Gödel provide a profound insight into the nominalist view of mathematics as the study of defined predicates, but do not preclude a formal development of mathematics.

With mathematics so described the set theory ZF can be seen to be a study of the first order derivability predicate defined using the axioms enumerated in [132]. The somewhat ad hoc motivation for the axioms as described in [151] and the necessity for an axiom of extensionality argue against their importance beyond their traditional use as a tool in the study of Cantor's transfinite number theory. But even this limited use may be better undertaken in ITT: By way of illustration a theorem of Cantor's is derived in Chapter 7.

Universal algebra and category theory can also be understood to be studies of defined predicates. They are particularly useful because the structures described by their axioms are widespread and they provide a more efficient way of reasoning about the consequences of their axioms than does a logic like ITT. They relate to logic in much the same way that higher order programming languages relate to say an assembly language - they find their justification in logic but are not necessarily the best tool in all situations. For a related point of view see [93] or [92].

MacLane has argued in [95] that traditional set theories and logics cannot formalize some arguments that can be developed in category theory. These arguments, at least in the form simplified by Feferman in [43], are shown to be formalizable in ITT; they make use of polytyping and secondary typing.

The three traditional philosophies of mathematics have been categorized as logicism, formalism, and intuitionism; see for example [18] or [91]. In this monograph a *nominalist* form of logicism has been described. The formalist view expressed by Curry in [35] and [36] can be understood as expressing that the intensions of predicates are defined by axioms and can therefore be understood as a form of the logicism of this monograph. Further, the nominalist motivation for the fully formalized HITT is compatible with an intuitionist/constructive point of view as expressed in [96]. Thus the three traditional philosophies of mathematics may be unified by the renewed *nominalist* logicism of this monograph.

Although Chapter 4 is in itself a contribution to theoretical computer science, a few topics that depend essentially on the secondary typing allowed in

ITT are discussed in Chapter 8. Examples are provided that illustrate how ITT can provide a logical foundation for the denotational semantics of programming languages as described by Mosses in [106]. They illustrate how the needed semantic domains can be defined in ITT using the methods described in Chapter 4.

In considering logical foundations for computer science it is essential, however, to distinguish carefully between the definition of predicates and the computation of values for arguments of given defined predicates. Thus, for example, although the definition of non-well-founded predicates presents no special problems within ITT, the computation of values for their arguments is beyond the scope of this book; see §8.2.

After a brief discussion of the role of formal specifications in the development of computer systems, Chapter 8 ends with the suggestion that meta-programs developed within a logic programming language as described in [81] can provide a programmable meta-language for ITT. With such a system available, ITT could be used to reason about the formal specification of a computer system. This proposal for a programming system built on ITT is modelled after the system HOL built on the simple theory of types using the programmable meta-language LCF [113]. Since the design and application of such a programming system will depend upon individuals familiar with the task of providing ITT derivations, many examples of derivations are offered throughout all the chapters and the reader is asked to supply others in exercises.

Acknowledgements. The book is dedicated to the memory of the four men who introduced me to the beauty, breadth and usefulness of mathematical logic: S.W.P. Steen at the University of Cambridge, England; Evert W. Beth and A. Heyting at the University of Amsterdam; and Abraham Robinson while he was at the University of Toronto.

During my last year at St John's College Cambridge, I participated in a reading and discussion of Ayer's [11]. This is the source of the nominalist view that has motivated the development of ITT. Although [53] was influenced by the reading, the first formal presentation of a view developed from [11] is in [55] which, with rewriting and extensions, became [58] and [59]. The route from the latter paper to the first presentation of ITT in [62] took an unfortunate detour when the nominalist motivation for the logics was abandoned for a bit of ad hocery. The resulting logic NaDSet is described and applied in the two papers [65, 66]. Although Girard proved the inconsistency of NaDSet in [67], most of the conclusions of the papers can be supported in ITT and some are described in Chapters 7 and 8. A return to the nominalist motivation and a change from the somewhat awkward notation of the earlier papers to the lambda notation resulted in the logic of [61]; this change of notation was suggested by correspondence with Thomas Forster and Nino Cocchiarella and [32, 33]. A closer study of the second order logic of [61] revealed that the technique it employed for maintaining the distinction between use and mention could be more simply employed in ITT.

However it should be noted that Jamie Andrews has succeeded in modifying the original NaDSet to maintain consistency. Cut elimination can be proved for the logic he describes in [6, 7], and Curry's Y combinator can be defined.

During the long gestation of ITT and the preparation of this monograph, I have been assisted and/or encouraged by a number of colleagues: Conversations with Jamie Andrews, Peter Apostoli, Jeff Joyce, Eric Borm, Michael Donat, Melvin Fitting, George Tsiknis, Alasdair Urquhart, and David Benson, and in addition to Thomas Forster and Nino Cocchiarella, correspondence with Henk Barendregt, Hendrik Boom, Per Martin-Löf, Jørgen Villadsen, Joachim Lambek and Peter Fletcher has helped in the clarification of some of the ideas and their sources. Michael Donat has been particularly helpful in the development of the polytyping algorithms, and Jesse Bingham in the final preparation of the manuscript for the publisher. Comments by a referee on two earlier drafts have resulted in the correction of some proofs and the removal of several deficiencies.

Correspondence with Chris Fox, Shalom Lappin, Carl Pollard, and Phil Scott on possible applications of intensional higher order logics to linguistics provided an opportunity to discuss in greater depth the subject shallowly treated in §1.1. This motivated in part the change from defining intensional identity $=$ in an earlier version of ITT to accepting it as a special primitive constant with its own semantic rules.

At the invitation of Roland Hinnion I gave three lectures on the contents of this monograph in Belgium. This opportunity to receive comments first hand from Roland, Marcel Crabbé, and others has been particularly helpful.

I am also most grateful for my wife's acceptance of the time I have spent in the company of Miss Mac during the preparation of this book.

Prior to my retirement financial support for my research was provided by the Natural Science and Engineering Research Council of Canada.

Paul Gilmore
Vancouver, Canada

CHAPTER 1

ELEMENTARY LOGIC

1.1. Language and Logic

The simplest sentences of a natural language have two components: a *subject* and a *predicate*. 'Africa is a continent' and 'One-half is a natural number' are examples. In the first 'Africa' is the subject and 'a continent' is the predicate, while in the second 'One-half' is the subject and 'a natural number' is the predicate. The verb 'is' asserts that the subject has the property described by the predicate; briefly that the predicate is *applied to* the subject, or that the subject is an *argument* for the predicate. The resulting sentence may or may not be true.

Some sentences, although in the subject-predicate form, have a predicate with an embedded subject. For example the sentence 'London is south of Paris' has 'London' as subject and 'south of Paris' as predicate. The subject 'Paris' is embedded in the predicate. Indeed the sentence can be understood to have two subjects 'London' and 'Paris' and a predicate 'south of', often called a *relation*. The number of subjects to which a predicate may be applied is called the *arity* of the predicate. Thus 'a continent' and 'a natural number' have arity one, while 'south of' has arity two.

From an arity two predicate an arity one predicate can be formed by applying the first of its two subjects. Thus for example 'south of Paris' is an arity one predicate. Later, notation will be introduced for the predicate suggested by 'London is south of'. In a slight generalization of the meaning of predicate, a sentence is understood to be a predicate of arity zero since a sentence results when it is applied to no subjects. Thus 'Africa is a continent' and 'London is south of Paris' are predicates of arity zero. Generalizing further, it will be assumed that there may be predicates of any given finite arity; that is, predicates which may be applied to any given finite number of subjects. A predicate of arity n, $1 \leq n$, when applied to n subjects forms an *elementary* sentence.

Consider the sentence 'One-half is a natural number' formed by applying an arity one predicate 'a natural number' to a single subject 'One-half'. Two names are involved, the name 'One-half' of the subject and the name 'a natural number' of the predicate. Each name is simply a *string* or finite sequence of letters and blanks. To understand the sentence, that is to know the circumstances under which it has a *truth value* true or false, it is necessary to know the meaning of each of the names, or to use the terminology of Carnap, the *intension* of the names [24].

This discussion can be generalized for a predicate of any arity. The intension of the predicate being south of is identified with its name 'south of'. The extension of the predicate is the set of ordered pairs of entities, the first of which is south of the second. In the sentence 'London is south of Paris' the subject names are 'London' and 'Paris' with intensions identified with the names. The extensions of these two names are respectively the city of London and the city of Paris. Since the ordered pair with first member the city of London and second member the city of Paris is not a member of the extension of the predicate 'south of', the sentence is false.

1.2. Types, Terms, and Formulas

In symbolic logic sentences such as 'Africa is a continent' and 'Three is a natural number' are represented by *formulas* of the form $C(a)$, where 'C' is a *constant* that is a name for a predicate and 'a' a constant that is a name for a subject to which the predicate may be applied. Sentences of the form 'London is south of Paris' are represented by formulas of the form $D(a, b)$, where 'D' is a constant that is a name for a binary predicate and 'a' and 'b' constants that are names of subjects. A type notation is introduced to distinguish between the names of subjects and of predicates; it is used to simplify and clarify the *grammar* or *syntax* of the logic.

DEFINITION 1. *First Definition of Type.*
Types are defined inductively as follows:

1. 1 *and* [] *are types*;
2. *For any sequence* 1^n *of* 1*'s*, $n \geq 1$, $[1^n]$ *is a type.*

Here 1 is the type of a name of an entity that may be a subject, for example a; the assertion that a is of type 1 is expressed by the *type assertion* a:1. [] is the type of a name of a formula, for example, $C(a)$:[] is a type assertion asserting that $C(a)$ has the type []. These types correspond to the types ι and o of Church's simple theory of types [28]. In addition to 1 and [], types include [1], [1,1], ... , $[1^n]$, ... The type [1], for example, is the type of a predicate name with one argument of type 1; for example C:[1] asserts that C is of type [1], the type of a predicate name with one argument. The type $[1^n]$ is the type of a predicate name with n arguments of type 1; for example D:[1, 1] asserts that D has the type [1,1], which is the type of a predicate with two arguments of type 1. The reason for the unconventional type notation will become apparent when the predicate abstraction operator λ is introduce in §1.5; the typing makes it more difficult to misapply λ.

It will be assumed that there are denumerably many constants and denumerably many variables available in the logic. The constants are sequences of upper and lower case Latin letters and numerals that begin with a letter other than u, v, w, x, y, or z; they will also include special symbols to be introduced later. These constants are potential names for subjects or predicates; that is, they may become an actual name when the intension of the name has been defined. The variables are sequences of upper and lower case Latin letters and numerals that begin with the letters u, v, w, x, y, or z. The role of the variables in the logic is explained in §1.5 with the introduction of the lambda abstraction operator. In the meantime it is not necessary to distinguish between constants and variables.

Each constant or variable in a given context has a particular type assigned to it. For the present, the variables will only be assigned the type 1.

To avoid a cluttered notation, the type assigned to a constant or variable is not explicitly indicated. Instead the types of variables and constants used in examples and applications will be declared within the context of their use by a *type declaration* such as C:[1] that asserts that the constant C has the type [1]. When such declarations are absent, it is assumed that the types are apparent from the context.

In contexts where it is necessary to assume that every constant and variable has a specific type, a function t[cv] from constants and variables to types is defined. For each constant or variable cv, t[cv] is the type of cv. Thus for the present when cv is a constant t[cv] is 1 or [1^n] for some $n \geq 0$, and is 1 when cv is a variable.

Although both upper and lower case latin letters can be the first letter of a constant or variable, upper case will generally be used in this chapter for names of predicates and lower case for names of subjects. However in preparation for the later chapters, where predicates may themselves be subjects, the practice is not always maintained.

Strings of upper and lower case latin letters have another distinct role to play in this monograph. This role is already evident in the discussion of types and their role. For example in the sentence "For each constant or variable cv, t[cv] is the type of cv" the pair of letters 'cv' is being used as a variable ranging over the union of the sets of constants and variables; that is over the set of all sequences of upper and lower case Latin letters and numerals that begin with letters. This role in the *metalogic* must be distinguished from the role in the logic itself; thus, for example, since 'cv' is a sequence of lower case Latin letters beginning with the letter 'c', it is a member of the set of constants of the logic and is therefore a potential name for a subject or predicate. In some expositions of formal logics distinct fonts or styles are used to distinguish these two roles. Here the reader is required to determine the role of a string from the context in which it is being used.

Although the definition of type will be extended several times, the two clauses of the first definition of term that follows remain essentially unchanged. However as special constants mentioned in the first clause are introduced in §1.3 and §1.5.1, the effect will be to introduce new terms.

DEFINITION 2. *First Definition of Term.*
A term is any non-empty sequence of constants, variables, special constants, and '(' and ')', typed as follows:

1. *Let cv be any constant, special or not, or variable. Then cv is a* term *of type* t$[cv]$;
2. *Let τ and τ_i be types for $0 \leq i \leq n$, where $n \geq 0$, and let P:$[\tau, \tau_1, \ldots, \tau_n]$ and Q:τ. Then (PQ) is a* term *of type* $[\tau_1, \ldots, \tau_n]$.

For example, if x,y:1 and C:[1,1] then (Cx):[1] and ((Cx)y):[ ].

This definition has been expressed in terms of any types τ and τ_i so as to accommodate extended definitions of types to come. A term of the form (PQ) is called an *application* term, since it will be understood to express the application of the predicate name P to the subject name Q. The more usual application notation can be introduced as "sugaring" by the following syntactic definitions:

DEFINITION 3. *Sugared Application.*
For terms P, Q, and $Q_1, \ldots, Q_n$,

1. $P(Q) \stackrel{\text{df}}{=} (PQ)$;
2. $P(Q_1, \ldots, Q_n, Q) \stackrel{\text{df}}{=} (P(Q_1, \ldots, Q_n)Q)$, $n \geq 1$.

Some explanation of the notation $\stackrel{\text{df}}{=}$ used here is necessary since this is but the first of many uses. Consider clause (1). The letters 'P' and 'Q' are metavariables ranging over terms as they have been defined above. For given terms P and Q, the intent of (1) is to allow any string of the form (PQ) to be replaced by the string $P(Q)$; that is, $P(Q)$ is to be understood as having the same intension as (PQ). For example, let C:[1, 1] and x, y:1. Then $((Cx)y)$ is a term of type [ ]. In this term the occurrence of (Cx) can, by clause (1), be replaced by $C(x)$. Thus $((Cx)y)$ can be replaced by $(C(x)y)$. Using clause (1) again, $(C(x)y)$ can be replaced by $C(x)(y)$, or alternatively using clause (2), by its more familiar form $C(x, y)$.

The "sugaring" of the application notation will be extensively used. But it will be essential at times to understand that the sugared notation is but a convenient way of expressing the fundamental application notation; the sugared notation, if you like, is for humans, while the unsugared is for machines, and for the convenience of some proofs.

An *elementary formula* is a term of type [ ] by the above definition of term. Elementary formulas are often called *atomic*; 'elementary' is used here instead because the definition of elementary formula will later be generalized in ways that result in formulas that are not atomic in the usual sense. As will be seen in §1.4, an elementary formula, like a sentence of a natural language, can be understood to have a *truth value* true or false when interpreted. By the above definition of term, an elementary formula has the sugared form $C(t_1, \ldots, t_n)$, where C:$[1^n]$, $n \geq 0$, and t_i:1 for $1 \leq i \leq n$. The rather complicated way in which this essentially very simple notion has been defined is to prepare the reader for later extensions of the definitions of type and term.

1.3. Logical Connectives

A *connective*, as the name suggests, is a linguistic device for connecting two or more sentences into a single more complex sentence. For example, both 'Africa is a continent and one-half is a natural number' and 'Africa is a continent or one-half is a natural number' are sentences; 'and' and 'or' are connectives of English. The connectives used in these sentences are called *logical* connectives since the truth value of each of the compound sentences is dependent only upon the truth values of its simpler components; that is upon the *extensions* of the simpler components. They are arity two connectives since they connect two sentences into one compound sentence. Although not strictly a connective, *negation* will be treated as an arity one logical connective, since the truth value of a sentence 'One half is not a natural number' depends only on the truth value of 'One half is a natural number'.

Not all connectives are logical. For example, the truth of a sentence such as 'If this computer program is started, then it will halt' is only seemingly dependent upon the truth values of its component sentences. Similarly, belief statements such as 'I believe that the program will stop' have a meaning going beyond that expressible in terms of the truth value of 'the program will halt'. For the present, study of such sentences will be put aside and the only sentences considered will be the sentences represented by elementary formulas, and compound sentences that can be formed from them using only logical connectives.

Compound sentences can themselves be understood to be in a subject predicate form. For example, the subjects of the sentence 'Africa is a continent and one-half is a natural number' are the sentences 'Africa is a continent' and 'One-half is a natural number' and the arity 2 predicate is 'and'. To accommodate such predicates, however, the first definition of type in §1.2 must be extended to admit the arity 1 type [[ ]], which is the type of negation, and the arity 2 type [[ ],[ ]], which is the type of conjunction. This could be accomplished by allowing clause (2) of the first definition to be applied not only to the type 1 but also to the type [ ]. But since the focus in this chapter is on a version of first order logic called Elementary Logic, or just EL, a less elegant method is chosen. The first definition of type in §1.2 is extended as follows:

DEFINITION 4. *Second Definition of Type.*

1. 1 *and* [ ] *are types.*
2. *For any sequence* 1^n *of* 1*'s*, $n \geq 1$, $[1^n]$ *is a type.*
3. [[ ]] *and* [[ ], [ ]] *are types.*

In a later definition of type, clause (3) becomes a special case of an extended clause (2).

1.3.1. Choice of connectives. The usual connectives of logic are negation $\neg$, conjunction $\wedge$, disjunction $\vee$, implication $\rightarrow$, and equivalence $\leftrightarrow$. Apart from negation which has the type [[ ]], each of the other connectives are of type [[ ],[ ]]. In classical elementary logic it is possible to define all of the connectives using only a pair consisting of negation and one of conjunction, disjunction, or implication, or using only implication and the primitive formula *contradiction* $\perp$ of type [ ]. But it is possible also to have only one binary connective, that of either joint denial $\downarrow$ or alternative denial $\uparrow$. A sentence formed from two simpler sentences using the joint denial connective is true if and only if each of the two sentences are false; one formed using alternative denial is true if and only if at least one of the two sentences is false.

The advantage of using a single connective is that there is only one case to be considered in many proofs about the logic. The disadvantage is that the sentences become incomprehensible to the human eye. But this disadvantage can be overcome, as will be done here, by defining all the usual connectives in terms of joint denial and using all the usual connectives in contexts where human comprehension is important. For contexts in which efficiency of proof is paramount, the single connective of joint denial can be used. Thus $\downarrow$ is introduced as a special constant of the logic for which t[$\downarrow$] is [[ ],[ ]] and is therefore included as a possible constant in clause (1) of Definition 2. Further in clause (2), τ may be the type [ ] whenever n is 1 or 0. The resulting definition of term is

DEFINITION 5. *Second Definition of Term.*
A term is any non-empty sequence of constants, variables, the special constant $\downarrow$, *and '(' and ')', typed as follows:*

1. *Let* cv *be any constant or variable. Then* cv *is a* term *of type* t[cv]. $\downarrow$ *is a term of type* [[ ], [ ]].
2. (a) *Let* P:$[1^{n+1}]$ *and* Q:1. *Then* (PQ) *is a* term *of type* $[1^n], 0 \leq n$.
   (b) *Let* F, G:[ ]. *Then* $(\downarrow F)$:[[ ]] *and* $(\downarrow F)G$:[ ].

A measure $ct[P]$ of the size of a term P is frequently needed in definitions and proofs. It is defined next.

DEFINITION 6. *First Definition of* $ct[P]$.

1. *For a constant, or variable* cv, $ct[cv]$ *is* 0; $ct[\downarrow]$ *is* 0.
2. *If* $ct[P]$ *is* m *and* $ct[Q]$ *is* n, *then* $ct[(PQ)]$ *is* $m + n + 1$.

Thus $ct[P]$, as defined here, is a count of the number of uses of application made in the construction of P. Later in §1.5 the definition of $ct[P]$ is extended to include a larger class of terms.

A *formula* is a term of type [ ]; thus an *elementary* formula as defined in §1.2 is a formula. Further, if F and G are formulas, then $(\downarrow F)$:[[ ]] and $((\downarrow F)G)$:[ ]; that is, $((\downarrow F)G)$ is a formula. The sugared application notation can also be used for $\downarrow$ so that $(\downarrow F)$ may be replaced by $\downarrow(F)$ and $((\downarrow F)G)$ by $\downarrow(F)(G)$ which in turn can be replaced by $\downarrow(F, G)$. But instead of this prefix notation for application, binary logical connectives like $\downarrow$ usually employ an infix notation:

DEFINITION 7. *Infix Notation for Connectives.*
For formulas F and G,

$$[F \downarrow G] \overset{\mathrm{df}}{=} \downarrow(F, G).$$

The change from '(' and ')' to '[' and ']' is to help a reader connect together matching pairs of parenthesis in complicated formulas. Note, however, that $[F \downarrow G]$ must be understood to be the unsugared term $((\downarrow F)G)$.

The definitions of sugared application in Notation 3, and the infix notation just defined, do not lead to a shortening of a string that is a term, but are intended to clarify the meaning of the term for a reader. Often definitions both shorten and clarify. For example, the definitions of the more usual logical connectives in terms of $\downarrow$ have these characteristics.

EXERCISES §1.3.

1. Determine $ct[[F \downarrow G]]$ as a function of $ct[F]$ and $ct[G]$.
2. For a formula F, the negated formula $\neg F$ can be defined $\neg F \overset{\mathrm{df}}{=} [F \downarrow F]$. Provide a similar definition for each of the usual binary connectives listed at the beginning of this section.
3. Prove that instead of choosing joint denial as the only primitive logical connective, it is possible to choose alternate denial $\uparrow$; that is the logical connective for which $[F \uparrow G]$ is true if and only if at least one of F and G is false.
4. Prove that all the usual connectives can be defined in terms of implication $\rightarrow$ and contradiction $\perp$.

1.4. Valuations

To understand whether a simple sentence such as 'Three is a natural number' is true or false it is necessary to know the intension of the subject name 'Three' and of the predicate name 'a natural number', so as to know whether the extension of the subject name is a member of the extension of the predicate name. For an elementary formula $C(c1)$ of the logic with subject name '$c1$' and predicate name 'C' no intension for either of these names is apparent so that it is impossible to know whether the formula is true or false. However if extensions were provided for the subject name and predicate name a truth value could be assigned: If the extension of the subject name is a member of the extension of the predicate name, then the formula is true, otherwise it is false.

Extensions for subject and predicate names are provided by *interpretations* that map each subject name to a member of a non-empty set of entities and each predicate name of arity n to a set of n-tuples of members. An interpretation necessarily assigns a truth value to each elementary formula $C(c_1, \ldots, c_n)$; the formula is assigned the truth value *true* if and only if the n-tuple of entities assigned to the subject names $c_1, \ldots, c_n$ by the interpretation is a member of the set of n-tuples of members assigned to C. Such an assignment of truth values to the elementary formulas defines a *valuation*. But clearly a valuation can be defined without first defining an interpretation. Further, from the truth values assigned to the elementary formulas, truth values can be determined for every formula by the extension of $\downarrow$; that is, a valuation of all formulas is determined.

If the only goal is to provide a valuation for the formulas of a logic, it is unnecessary to provide an interpretation to a set of entities. Instead a valuation of the formulas can be provided directly in terms of the constants of the logic as will be done next.

DEFINITION 8. *First Definition of Valuation Domain.*
A domain for a valuation is a function D from the types of the logic to the following sets:

1. $D(1)$ *is the set of constants of type* 1.
2. $D([\,])$ *is* $\{+, -\}$.
3. (a) $D([1^n])$, $1 \leq n$, *is a non-empty subset of* $\wp(D(1) \times \cdots \times D(1))$.
   (b) $D([[\,]])$ *is a non-empty subset of* $\wp D([\,])$.
   (c) $D([[\,], [\,]])$ *is a non-empty subset of* $\wp(D([\,]) \times D([\,]))$.

The notation $\wp$ used here denotes the power set operator. Thus, for example, that $D([1^n])$, $1 \leq n$, is a non-empty subset of $\wp(D(1) \times \cdots \times D(1))$ means that

$$D([1^n]) \subseteq \{\mathsf{S}|\mathsf{S} \subseteq D(1) \times \cdots \times D(1)\}$$

where $D(1) \times \cdots \times D(1)$ is the set

$$\{\langle e_1, \ldots, e_n\rangle | e_1 \ldots e_n \in D(1)\}.$$

Thus each member of $D([1^n])$ is a possibly empty set of n-tuples $\langle e_1, \ldots, e_n\rangle$ for which $e_1 \ldots e_n \in D(1)$; that is, a possible extension for an n-ary predicate.

Since the constants of type 1 are the subject names of the logic, $D(1)$ is the set of subject names of which there is assumed to be denumerably many. The arithmetic signs $+$ and $-$ that are members of $D([\,])$ represent the truth values *true* and *false* respectively.

The valuation domain D for which

$$D([1^n]) \text{ is } \wp(D(1) \times \cdots \times D(1)),$$
$$D([[\,]]) \text{ is } \wp D([\,]) \text{ and}$$
$$D([[\,],[\,]]) \text{ is } \wp(D([\,]) \times D([\,]))$$

is called the *standard* domain. Domains other than the standard may be considered for special purposes, as will be done later, although for first order logic the standard domain is commonly the only one considered.

DEFINITION 9. *First Definition of Valuation to a Domain.*
A valuation Φ *to a domain* D *is a function for which* $\Phi(T)$ *is defined for each term* T *by induction on* $ct[T]$ *as follows*:

V.1.1: $\Phi(cv) \in D(\mathrm{t}[cv])$, *for each constant or variable* cv *other than* $\downarrow$.
V.1.2: $\Phi(\downarrow)$ *is* $\{\langle -, - \rangle\}$.
V.2: *Let* $\Phi(P)$ *and* $\Phi(Q)$ *be defined,* $P{:}[\tau^{n+1}]$ *and* $Q{:}\tau$, *where*
τ *is* $1 \Rightarrow 0 \leq n$; *and* τ *is* $[\,] \Rightarrow 0 \leq n \leq 1$. *Then*
when $n \geq 1$, $\Phi((PQ))$ *is the set of n-tuples* $\langle e_1, \ldots, e_n \rangle$ *for which*
$\langle \Phi(Q), e_1, \ldots, e_n \rangle \in \Phi(P)$.
when $n = 0$, $\Phi((PQ))$ *is*
$+$ *if* $\Phi(Q) \in \Phi(P)$ *and is* $-$ *otherwise.*

Consider first the clause (V.1.1). When $c{:}1$, $\Phi(c)$ is some member $c1{:}1$ of $D(1)$. This is characteristic of valuations. That $\Phi(v)$ is a member of $D(\mathrm{t}[v])$ when v is a variable, means that the extension of a variable v of type τ may be any member of $D(\tau)$.

Clause (V.2) defines an extension for application terms (PQ), when extensions have been defined for both P and Q.

Consider now clause (V.1.2). That $\Phi(\downarrow)$ is $\{\langle -, - \rangle\}$ expresses that $\downarrow$ is joint denial. Consider the effect of the clause. Let $F, G{:}[\,]$, and let $\Phi(F)$ and $\Phi(G)$ be defined and therefore either $+$ or $-$. The possible values for $\Phi(\downarrow F)$ and $\Phi((\downarrow F)G)$ determined by (V.2) are listed here:

a) $\Phi(\downarrow F)$ is the set of e for which $\langle \Phi(F), e \rangle \in \Phi(\downarrow)$.

b) $\Phi(F)$ is $+ \Rightarrow \Phi(\downarrow F)$ is $\oslash$, the empty subset of $\{+, -\}$;
$\Phi(F)$ is $- \Rightarrow \Phi(\downarrow F)$ is $\{-\}$.

c) $\Phi(F)$ is $+$, $\Phi(G)$ is $\pm \Rightarrow \Phi((\downarrow F)G)$ is $-$ since $\Phi(G) \notin \oslash$.
$\Phi(F)$ is $-$, $\Phi(G)$ is $+ \Rightarrow \Phi((\downarrow F)G)$ is $-$ since $\Phi(G) \notin \{-\}$.
$\Phi(F)$ is $-$, $\Phi(G)$ is $- \Rightarrow \Phi((\downarrow F)G)$ is $+$ since $\Phi(G) \in \{-\}$.

1.4.1. Domains, valuations, and models. An important property of valuations to the standard valuation domain is expressed in the following theorem.

THEOREM 1. *Standard Valuation Domain.*
Let D *be the standard valuation domain and let* Φ *be a valuation to* D. *Then* $\Phi(T)$ *is defined and a member of* $D(\tau)$, *for each term* $T{:}\tau$.

PROOF. The theorem is proved by induction on $ct[T]$. Should $ct[T]$ be 0, then T is a constant, a variable, or $\downarrow$; the conclusion of the theorem follows from (V.1.1) or (V.1.2).

Assume that the theorem is proved whenever $ct[T] < ct$, and consider those T for which $ct[T]$ is ct. Necessarily T is an application term (PQ), where $P{:}[\tau^{n+1}]$ and $Q{:}\tau$. Necessarily $ct[P], ct[Q] < ct$, so that by the induction assumption both $\Phi(P)$ and $\Phi(Q)$ are defined and members of $D([\tau^{n+1}])$ and $D(\tau)$, respectively. By (V.2) therefore $\Phi(PQ)$ is defined and a member of $\wp(D(\tau) \times \cdots \times D(\tau))$. Hence $\Phi(PQ) \in D([\tau^n])$, since D is the standard domain. ⊣

Note that the theorem cannot be proved without some condition on the valuation domain D, although not necessarily that it is standard. Let C and c be constants of type $[1^2]$ and 1 respectively. Let $D([1^2])$ be $\wp(D(1) \times D(1))$ and let $\Phi(C)$ be $(D(1) \times D(1))$. $\Phi(c) \in D(1)$ so that $C(c){:}[1]$ and $\Phi(C(c))$ is $\{D(1)\}$. But $\{D(1)\} \in D([1])$ is not required.

DEFINITION 10. *Definition of Model.*
A model with domain D is a valuation to D for which $\Phi(T)$ is defined and a member of $D(\tau)$ for each term $T{:}\tau$.

This definition will continue to be valid through all extensions of types, terms, domains and valuations considered in this and all subsequent chapters as will a fundamental theorem for valuations of application terms stated next.

THEOREM 2. *Fundamental Theorem for Application.*
Let P and t_i be terms for which $P{:}[\tau_1, \ldots, \tau_n]$ and $t_i{:}\tau_i$, $1 \le i \le n$. Let Φ be any valuation. Then

$$\Phi(P(t_1, \ldots, t_n)) \text{ is } + \Leftrightarrow \langle \Phi(t_1), \ldots, \Phi(t_n)\rangle \in \Phi(P).$$
$$\Phi(P(t_1, \ldots, t_n)) \text{ is } - \Leftrightarrow \langle \Phi(t_1), \ldots, \Phi(t_n)\rangle \notin \Phi(P).$$

PROOF. $P(t_1, \ldots, t_n)$ is $(((\ldots((Pt_1)t_2)\ldots)t_{n-1})t_n)$. Hence by (V.2) of Definition 9,

$$\begin{aligned}\Phi(P(t_1, \ldots, t_n)) \text{ is } + &\Leftrightarrow \Phi(t_n) \in \Phi(((\cdots((Pt_1)t_2)\cdots)t_{n-1}))\\ &\Leftrightarrow \langle \Phi(t_{n-1}), \Phi(t_n)\rangle \in \Phi((\cdots((Pt_1)t_2)\cdots)))\\ &\cdots\\ &\Leftrightarrow \langle \Phi(t_1), \cdots, \Phi(t_n)\rangle \in \Phi(P).\end{aligned}$$

The proof of the case $\Phi(P(t_1, \ldots, t_n))$ is $-$ is similar. ⊣

Since the notation $[F{\downarrow}G]$ is a syntactic sugaring of $((\downarrow F)G)$, the following is an immediate corollary:

COROLLARY 3. *Let $F, G{:}[\,]$. Then*

$$\Phi([F{\downarrow}G]) \text{ is } + \Rightarrow \Phi(F) \text{ is } - \text{ and } \Phi(G) \text{ is } -.$$
$$\Phi([F{\downarrow}G]) \text{ is } - \Rightarrow \Phi(F) \text{ is } + \text{ or } \Phi(G) \text{ is } +.$$

EXERCISES §1.4.

1. An exercise in §1.3 asked for definitions of the usual binary logical connectives in terms of $\downarrow$. Properties of valuations for these connectives similar to those of the corollary can be stated. For example, properties for $\neg$ are:

$$\Phi(\neg F) \text{ is } + \Rightarrow \Phi(F) \text{ is } -.$$
$$\Phi(\neg F) \text{ is } - \Rightarrow \Phi(F) \text{ is } +.$$

Prove these properties for $\neg$ and state and prove similar properties for the binary logical connectives.

1.5. The Lambda Abstraction Operator

The abstraction operator λv is of such importance for logic that it will be introduced and used now in a simple application so that readers may become familiar with it before it is more extensively used. Although the operator is not usually part of first order logic, it may be added to EL with few complications. In [46] it is shown to be valuable for making subtle distinctions of meaning in first order modal logic. The operator makes use of variables, and it is this use that is the only role for variables in the logics described in this book.

The use of the lambda operator can be illustrated with a simple example. Let *Emp* be the predicate name for the predicate of being a current employee of a particular corporation, and let *Male* be the predicate name for the predicate of being a male person. Thus *Emp*:[1] and *Male*:[1]. Let the constants of type 1, the subject names, be potential names of persons. Assume that a valuation Φ has been defined consistent with the predicates *Emp* and *Male*; that is, $\Phi(Emp)$ is the set of constants that are names of employees of the corporation, and $\Phi(Male)$ is the set of constants that are names of male persons. Let x be a variable of type 1 and consider the expression

$$(\lambda x.[Emp(x) \wedge Male(x)]).$$

It is the name for the predicate of being a male employee of the corporation; that is, the truth value of the expression

$$(\lambda x.[Emp(x) \wedge Male(x)])(c),$$

where c:1, is the truth value of the formula $[Emp(c) \wedge Male(c)]$. Thus the lambda abstraction operator λ together with variables of type 1 permit the definition of new predicate names of type 1^n, $n \geq 1$, from given predicate names.

There is no reason to restrict variables to being only of type 1, although for EL or first order logic that is the only type of variable needed. Consider for example the expression $(\lambda x.((\downarrow x)x)$ where x is a variable of type [ ]. By application sugaring it can be written $(\lambda x.[x{\downarrow}x])$. The truth value of the expression $(\lambda x.[x{\downarrow}x])(F)$, where F:[ ], is the truth value of the formula $[F{\downarrow}F]$ which, by the exercise of §1.4.1, is the truth value of $\neg F$. Thus the use of λ in the expression $(\lambda x.((\downarrow x)x)$ provides a means of defining $\neg$ without using the metavariable F as in §1.3.1.

1.5.1. Syntax for quantification. A predicate may itself be the subject of a sentence; that is, predicates of predicates do exist. An important example is the existential quantifier $\exists$ over the type 1 domain $D(1)$. The first use of λ is to define subjects for this predicate. But in order to do this it is necessary to extend the set of types to include a type for $\exists$. Since $\exists$ is a predicate of unary predicates, and since [1] is the type of predicates with one argument, the type of $\exists$ must be [[1]]; that is, t[$\exists$] for the special constant $\exists$ is [[1]]. In the second definition of types, Definition 4 in §1.3, the types [[ ]] and [[ ], [ ]] needed for terms in which $\downarrow$ occurs are added as a special case; the type [[1]] needed for terms in which $\exists$ occurs is added in the same inelegant way:

DEFINITION 11. *Third Definition of Type.*

1. 1 *and* [ ] *are types.*
2. *For any sequence* 1^n *of* 1*'s*, $n \geq 1$, $[1^n]$ *is a type.*
3. [[ ]] *and* [[ ], [ ]] *and* [[1]] *are types.*

The clause (1) of the first definition of term, Definition 2 in §1.2, allows for special constants; the special constant $\downarrow$ was added in §1.3. Now the special constant $\exists$ is added; but further a third clause must be added to accommodate the untyped symbol λ.

DEFINITION 12. *Third Definition of Term.*
A string may now include λ *and* $\exists$, *but unlike* $\exists$, λ *is not a special constant and is not assigned a type. A term is any non-empty sequence of constants, variables, the special constant* $\downarrow$ *and* $\exists$, *and '(' and ')', typed as follows*:

1. *Let* cv *be any constant or variable. Then* cv *is a* term *of type* t[cv]. $\downarrow$ *is a term of type* [[ ], [ ]]. $\exists$ *is a term of type* [[1]].
2. (a) *Let* P:$[1^{n+1}]$ *and* Q:1. *Then* (PQ) *is a* term *of type* $[1^n], 0 \leq n$.
   (b) *Let* F_1, F_2:[ ]. *Then* $(\downarrow F_1)$:[[ ]] *and* $(\downarrow F_1)F_2$:[ ].
   (c) *Let* P:[1]. *Then* $(\exists P)$:[ ].
3. *Let* P:$[1^n], 0 \leq n$, *and* v:1 *be a variable. Then* $((\lambda v)P) : [1^{n+1}]$.

To eliminate the use of extra pairs of parentheses the following abbreviating notation will be used:

DEFINITION 13. $\lambda v.P$ *abbreviates* $(\lambda v.P)$ *which in turn abbreviates* $((\lambda v)P)$.

The terms of the form (PQ) have been called application terms; those of the form $(\lambda v.P)$ will be called *abstraction* terms.

The Definition 6 of $ct[P]$ in §1.3.1 can now be given its final form; the third clause added to the definition is: If $ct[P]$ is m, then $ct[(\lambda v.P)]$ is $m + 1$.

The prefix λv of an abstraction term $(\lambda v.P)$ is an instance of an *abstraction* operator; the term P is called the *scope* of the abstraction operator λv. Note that the type of the scope of an abstraction operator is never a term of type 1. The abstraction operator is only used in this book as a device for defining predicate names; expressions such as $\lambda v.v$, where v:1, are not typed and are therefore not terms; that is, they are not well formed.

Again a *formula* is a term of type [ ]. Among the formulas are ones of the form $(\exists P)$, where P:[1], since $\exists$:[[1]].

$\exists$ is an example of an *extensional* predicate of predicates in the following sense: Whether $\Phi(\exists P)$ is $+$ or $-$ depends only on the extension of the subject predicate P, and not on its intension. Thus if P and Q are both of type [1] and have the same extension, then $\Phi(\exists P)$ is $\Phi(\exists Q)$. The study of predicates of predicates will be continued in greater depth in chapters 2 and 3.

When there is no risk of confusion the outer brackets of application and abstraction terms will be dropped. Application sugaring can be used for these extended terms. Further, a sequence of abstraction operators can be abbreviated with a single λ followed by several variables:

$$(\lambda x_1, \ldots, x_n.P) \stackrel{\text{df}}{=} (\lambda x_1(\ldots.(\lambda x_n.P)\ldots)).$$

A *subterm* of a term P is a substring of P that is itself a term. Such a subterm has a depth in P defined as follows:

DEFINITION 14. *Depth of a Subterm of a Term.*

1. *A term P is a subterm of itself of depth* 0.
2. *Let P' be a subterm of P of depth d. Then*
   (a) *P' is a subterm of PQ and QP of depth $d+1$;*
   (b) *P' is a subterm of $(\lambda x.P)$ of depth $d+1$.*

P' is said to be a *proper* subterm of P if it is a subterm of depth $d \geq 1$.

1.5.2. Semantics for abstraction terms. Before extending the first definition of valuation, Definition 9 in §1.4, for the new terms it is necessary to extend first Definition 8 of a domain function for the new type [[1]]; this is accomplished by adding to clause (3b) of Definition 8. The resulting definition is:

DEFINITION 15. *Second Definition of Valuation Domain.*
A domain *for a valuation is a function D from the types of the logic to the following sets*:

1. *$D(1)$ is the set of constants of type* 1.
2. *$D([\,])$ is $\{+, -\}$.*
3. (a) *$D([1^n])$, $1 \leq n$, is a non-empty subset of $\wp(D(1) \times \cdots \times D(1))$.*
   (b) *$D([[\,]])$ is a non-empty subset of $\wp D([\,])$.*
   (c) *$D([[\,],[\,]])$ is a non-empty subset of $\wp(D([\,]) \times D([\,]))$.*
   (d) *$D([[1]])$ is a non-empty subset of $\wp D([1])\}$.*

Definition 9 of valuation in §1.4 must be extended in several ways. Clause (V.1.2) of the definition must be extended by defining a value for $\Phi(\exists)$, and clause (V.2) extended by defining a value for (PQ) when P:[[1]] and Q:[1]. Finally a new clause (V.3) is required to define a value for $\Phi(\lambda v.P)$. This requires the prior definition of a valuation that is a v-variant of a given valuation.

DEFINITION 16. *Variable variant of a valuation.*
Let v be a variable of type 1 *and let Φ be a valuation to a given domain. A valuation Φ^v to the same domain is a* v-variant *of Φ if $\Phi^v(u)$ differs from $\Phi(u)$ at most when the variable u is v.*

Thus $\Phi^v(v)$ may be different from $\Phi(v)$, but $\Phi^v(u)$ is $\Phi(u)$ if u is not v. The new definition of valuation is:

DEFINITION 17. *Second Definition of Valuation to a Domain.*
A valuation Φ *to a domain D is a function for which $\Phi(T)$ is defined for each term T by induction on $ct[T]$ as follows*:

*V.*1.1: $\Phi(cv) \in D(\mathrm{t}[cv])$, *for each constant or variable cv other than $\downarrow$ or $\exists$.*
*V.*1.2: $\Phi(\downarrow)$ *is* $\{\langle -, -\rangle\}$ *and*
$\Phi(\exists)$ *is a non-empty set of non-empty subsets of* $D([1])$.
*V.*2: *Let $\Phi(P)$ and $\Phi(Q)$ be defined, P:$[\tau^{n+1}]$ and Q:τ, where*

τ *is* $1 \Rightarrow 0 \leq n$; τ *is* $[\,] \Rightarrow 0 \leq n \leq 1$; *and* τ *is* $[1] \Rightarrow n = 0$. *Then*
when $n \geq 1$, $\Phi((PQ))$ *is the set of n-tuples* $\langle e_1, \ldots, e_n\rangle$ *for which*
$\langle \Phi(Q), e_1, \ldots, e_n\rangle \in \Phi(P)$;
when $n = 0$, $\Phi((PQ))$ *is*
$+$ *if* $\Phi(Q) \in \Phi(P)$ *and is* $-$ *otherwise.*

*V.*3: *Let $\Phi(P)$ be defined, P:$[1^n]$. Then*

when $n \geq 1$, $\Phi(\lambda v.P)$ *is the set of tuples*
$\langle \Phi^v(v), e_1, \ldots, e_n\rangle$
for which
$\langle e_1, \ldots, e_n\rangle \in \Phi^v(P)$ *for some v-variant Φ^v of Φ;*
when $n = 0$, $\Phi(\lambda v.P)$ *is the set of values* $\Phi^v(v)$ *for which*
$\Phi^v(P)$ *is* $+$.

Note that the addition to clause (V.2) is a special case of Theorem 2 of §1.4.1 for application.

The definition of model given in §1.4.1 is unchanged: A valuation to a domain D for which $\Phi(P) \in D(\tau)$ holds for each term P of type τ is called a *model.*

EXAMPLE. It was stated in the introduction to §1.5 that the truth value of the expression

$$(\lambda x.[Emp(x) \wedge Male(x)])(c)$$

is the truth value of the formula

$$[Emp(c) \wedge Male(c)].$$

That is, it is claimed that for any valuation Φ, that

$$\Phi((\lambda x.[Emp(x) \wedge Male(x)])(c)) \text{ is } \Phi([Emp(c) \wedge Male(c)])$$

This can be verified as follows:

(a) $\Phi((\lambda x.[Emp(x) \wedge Male(x)])(c))$ is $\pm \Leftrightarrow$
$\Phi(c) \in \Phi((\lambda x.[Emp(x) \wedge Male(x)]))$, respectively
$\Phi(c) \notin \Phi((\lambda x.[Emp(x) \wedge Male(x)]))$.

(b) $\Phi((\lambda x.[Emp(x) \wedge Male(x)]))$ is
$\{\Phi^x(x) \parallel \Phi^x([Emp(x) \wedge Male(x)])$ is $+$ and Φ^x is an x-variant of $\Phi\}$

(c) Consider the x-variant Φ^x of Φ for which $\Phi^x(x)$ is $\Phi(c)$.
The values of $\Phi^x(\wedge)$, $\Phi^x(Emp)$, $\Phi^x(Male)$, and $\Phi^x(x)$ are respectively the values of $\Phi(\wedge)$, $\Phi(Emp)$, $\Phi(Male)$, and $\Phi(c)$,while the value of the latter is c by (V.1).

(d) $\Phi^x([Emp(x) \wedge Male(x)])$ is $\Phi([Emp(c) \wedge Male(c)])$.

(a) is justified by (V.2), (b) by (V.3), and (d) by repeated uses of (V.2) and by the value of $\Phi(\wedge)$ determined from the definition of $\wedge$ and (V.1.2).

The role of the variable x in the abstraction term $(\lambda x.[Emp(x) \wedge Male(x)])$ is that of a *placeholder*. It indicates in what places in the scope $[Emp(x) \wedge Male(x)]$ of the abstraction operator λx the subject c of the predicate name is to be inserted. Any variable of type 1 can play the same role. For example, if y:1 then $(\lambda y.[Emp(y) \wedge Male(y)])$ also names the predicate of being a male employee of the corporation; that is, the predicate names $(\lambda x.[Emp(x) \wedge Male(x)])$ and $(\lambda y.[Emp(y) \wedge Male(y)])$ have the same intensions, and therefore the same extensions.

1.5.3. Free and bound variables. In the term $(\lambda x.[Emp(y) \wedge Male(x)])$ the occurrence of x in $Male(x)$ is *bound by* the abstraction operator λx while the occurrence of y in $Emp(y)$ is *free*. These concepts are defined next by induction on Definition 12 of §1.5.1:

DEFINITION 18. *Free and Bound Occurrences.*

1. *The occurrence of a variable in itself is a* free *occurrence.*
2. *A free occurrence of a variable in P or Q is a free occurrence in (PQ).*
3. *A free occurrence of a variable v in P is a* bound *occurrence in $(\lambda v.P)$. An occurrence of a variable u other than v is free or bound in $(\lambda v.P)$ if it is free, respectively bound, in P.*

The term $(\lambda y.[Emp(y) \wedge Male(y)])$ results from $(\lambda x.[Emp(x) \wedge Male(x)])$ by replacing the abstraction operator λx by λy and each occurrence of x bound by λx in $(\lambda x.[Emp(x) \wedge Male(x)])$ by y. Such a transformation is called a *change of bound variable* and $(\lambda y.[Emp(y) \wedge Male(y)])$ is said to be a bound variable variant of $(\lambda x.[Emp(x) \wedge Male(x)])$. The transformation is permitted whenever the new variable y has no free occurrence in the scope of the abstraction operator λx of the original variable. This restriction is necessary if the intension of the abstraction term is to be maintained. For example, changing the bound variable x in $(\lambda x.[Emp(y) \wedge Male(x)])$ to y without regard to the restriction, results in the term $(\lambda y.[Emp(y) \wedge Male(y)])$. Whatever the intension of $(\lambda x.[Emp(y) \wedge Male(x)])$ may be, it clearly must be different from the intension of $(\lambda y.[Emp(y) \wedge Male(y)])$, since for $c{:}1$, $(\lambda x.[Emp(y) \wedge Male(x)])(c)$ is $[Emp(y) \wedge Male(c)]$ while $(\lambda y.[Emp(y) \wedge Male(y)])(c)$ is $[Emp(c) \wedge Male(c)]$.

A variable u is said to be *free to replace* a variable v of the same type in a term if no occurrence of v in the term is within the scope of an abstraction operator λu. Thus x is not free to replace y in $(\lambda x.[Emp(y) \wedge Male(x)])$ since the single occurrence of y is within the scope of λx.

A term in which no variable has a free occurrence is sometimes said to be a *closed* term.

1.5.4. Substitutions and contractions. The transformation of $(\lambda y.[Emp(y) \wedge Male(y)])(c)$ to $[Emp(c) \wedge Male(c)]$ is an example of a *β-contraction*. The definition of such contractions requires the prior definition of a substitution operator $[t/v]$, where t is a term of the same type as the variable v.

Many different notations have been employed in mathematics and computer science to express the operation of substitution. The notation $[t/v]$ defined next is used throughout the book as a prefix operator. A simple mnemonic device for understanding the difference between the roles of t and v in the operator is to think of it as a fraction with numerator t and denominator v, and to think of its application to a term as multiplication; the effect of $[t/v]$ on a term P is to "cancel out" each free occurrence of v in P and replace it with t. However bound variables introduce a complication that is dealt with in the following definition.

DEFINITION 19. *Definition of Substitution Operator.*
Let t be a term of the same type as the variable v. Then

1. $[t/v]v$ *is* t; $[t/v]cw$ *is* cw *for any constant or variable* cw *distinct from* v.
2. $[t/v](PQ)$ *is* $([t/v]P[t/v]Q)$.
3. (a) $[t/v](\lambda w.P)$ *is* $(\lambda w.P)$ *if* w *is* v;
   (b) $[t/v](\lambda w.P)$ *is* $(\lambda w.[t/v]P)$ *if* w *is not* v *and* w *has no free occurrence in* t;
   (c) $[t/v](\lambda w.P)$ *is* $(\lambda u.[t/v][u/w]P)$ *if* w *is not* v, w *has a free occurrence in* t, *and* u *is free to replace* w *in* P *and has no free occurrence in* t.

There are no restrictions on the application of the substitution operator to a term; however, because of clause (3c), an application of $[t/v]$ to a term can result in a change of a bound variable in the term; for example, the application of $[x/y]$ to the term $(\lambda x.[Emp(y) \wedge Male(x)])$ may be $(\lambda z.[Emp(x) \wedge Male(z)])$, where z is not x. Since there will be denumerably many variables u free to replace w in a term P, $[t/v](\lambda w.P)$ as defined in clause (3c) is not unique. However by always choosing u to be the first variable in an enumeration of the variables of type t[w] free to replace w in P, the result of applying the substitution operator can be assumed to be unique. In any case it is unique up to bound variable variants.

LEMMA 4. *Substitution.*
$[t/v]P$ has the same type as P, for any terms P and t, and variable v.

A proof of the lemma is left as an exercise.

Occasionally a simultaneous substitution operator

$$[t_1, \ldots, t_n/v_1, \ldots, v_n]$$

will be used, where $n \geq 2$. The term

$$[t_1, \ldots, t_n/v_1, \ldots, v_n]P$$

differs from

$$[t_1/v_1][t_2/v_2] \cdots [t_n/v_n]P$$

in the following way: The latter results from P by first applying $[t_n/v_n]$ to P, and then applying $[t_{n-1}/v_{n-1}]$ to $[t_n/v_n]P$, and so on. As a result each free occurrence of v_{n-1} in t_n is replaced by t_{n-1}, not just those in P; and so on. The result $[t_1, \ldots, t_n/v_1, \ldots, v_n]P$, on the other hand, is obtained without these additional substitutions; the substitution operators $[t_1/v_1][t_2/v_2] \cdots [t_n/v_n]$ are applied simultaneously to P. A sequence $[t_1/v_1] \cdots [t_n/v_n]P$ of substitutions does define a simultaneous substitution operator $[t_1', \ldots, t_n'/v_1, \ldots, v_n]$, where t_i' is $[t_1/v_1] \cdots [t_{i-1}/v_{i-1}]t_i$.

The most important use of substitution is in the definition of β-contractions defined next along with η-contractions.

DEFINITION 20. *Contractions.*

1. *$[Q/v]P$ is a β-contraction of $((\lambda v.P)Q)$.*
2. *Let the variable v have no free occurrence in P. Then P is an η-contraction of $(\lambda v.P(v))$.*

Contractions are sometimes called *reductions* in the literature. Note that the contraction of a term has the same type as the term.

The following lemma describes the effect of substitution on the semantics of terms.

LEMMA 5. *Semantic Substitution.*
Let Φ be a valuation to a domain D. Let x be a variable and t a term of type 1, and T a term of any type. Let Φ^x be the x-variant of Φ for which $\Phi^x(x)$ is $\Phi(t)$. Then $\Phi^x(T)$ is $\Phi([t/x]T)$.

PROOF. It may be assumed that x has a free occurrence in T, since otherwise the conclusion is immediate. The proof is by induction on $ct[T]$ as defined in §1.5.1.

Let $ct[T]$ be 0. Then T is x, $[t/x]T$ is t, and therefore $\Phi^x(T)$ is $\Phi([t/x]T)$. Assume the conclusion of the lemma whenever $ct[T] < ct$. Let $ct[T] = ct$, so that T is either an application or abstraction term.

• T is PQ. Then either

(a) For some τ that is 1 or [] and some n, $n \geq 0$, $P{:}[\tau^{n+1}]$ and $Q{:}\tau$, or
(b) $P{:}[[1]]$ and $Q{:}[1]$.

Consider (a) with $n > 0$. Then $\Phi^x(PQ)$ is the set of n-tuples $\langle e_1, \ldots, e_n\rangle$ for which

$$\langle \Phi^x(Q), e_1, \ldots, e_n\rangle \in \Phi^x(P).$$

But by the induction assumption the latter is

$$\langle \Phi([t/x]Q), e_1, \ldots, e_n\rangle \in \Phi([t/x]P).$$

Therefore $\Phi^x(T)$ is $\Phi([t/x]T)$. A similar argument deals with the case $n = 0$.

Consider (b). Necessarily P is $\exists$, $\Phi(\exists) \in D([[1]])$, and $\Phi(Q) \in D([1])$. Hence

$$\begin{aligned}
\Phi^x((\exists Q)) \text{ is } + &\Leftrightarrow \Phi^x(Q) \in \Phi^x(\exists) && \text{(V.2)}\\
&\Leftrightarrow \Phi([t/x]Q) \in \Phi(\exists) && \text{by induction}\\
&\Leftrightarrow \Phi((\exists[t/x]Q)) \text{ is } + && \text{(V.2)}\\
&\Leftrightarrow \Phi(([t/x]\exists Q)) \text{ is } +.
\end{aligned}$$

• T is $\lambda y.P$, where $P{:}[1^n]$, $n \geq 0$. It may be assumed that y has no free occurrence in t and is not x so that $[t/x](\lambda y.P)$ is $(\lambda y.[t/x]P)$.

Let $n \geq 1$. Then $\Phi^x(\lambda y.P)$ is the set of n-tuples

$$\langle \Phi^{yx}(y), e_1, \ldots, e_n\rangle$$

for which

$$\langle e_1, \ldots, e_n\rangle \in \Phi^{yx}(P).$$

Since x is not y, $\Phi^{yx}(y)$ is $\Phi^y(y)$. Hence

$$\langle \Phi^y(y), e_1, \ldots, e_n\rangle$$

for which

$$\langle e_1, \ldots, e_n\rangle \in \Phi^y([t/x]P).$$

So that $\Phi^x(\lambda y.P)$ is $\Phi(\lambda y.[t/x]P)$ as required. A similar argument deals with the case $n = 0$. ⊣

EXERCISES §1.5.

1. Determine $ct[(\lambda x.[Emp(x) \wedge Male(x)])]$. Use the definition of $\wedge$ provided in the exercise of §1.3.1.
2. Complete the details required in the last step of the example of §1.5.2.
3. Provide a proof for Lemma 4 by induction on $ct[P]$.

1.6. Syntax and Semantics for EL

This chapter has been written to provide motivation and foundations for the higher order logics described in later chapters. The syntax and semantics for an unconventional form EL of first order logic have been described in an inelegant way that is the usual very elegant way of describing the syntax and semantics for higher order logics. In this section the unconventional presentation of EL is related to a more conventional that at the same time provides a semantic basis for its proof theory.

The conventional notation for existential quantification is defined in terms of $\exists$ and the abstraction operator

$$\exists v.F \stackrel{\text{df}}{=} \exists(\lambda v.F),$$

where F:[ ] and v:1 so that $(\lambda v.F)$:[1] and hence $\exists v.F$:[ ].

DEFINITION 21. *Formulas of EL.*

1. $C(t_1, \ldots, t_n)$ *is a formula if* C:[1^n] *and* t_i:1, $0 \leq i \leq n$.
2. *If* F *and* G *are formulas, then* $[F \downarrow G]$ *is a formula.*
3. *If* F *is a formula, then* $\exists v.F$ *is a formula for any variable* v *of type* 1.
4. *If* $([t/v]P)(t_1, \ldots, t_n)$ *is a formula, then* $(\lambda v.P)(t, t_1, \ldots, t_n)$ *is a formula.*

The formulas of clause (1) are called *elementary* formulas in this monograph. In first order logic they are generally called *atomic* formulas since for any subterm t of them $ct[t]$ is 0. However because of abstraction terms, an elementary formula of higher order logic may have a subterm t for which $ct[t] > 0$.

EXAMPLE. $(\lambda x.[Emp(x) \wedge Male(x)])(c)$ is a formula when Emp and $Male$ are constants of type [1], x is a variable and c is a constant of type 1. This can be seen as follows:

a) $Emp(x)$ and $Male(x)$ are formulas by (1) of Definition 21;
b) $[Emp(x) \wedge Male(x)]$ is a formula by (a) and the definition of $\wedge$ in terms of $\downarrow$, and repeated use of (2) of Definition 21;
c) $[c/x][Emp(x) \wedge Male(x)]$ is a formula by (b) and the definition of the substitution operator $[c/x]$;
d) $(\lambda x.[Emp(x) \wedge Male(x)])(c)$ is a formula by (b) (4) of Definition 21.

Clause (d) of the definition of formula does not appear in most formulations of first order logic because abstraction terms are not admitted. But such terms can be useful in applications of logic to databases as suggested by the term $(\lambda x.[Emp(x) \wedge Male(x)])$; the term can be seen as a request to a database to return the extension of the predicate name.

The semantics of EL are described in §1.5.2 in terms of a domain D and valuations Φ to the domain. Recall that a model of EL is a valuation Φ to D for which

$$\Phi(T) \in D(\tau)$$

for each term $T{:}\tau$ of EL. The following theorem provides the semantic basis for the proof theory of EL.

THEOREM 6. *Semantic Basis for the Proof Theory of EL.*
Let Φ be a valuation that is a model. Let C be a constant of type $[\tau_1, \dots, \tau_n]$, $t_i{:}\tau_i$ terms for $1 \leq n$, and F and G be formulas.

1. $\langle\Phi(t_1), \dots, \Phi(t_n)\rangle \in \Phi(C) \Rightarrow \Phi(C(t_1, \dots, t_n))$ *is* $+$;
   $\langle\Phi(t_1), \dots, \Phi(t_n)\rangle \notin \Phi(C) \Rightarrow \Phi(C(t_1, \dots, t_n))$ *is* $-$.
2. $\Phi([F \downarrow G])$ *is* $+ \Rightarrow \Phi(F)$ *is* $-$ *and* $\Phi(G)$ *is* $-$;
   $\Phi([F \downarrow G])$ *is* $- \Rightarrow \Phi(F)$ *is* $+$ *or* $\Phi(G)$ *is* $+$.
3. $\Phi(\exists u.F)$ *is* $+ \Rightarrow \Phi([t/u]F)$ *is* $+$ *for some* $t{:}1$;
   $\Phi(\exists u.F)$ *is* $- \Rightarrow \Phi([t/u]F)$ *is* $-$ *for every* $t{:}1$.
4. *Let F be $([t/v]P)(t_1, \dots, t_n)$ and let G be $(\lambda v.P)(t, t_1, \dots, t_n)$. Then* $\Phi(G)$ *is* $\pm \Rightarrow \Phi(F)$ *is* $\pm$*, respectively.*

PROOF. All references here to clauses (V.n) is to the clauses of Definition 17 in §1.5.2.
(1) is a special case of Theorem 2 of §1.4 and (2) repeats its corollary. To prove (3) consider the value of $\exists$ for an argument $(\lambda u.F)$.

$$\begin{array}{lll}
\Phi(\exists(\lambda u.F)) \text{ is } + & \Rightarrow \Phi(\lambda u.F) \in \Phi(\exists) & \text{(V.2)} \\
& \Rightarrow \{\Phi^u(u) \| \Phi^u(F) \text{ is } +\} \in \Phi(\exists) & \text{(V.3)} \\
& \Rightarrow \{\Phi^u(u) \| \Phi^u(F) \text{ is } +\} \text{ is non-empty} & \text{(V.1.2)} \\
& \Rightarrow [t/u]F \text{ is } + \text{ for some } t{:}1. &
\end{array}$$

The second half of (3) is left as an exercise.

Consider now (4).

$$\begin{array}{lll}
\Phi(G) \text{ is } + & \Rightarrow \langle\Phi(t), \Phi(t_1), \dots, \Phi(t_n)\rangle \in \Phi(\lambda v.P) & \text{Thm 2, §1.4.1} \\
& \Rightarrow \Phi(t) \text{ is } \Phi^v(v) \text{ and } \langle\Phi(t_1), \dots, \Phi(t_n)\rangle \in \Phi^v(P) & \text{(V.3)} \\
& \Rightarrow \langle\Phi(t_1), \dots, \Phi(t_n)\rangle \in \Phi([t/v]P) & \text{Lemma 5, §1.5.4} \\
& \Rightarrow \Phi(F) \text{ is } + & \text{Thm 2, §1.4.1}
\end{array}$$

Similarly $\Phi(G)$ is $- \Rightarrow \Phi(F)$ is $-$. ⊣

EXERCISES §1.6.

1. A formula has been defined to be a term of type []. Prove that an EL formula is a term of type []. Is the converse true; that is, is a term of type [] necessarily an EL formula as defined in Definition 21?.
2. Complete the proofs of clauses (3) and (4) of Theorem 6.
3. Clause (3) of Theorem 6 states a property of valuations relating to the quantifier $\exists$. Using your definition of the quantifier $\forall$ from exercise (1), state and prove a similar property of valuations for $\forall$.

1.7. Elementary Proof Theory

A semantic tree proof theory for the variant of first order logic referred to as EL, Elementary Logic, is presented here. The theory has been derived from the semantic tableaux method of Beth [19]. It is introduced here and is used in one form or another throughout the remainder of the book. More recent expositions of similar proof theories are in [85], [134], and [47].

The proof theory for EL defines derivations for sequents of formulas, not individual formulas. A sequent consists of two finite, possibly empty, sets

$$\{F_1, \ldots, F_m\} \text{ and } \{G_1, \ldots, G_n\}$$

of formulas, the *antecedent* and the *succedent* respectively. Sequents are displayed by listing the members of the antecedent, followed by the symbol ⊢ followed by a listing of the members of the succedent

(Seq) $$F_1, \ldots, F_m \vdash G_1, \ldots, G_n.$$

Because the antecedent and succedent are *sets* of formulas, duplicates and the order of the listings of the formulas are irrelevant.

A proof theory of sequents, rather than of formulas, is being presented here since the proof theory can then be given a very simple form closely related to the semantics. This was first demonstrated by Gentzen [52]. There a sequent calculus was defined for both the classical and the intuitionist first order logics, the latter formalizing the logic defined by Heyting [77]. It should be noted, however, that the sequents used here are different from the sequents introduced by Gentzen. The antecedent and succedent of a Gentzen sequent are *sequences* of formulas displayed in the form (Seq) for which duplicates and order are significant. Structural rules are needed in the Gentzen sequent calculus to introduce new formulas into the antecedent of succedent, to change the order of the formulas, and to remove duplicates. By understanding the antecedent and succedent of a sequent to be a listing of the members of a set, rules for changing the order of formulas or for removing duplicates are not necessary. In addition, the manner in which the proof theory is defined below removes the need for adding formulas to the antecedent and succedent of a sequent, except for the succedent of an intuitionist logic as will be seen in Chapter 6. But until then the exposition will be of classical logics and no structural rules are needed.

A sequent is said to be *satisfied* by a valuation Φ if $\Phi(F)$ is $-$ for a member F of the antecedent or is $+$ for a member F of the succedent. As a consequence, an *empty* sequent, that is one in which both the antecedent and succedent are empty, cannot be satisfied by any valuation. A valuation Φ is a *counter-example* for a sequent if $\Phi(F)$ is $+$ for every F in the antecedent and $-$ for every F in the succedent. A sequent is *valid* if it is satisfied by every valuation that is a model; thus a valid sequent can have no counter-example.

The proof theory provides an implicit systematic search procedure for a counter-example for a given sequent $\Gamma \vdash \Theta$. Should the procedure fail to find such a valuation, and if it does fail it will fail in a finite number of steps, then that $\Gamma \vdash \Theta$ is valid follows from the completeness theorem to be proved in §1.9. The steps resulting in a failure are recorded as a *derivation*. Thus a derivation for a sequent is constructed by an attempt to find a counter-example for the sequent, and the existence of a derivation establishes that no such counter-example can exist. *Signed* formulas are introduced to abbreviate assertions about the truth value assigned to a formula by a conjectured valuation Φ. Thus $+F$ is to be understood as an abbreviation for "$\Phi(F)$ is $+$" and $-F$ for "$\Phi(F)$ is $-$", for some conjectured counter-example Φ.

Note that $\Gamma \vdash \Theta$ has no counter-example if $\Gamma' \vdash \Theta'$ has no counter-example, where $\Gamma' \subseteq \Gamma$ and $\Theta' \subseteq \Theta$, and $\Gamma' \cup \Theta'$ is not empty.

The proof theory is presented in a semantic tree formulation that has evolved from [19] but is better known in the form described in [134] or [47]. Another version is presented in [85]. The semantic tree formulation will be seen to be a condensed notational form of Gentzen's sequent calculus formulation. A derivation of a sequent is a finite binary tree with nodes consisting of signed formulas that are related by semantic rules and that satisfy special conditions. The semantic rules are described in §1.7.1 for the single logical connective $\downarrow$, the quantifier $\exists$, and the abstraction operator λ; the semantic rules are suggested by the clauses (2), (3) and (4) of Theorem 6. The special conditions for trees to be derivations are stated in §1.7.2, along with some properties of derivations.

1.7.1. Semantic rules for EL.

$$+\downarrow \quad \frac{+[F\downarrow G]}{-F} \quad \frac{+[F\downarrow G]}{-G} \qquad\qquad -\downarrow \quad \frac{-[F\downarrow G]}{+F \quad +G}$$

$$+\exists \quad \frac{+\exists P}{P(x)} \qquad\qquad -\exists \quad \frac{-\exists P}{-P(t)}$$

$P{:}[\tau]$ and $x, t{:}\tau$; x is new

$$+\lambda \quad \frac{+(\lambda v.T)(t, s_1, \ldots, s_n)}{+[t/v]T(s_1, \ldots, s_n)} \qquad\qquad -\lambda \quad \frac{-(\lambda v.T)(t, s_1, \ldots, s_n)}{-[t/v]T(s_1, \ldots, s_n)}$$

The requirement that x be new for the conclusion of the $+\exists$ rule means that x has no free occurrence in a node above the conclusion of the rule.

The variable x is called the *eigen* variable of an application of $+\exists$, and the term t the *eigen* term of an application of $-\exists$. The terminology *eigen* variable was used in a similar context by Hilbert and Bernays in [79]; the German word 'eigen' means 'own' or 'characteristic' in English.

The last rule has a character different from these *logical* rules. It has no premiss and two conclusions:

$$Cut \quad \frac{}{+F \qquad -F}$$

The Cut rule is *eliminable* in the sense that any derivation of a sequent in which it is used can be replaced with a derivation in which it is not used. A proof of this result follows from Theorem 11 of §1.9. Nevertheless, Cut is a useful rule since it makes possible the reuse of derivations as will be described in §1.7.4.

The name "Cut" of this rule is a translation of the German word "Schnitte" that was the name given to the rule in Gentzen [52]. In Kleene [88], one of the first widely available English descriptions of Gentzen's work in logic, as well as in [140], the name of the rule is translated as "Cut". Later "Cut" was used in the Prolog literature as a name for a tree pruning technique; see for example, Chapter 11 of [138]. These two uses of "Cut" should not be confused.

1.7.2. Derivation of a sequent. A derivation of a sequent is a form of semantic tree based on the sequent. Such trees are defined next:

DEFINITION 22. *Given a sequent* $\Gamma \vdash \Theta$ *a semantic tree* based on *the sequent consists of a tree of signed formulas defined inductively as follows*:

1. *Any tree* $\Im$ *with a single branch consisting of nodes* $+F$ *and* $-G$, *where* $F \in \Gamma$ *and* $G \in \Theta$ *is a tree based on the sequent.*
2. *Let* $\Im$ *be a tree based on the sequent, and let* $\Im'$ *be obtained from* $\Im$ *by adding to the end of a branch of* $\Im$ *either*
   (a) *a signed formula that is the single conclusion of an application of one of the rules* $+\downarrow$, $\pm\exists$, *or* $\pm\lambda$ *with premiss a signed formula on the given branch, provided the eigen variable* x *introduced in the conclusion of an application of* $+\exists$ *does not occur free in any node above the conclusion, or*
   (b) *two signed formulas on separate branches that are the conclusions of Cut, or of an application of* $-\downarrow$ *with premiss a signed formula on the given branch.*

   Then $\Im'$ *is a tree based on the sequent.*

Note that not all the formulas in the antecedent or succedent of $\Gamma \vdash \Theta$ need be signed and added as nodes of a semantic tree based on the sequent. The nodes that are so added are called the *initial* nodes of the semantic tree.

A *closing* pair of nodes on a branch are nodes $+F$ and $-F$. A branch of a semantic tree is *closed* if there is a closing pair of nodes on it; otherwise it is *open*. A semantic tree is *closed* if each of its branches is closed. A *derivation* of a sequent is a closed semantic tree based on the sequent.

EXAMPLE DERIVATION. A derivation follows for the sequent

$$\exists x.[\neg C(x)\downarrow\neg B(x,a)] \vdash \neg[\exists x.C(x)\downarrow\neg\exists x.B(x,a)],$$

where $\neg \stackrel{\text{df}}{=} \lambda X.[X \downarrow X]$ and the following rules for $\neg$ are assumed to be derivable:

$$+\neg \quad \frac{+\neg F}{-F} \qquad\qquad -\neg \quad \frac{-\neg F}{+F}$$

$+\exists x.[\neg C(x) \downarrow \neg B(x,a)]$	(1)	initial node
$-[\neg\exists x.C(x) \downarrow \neg\exists x.B(x,a)]$	(2)	initial node
$+[\neg C(x) \downarrow \neg B(x,a)]$	(3)	$+\exists$(1)
$-\neg C(x)$	(4)	$+\downarrow$(3)
$+C(x)$	(5)	$-\neg$(4)
$-\neg B(x,a)$	(6)	$+\downarrow$(3)
$+B(x,a)$	(7)	$-\neg$(6)
L R		$-\downarrow$(2)
$+\neg\exists x.C(x)$	(8)	
$-\exists x.C(x)$	(9)	$+\neg$(8)
$-C(x)$	(10)	$-\exists$(9)
		(5)&(10)
R		
$+\neg\exists x.B(x,a)$	(11)	
$-\exists x.B(x,a)$	(12)	$+\neg$(11)
$-B(x,a)$	(13)	$-\exists$(12)
		(6)&(13)

The initial nodes are (1) and (2). Each of the other nodes is a conclusion of the rule cited to its right with the cited premiss. The thicker lines at the bottom of the two branches indicate that the branches are closed. The nodes $+C(x)$ and $-C(x)$ are a closing pair for the L branch of the derivation, and the nodes $+B(x,a)$ and $-B(x,a)$ for the R branch.

1.7.3. Terminology for and properties of derivations. The above derivation has been illustrated with its root at the top and with its branches spreading downward, each branch ending in a different *leaf* node. The terminology used in discussing trees reflects this orientation. Thus a node $\eta 1$ is *above* a node $\eta 2$ if they are both on the same branch and $\eta 1$ is closer to the root of the tree than $\eta 2$; it is *below* $\eta 2$ if $\eta 2$ is above it. The *height* of a node on a given branch is the number of nodes below it on the branch; the *height* of a node in a tree is the maximum of its heights on the branches on which it occurs. A leaf node of a tree is a node of height zero.

A subbranch of a branch may be the entire branch, or be a node that is a conclusion of an application of the $-\downarrow$ or Cut rules and all nodes below it on the branch. The conclusion is said to be the *head* of the subbranch. The head of a subbranch is labelled with a finite sequence of the letters L and R, known as the *signature* of the head. The node that is the root of the tree is the head of all the branches of the tree; its signature is the empty sequence. The two conclusions of an application of the $-\downarrow$ or Cut rules are labelled $sigL$ and $sigR$ respectively where sig is the signature of the head of least height above the conclusions. Thus in the example derivation, the empty signature labels the node (1) that is head of each of the two branches, L is the signature of node (3) that is the head of the subbranch consisting of the nodes (3) - (6), and R is the signature of the node (7) that is the head of the subbranch consisting of the nodes (7) - (10). The head of least height on a branch determines the branch. As a consequence of this labelling, a node with signature $sig1$ is above every node with signature $sig1sig2$, where $sig2$ is not empty.

A node $\eta 3$ of a semantic tree is said to be a *descendant* of a node $\eta 1$ if $\eta 3$ is $\eta 1$, or if there is a descendant $\eta 2$ of $\eta 1$ which is the premiss of an application of a rule of deduction with a conclusion $\eta 3$.

The properties of derivations described in the next lemma can be useful when attempting to find a derivation for a given sequent since they can narrow the choices to be considered.

LEMMA 7. *Properties of Derivations.*
If a sequent $\Gamma \vdash \Theta$ has a derivation, then it has a derivation with the following properties:

1. *The same signed formula does not occur twice as a node of the same branch.*
2. *Each branch of the derivation has exactly one closing pair.*
3. *Each variable with a free occurrence in the eigen term t of an application of $-\exists$ has a free occurrence in a node above the conclusion of the rule. With the possible exception of a single constant for each type, each constant occurring in the eigen term t of an application of $-\exists$ has an occurrence in a node above the conclusion of the rule.*
4. *Let the sequent $\Gamma \vdash \Theta$ have a derivation. Then there is a derivation for a sequent $\Gamma' \vdash \Theta'$, where $\Gamma' \subseteq \Gamma$ and $\Theta' \subseteq \Theta$, for which each node of the derivation has a descendant that is a closing node.*

PROOF. Consider property (1). Let $\eta 1$ and $\eta 3$ be distinct nodes of the same branch that are identical signed formulas, and for which $\eta 1$ is above $\eta 3$. Let $\eta 2$ be immediately above $\eta 3$; it is possible that it is the node $\eta 1$. If $\eta 3$ is an initial node then it can be removed. If $\eta 3$ is the single conclusion of a rule then it can be dropped from the branch since its descendants are descendants of $\eta 1$. Similarly if $\eta 3$ and $\eta 4$ are the two conclusions of a rule, then $\eta 3$, $\eta 4$, and all nodes below $\eta 4$ can be removed from the tree and the branch remains closed. In each case $\eta 2$ takes the place of $\eta 3$ in the tree and the tree remains a derivation.

Consider property (2). Let there be a branch of a derivation with more than one closing pair. Assume that the derivation satisfies condition (1). Define the height of a closing pair to be the height of the lowest node of the pair. Consider the closing pair of the branch of greatest height, and consider its lowest node $\eta 2$. All nodes below $\eta 2$ in the tree can be removed from the tree with the branch and tree remaining closed. There cannot remain more than one closing pair on the branch since that would require there to be a signed formula occurring more than once as a node on the branch violating (1).

Consider property (3). Let v be a variable with a free occurrence in the eigen term t that does not have a free occurrence in a node above the conclusion of the application of the rule. Let w be a variable with a free occurrence in a node above the conclusion of the rule; if no such variable exists, let c be a constant of type $\mathrm{t}[v]$ occurring in a node above the conclusion of the rule; if no such constant exists then let c be any constant of type $\mathrm{t}[v]$. Replace each free occurrence of v in t and in any descendant of the conclusion of the rule by w, respectively c. The resulting tree is a derivation since each application of a rule remains an application under the change. Similarly a constant occurring in the eigen term t but not in a node above the conclusion of the application of the rule can be replaced by a constant of the same type that does if such a constant exists.

Consider property (4). Let $\eta 2$ be a node of the derivation of $\Gamma \vdash \Theta$ that has no descendant which is a member of a closing pair. Then $\eta 2$ can be removed from the derivation in the same manner as $\eta 2$ was removed in the proof establishing property (2). After all such nodes have been removed, the initial nodes remaining are those of a sequent $\Gamma' \vdash \Theta'$, for which $\Gamma' \subseteq \Gamma$ and $\Theta' \subseteq \Theta$. ⊣

1.7.4. Reusing derivations. As defined in §1.7.2, a branch of a derivation is closed when both $+F$ and $-F$ appear as nodes on the branch for some F. More general methods of closing a branch are justified here; they amount to the reuse of derivations through applications of the Cut rule.

Let the following sequent have a derivation

$$F_1, \ldots, F_m \vdash G_1, \ldots, G_n \tag{1}$$

where $m, n \geq 1$. A number of derivable rules of deduction can then be justified by applications of the Cut rule. For example, if n is 1 then the following is a derivable rule:

$$\frac{+F_1, \ldots, +F_m}{+G_1}.$$

Thus a branch of a semantic tree can be extended with the node $+G_1$ provided each of $+F_1, \ldots, +F_m$ is a node of the branch. The extended semantic tree is justified by applying Cut with Cut formula G_1 below the leaf node η of the branch:

$$\frac{\begin{matrix}\eta \\ \cdots\end{matrix}}{+G_1 \qquad -G_1}$$

The extended branch with leaf node $-G_1$ can then be closed by attaching below this leaf the nodes of the derivation of (1) below its initial nodes.

EXAMPLE. Derivations can be found for each of the sequents

a) $F, [F \to G] \vdash G$

b) $G, [\neg H \to \neg G] \vdash H$

Consider now the sequent

c) $[F \to G], [\neg H \to \neg G] \vdash [F \to H]$

A derivation can be given as follows

$+[F \to G]$
$+[\neg H \to \neg G]$
$-[F \to H]$
$+F$
$-H$
$+G$ from (a)
$+H$ from (b)

A demonstration that a derivation can be constructed for (c) without using Cut will be left to the reader.

Similarly, if n is 2 then the following is a derivable rule:

$$\frac{+F_1, \ldots, +F_m}{+G_1 \qquad +G_2}.$$

Thus a branch of a semantic tree can be split and extended with the nodes $+G_1$ and $+G_2$ provided each of $+F_1, \ldots, +F_m$ is a node of the branch. The extended semantic tree is justified by applying Cut with Cut formula G_1 below the leaf node η of the branch, and then applying Cut with Cut formula G_2 below the node $-G_1$:

$$\frac{\eta}{+G_1 \qquad \dfrac{-G_1}{+G_2 \qquad -G_2}}$$

Again the extended branch with leaf node $-G_2$ can then be closed by attaching below this leaf the nodes of the derivation of (1) below the initial nodes.

Because of the symmetric form of sequents, similar rules can be derived for formulas of the antecedent. For example, if m is 2 then

$$\frac{-G_1, \ldots, -G_m}{-F_1 \qquad -F_2}$$

can be similarly justified as a derived rule. But clearly many other derived rules can be justified as well.

1.7.5. An alternative proof theory. The proof theory described in §1.7.1 and §1.7.2 makes use of signed formulas. An alternative proof theory that uses only formulas is described here. It is the form of the proof theory presented in the very thorough introduction to logic provided in [85].

The proof theory that has been described uses only the single logical connective $\downarrow$, since all other logical connectives can be defined in terms of it. The new proof theory uses negation and conjunction as the only logical connectives, although disjunction or implication would do just as well in place of conjunction. A node $+F$ in a derivation is replaced with just F, while a node $-F$ is replaced with $\neg F$. The removal of double negation is the only rule needed for negation. The rules are

$$\neg \quad \frac{\neg\neg F}{F}$$

$$\wedge \quad \frac{[F \wedge G]}{F} \qquad \frac{[F \wedge G]}{G} \qquad\qquad \neg\wedge \quad \frac{\neg[F \wedge G]}{\neg F \quad \neg G}$$

$$\exists \quad \frac{\exists x.F}{[y/x]F} \qquad\qquad \neg\exists \quad \frac{\neg\exists x.F}{\neg[t/x]F}$$

$$\lambda \quad \frac{(\lambda x.F)(t)}{[t/x]F} \qquad\qquad \neg\lambda \quad \frac{\neg(\lambda x.F)(t)}{\neg[t/x]F}$$

$$Cut \quad \frac{}{F \qquad \neg F}$$

The $\exists$ rule has the usual restriction on the eigen variable y; it may not occur free in any node above the conclusion of the rule

1.7.6. Derivations as binary trees of sequents. In the introduction to this section it was noted that the semantic tree presentation of EL is closely related to the Gentzen sequent calculus formulation of first order logic described in [1], [88] or [140]. Here a variant of Gentzen's formulation, referred to as ELG [EL*Gentzen*], is described and a sketch is provided of its connection to the semantic tree formulation of EL in §1.7.1.

In Chapter 6 a sequent calculus formulation ITTG [ITT*Gentzen*] of the semantic tree formulation of the logic ITT described in Chapter 3 is provided.

Apart from the different primitive logical symbols used by Gentzen, the sequent calculus ELG presented here differs from Gentzen's in the absence of any structural rules used to change the order of the listing of formulas in Γ and Θ; only *thinning* rules are needed that can add a formula to either the antecedent or succedent of a sequent.

Corresponding to a closing pair of nodes in the semantic tree formulation is an infinite collection of axioms in the sequent calculus formulation:

Axioms $F \vdash F$

for each formula F. The rules of deduction are:

$$\frac{\Gamma \vdash \Theta, F}{\Gamma, [F\downarrow G] \vdash \Theta} \qquad \frac{\Gamma \vdash \Theta, G}{\Gamma, [F\downarrow G], \vdash \Theta} \qquad \frac{\Gamma, F \vdash \Theta 1 \qquad \Gamma, G \vdash \Theta 2}{\Gamma \vdash \Theta, [F\downarrow G]}$$

$$\frac{\Gamma, P(x) \vdash \Theta}{\Gamma, \exists P \vdash \Theta} \qquad \frac{\Gamma \vdash \Theta, P(t)}{\Gamma \vdash \Theta, \exists P}$$

$P{:}[\tau]$ and $x, t{:}\tau$; x is new

$$\frac{\Gamma, [t/x]T(\overline{s}) \vdash \Theta}{\Gamma, (\lambda v.T)(t, \overline{s}) \vdash \Theta} \qquad \frac{\Gamma \vdash \Theta, [t/x]T(\overline{s})}{\Gamma \vdash \Theta, \lambda v.T)(t, \overline{s})}$$

Here $\overline{s}$ abbreviates $s_1, \ldots, s_n$, where $0 \leq n$.

The two thinning rules are:

$$\frac{\Gamma \vdash \Theta}{\Gamma, F \vdash \Theta} \qquad \frac{\Gamma \vdash \Theta}{\Gamma \vdash \Theta, F}$$

The Cut rule has the following form:

$$\frac{\Gamma \vdash \Theta, F \qquad \Gamma, F \vdash \Theta}{\Gamma \vdash \Theta}$$

The rules are expressed using the following notation: 'Γ, F' and 'Θ, G' are understood to be '$\Gamma \cup \{F\}$' and '$\Theta \cup \{G\}$'.

The rules are referred to as follows. The first two rules for $\downarrow$ are the $\downarrow\vdash$ rules and the third is the $\vdash\downarrow$ rule; the two rules for $\exists$ are $\exists\vdash$ and $\vdash\exists$; the two rules for λ are $\lambda\vdash$ and $\vdash\lambda$; the two thinning rules are *Thin* $\vdash$ and $\vdash$ *Thin*; and the Cut rule is *Cut*.

THEOREM 8. *Equivalence of EL and ELG.*
A sequent is derivable in EL if and only if it is derivable in ELG.

PROOF. It will be proved first that a sequent derivable in EL is derivable in ELG.

Consider an EL derivation for a sequent $\Gamma \vdash \Theta$. By the *size* of the derivation is meant the number of applications of rules made in the derivation. The conclusion will be proved by induction on the size of the EL derivation of the sequent.

Should the size of the derivation be 0, then some formula is common to Γ and Θ. Therefore the sequent is derivable from an axiom of ELG by zero or more applications of the thinning rules.

Assume that a derivation of a sequent in EL of size k, where $k < n$, can be transformed into a derivation of the same sequent in ELG. Consider now a derivation of a sequent in EL of size n. Necessarily the derivation has at least one application of a semantic rule. Consider the possibilities for the rule first applied; that is for the rule with conclusion, or conclusions, immediately below the initial nodes. The argument will be illustrated for the one case $-\downarrow$; the remaining cases are left for the reader.

The derivation of the sequent $\Gamma \vdash \Theta$ in EL is assumed to take the form on the left, where s_i is $+$ if $F_i \in \Gamma$ and is $-$ if $F_i \in \Theta$, and $-[F \downarrow G]$ is some $s_i F_i$. The derivation on the left is transformed into the two derivations on the right each with size smaller than n:

$$\begin{array}{ll} s_1 F_1 & \\ \ldots & \\ -[F \downarrow G] & \\ \ldots & \\ s_n F_n & \\ \hline +F & +G \\ \ldots & \ldots \end{array} \qquad\qquad \begin{array}{l} s_1 F_1 \\ \ldots \\ s_n F_n \\ +F \\ \ldots \end{array} \qquad\qquad \begin{array}{l} s_1 F_1 \\ \ldots \\ s_n F_n \\ +G \\ \ldots \end{array}$$

From the induction assumption, there exist derivations in ELG for the two sequents

$$\Gamma, F \vdash \Theta \text{ and } \Gamma, G \vdash \Theta$$

and therefore, by one application of the $\vdash\downarrow$ rule, a derivation for the sequent

$$\Gamma \vdash \Theta, [F \downarrow G]$$

as required.

Consider now an ELG derivation of $\Gamma \vdash \Theta$. By the *size* of the derivation is meant the number of applications of rules of deduction of ELG made in the derivation. It will be proved by induction on the size of the derivation of $\Gamma \vdash \Theta$ that the ELG derivation can be transformed into an EL derivation.

Should the size be 0 then the derivation consists of an axiom of ELG so that $\Gamma \vdash \Theta$ is $F \vdash F$ and the initial branch of a semantic tree based on $\Gamma \vdash \Theta$ is closed.

Assume now that each ELG derivation with fewer than n applications of rules of ELG can be transformed into an EL derivation of the same sequent. Let $\Gamma \vdash \Theta$ have an ELG derivation with n applications. Consider the possibilities for the last rule applied. Again only one case will be demonstrated, that in which the last rule applied is $\vdash\downarrow$; the remaining cases will be left to the reader.

$$\dfrac{\begin{array}{cc} \cdots & \cdots \\ \Gamma, F \vdash \Theta & \Gamma, G \vdash \Theta \end{array}}{\Gamma \vdash \Theta, [F \downarrow G]} \qquad \begin{array}{cc} +F & +G \\ s_1 F_1 & s_1' G_1 \\ \cdots & \cdots \\ s_m F_m & s_n' G_n \\ \cdots & \cdots \end{array} \qquad \begin{array}{cc} \multicolumn{2}{c}{-[F \downarrow G]} \\ \hline +F & +G \\ s_1 F_1 & s_1' G_1 \\ \cdots & \cdots \\ s_m F_m & s_n' G_n \\ \cdots & \cdots \end{array}$$

The premises of the application of $\vdash\downarrow$ in the ELG derivation on the left have derivations of size less than n. By the induction assumption there are EL derivations for them; these derivations have their initial formulas listed to the right of the ELG derivation. The EL derivation on the right is therefore a derivation of the conclusion of the ELG derivation. ⊣

As can be seen from the proof of the lemma, a derivation in the formulation EL of elementary logic can be seen to be an abbreviation of a derivation in the formulation ELG that loses none of the essential structure of the derivation.

1.7.7. Other formulations of first order logic. The intuitionist version of first order logic was first formulated by Heyting in [77]. Gentzen proved that a sequent calculus formulation of Heyting's intuitionist logic can be obtained from the sequent calculus formulation for classical logic by restricting the succedent of sequents in the intuitionist formulation to at most one formula. In Chapter 6 this same restriction on sequents is applied to the sequent calculus formulation ITTG of ITT to define the logic HITTG [*Heyting*ITTG]. A semantic tree version HITT [*Heyting*ITT] of HITTG is described there.

Gentzen also described in [52] what are called natural deduction formulations of both the classical and intutionistic logics; these formulations are intended to more closely mimic the form of reasoning used in proofs in mathematics. A widely used reference for natural deduction is [119] and a more recent and accessible one is [1].

Exercises §1.7.

1. The rules for the connective $\downarrow$, the quantifier $\exists$, and the abstraction operator λ are suggested by clauses (2) - (4) of Theorem 6 of §1.6. So also are derivable rules for the traditional connectives suggested by the exercise of §1.4 and the universal quantifier $\forall$ by its usual definition in terms of $\neg$ and $\exists$. State these rules and prove that they are derivable from the rules for $\downarrow$ and $\exists$ given in §1.7.1.
2. Provide derivations for the following sequents:

$$\vdash [\forall y.[[\exists x{:}H].F \rightarrow G] \leftrightarrow [\forall x{:}H].\forall y.[F \rightarrow G]],$$

when x has no free occurrence in G and y no free occurrence in H, and

$$\vdash [\forall y.[F \rightarrow [\forall x{:}H].G] \leftrightarrow [\forall x{:}H].\forall y.[F \rightarrow G]],$$

when x has no free occurrence in F and y no free occurrence in H. Here

$$[\exists x{:}H].F \stackrel{\mathrm{df}}{=} \exists x.[H \wedge F] \text{ and } [\forall x{:}H].G \stackrel{\mathrm{df}}{=} \forall x.[H \rightarrow G].$$

3. Prove that a sequent is derivable using signed formulas as nodes in a derivation if and only if it is derivable using only formulas as nodes.
4. Complete the omitted cases of the proof of Theorem 8.

1.8. Consistency of EL

The logic EL is the logic with the rules of deduction of §1.7.1. It is convenient to use the unsugared form of the two $\exists$ rules with premiss $\pm\exists P$ and conclusion respectively $+P(y)$ and $-P(t)$.

The logic is inconsistent if for some formula F both the sequents $\vdash F$ and $F \vdash$ are derivable. Such a logic would be of little value since if $F \vdash$ is derivable then so is $F \vdash G$ for any formula G, and therefore by the Cut rule $\vdash G$ would be derivable for any formula G.

There are several ways in which the logic can be proved consistent. The method used in [52] was to prove that the Cut rule is eliminable in the following sense: If a sequent has a derivation, then it has a derivation in which Cut is not used. Consistency follows from this result because if both $\vdash F$ and $F \vdash$ are derivable then the empty sequent $\vdash$ is also derivable. But that is obviously not possible using the rules of §1.7.1 other than Cut.

The proof of the eliminability of Cut provided in [52] was a purely syntactic one not involving the semantics of the logic; other such proofs are also available in [88] and [47]; as stressed in the latter the eliminability of Cut is important for theorem proving in first order logic. However, no purely syntactic proof of the eliminability of Cut is known for the logic ITT to be described in Chapter 3. A semantic proof is provided but it requires a prior proof of consistency. To prepare the reader for this proof, a semantic proof of the consistency of EL is provided next that is similar to the proof provided for ITT.

THEOREM 9. *Consistency.*
A derivable sequent of first order logic is valid.

PROOF. Consider a derivation for a sequent $\Gamma \vdash \Theta$. Let η be any node of the derivation which does not have an initial node below it. The sets $\Gamma[\eta]$ and $\Theta[\eta]$ are defined as follows: $\Gamma[\eta]$ is the set of all nodes $+F$ and $\Theta[\eta]$ the set of all nodes $-F$ that are either η or above it. $\Gamma[\eta] \vdash \Theta[\eta]$ will be shown to be valid by induction on $h(\eta)$, the height of η as defined in §1.7.3.

If $h(\eta)$ is 0, then η is a leaf node of a branch of the derivation. Since the branch is closed, $\Gamma[\eta] \vdash \Theta[\eta]$ is valid. Assume therefore that $h(\eta) > 0$, and that there is a model Φ that is a counter-example for $\Gamma[\eta] \vdash \Theta[\eta]$. Necessarily η is immediately above a conclusion $\eta 1$ or conclusions $\eta 1$ and $\eta 2$ of one of the rules of deduction. There are therefore two main cases to consider corresponding to the single conclusion rules $+\downarrow$ and $\pm\exists$ and to the two conclusion rules $-\downarrow$ and Cut.

• For the single conclusion rules it is sufficient to illustrate the argument with the $\pm\exists$ rules with premiss $\pm\exists P$ and conclusion respectively $+P(y)$ and $-P(t)$, where P:[1] and t, y:1 with y not occurring free in any node above the conclusion of the $+\exists$ rule.

For the $+$ case, $\exists P \in \Gamma[\eta]$, $\Gamma[\eta 1]$ is $\Gamma[\eta] \cup \{P(y)\}$, and $\Theta[\eta 1]$ is $\Theta[\eta]$. Since Φ does not satisfy $\Gamma[\eta] \vdash \Theta[\eta]$ it follows that

1. $\Phi(\exists P)$ is $+$.
2. $F \in \Gamma[\eta 1] \Rightarrow \Phi(F)$ is $+$.
3. $F \in \Theta[\eta 1] \Rightarrow \Phi(F)$ is $-$.

By (1) and clause (3) of Theorem 6 of §1.6, it follow that there is a term t:1 for which $\Phi(P(t))$ is $+$. Define Φ^y to be the y-variant of Φ for which $\Phi^y(y)$ is t. Since $\Gamma[\eta 1] \vdash \Theta[\eta 1]$ is valid Φ^y must satisfy it. By Lemma 5 of §1.5.4, semantic substitution, (2) and (3) hold when Φ is Φ^y since y does not occur free in any F. Thus for Φ^y to satisfy $\Gamma[\eta 1] \vdash \Theta[\eta 1]$ it is necessary that $\Phi^y(P(y)$ be $-$. By Theorem 2 of §1.4.1, the fundamental theorem for application, it follows that $\Phi^y(y) \notin \Phi^y(P)$ and therefore by Lemma 5 that $\Phi^y(y) \notin \Phi(P)$ so that $t \notin \Phi(P)$ and therefore $\Phi(P(t))$ is $-$ contradicting $\Phi(P(t))$ is $+$.

For the $-$ case, $\Gamma[\eta 1]$ is $\Gamma[\eta]$, $\exists P \in \Theta[\eta]$, and $\Theta[\eta 1]$ is $\Theta[\eta] \cup \{P(t)\}$. Since Φ satisfies $\Gamma[\eta 1] \vdash \Theta[\eta 1]$ but does not satisfy $\Gamma[\eta] \vdash \Theta[\eta]$ it follows that $\exists P$ is $-$ and $\Phi(P(t))$ is $+$ contradicting clause (3) of Theorem 6.

• Let the premiss of an application of $-\downarrow$ be $-[F\downarrow G]$ and the conclusions $+F$ and $+G$. Thus $[F\downarrow G] \in \Theta[\eta]$, $\Gamma[\eta 1]$ is $\Gamma[\eta] \cup \{F\}$, $\Gamma[\eta 2]$ is $\Gamma[\eta] \cup \{G\}$, $\Theta[\eta 1]$ is $\Theta[\eta]$ and $\Theta[\eta 2]$ is $\Theta[\eta]$. As before it follows that $\Phi(F)$ is $\Phi(G)$ is $-$ so that $\Phi([F\downarrow G])$ is $+$. This contradicts clause (2) of the theorem.

For the case of Cut, let the Cut formula be F. In this case $\Gamma[\eta 1]$ is $\Gamma[\eta] \cup \{F\}$, $\Theta[\eta 1]$ is $\Theta[\eta]$, $\Gamma[\eta 2]$ is $\Gamma[\eta]$, and $\Theta[\eta 2]$ is $\Theta[\eta 2] \cup \{F\}$. It follows therefore that $\Phi(F)$ is both $+$ and $-$, which is impossible. $\dashv$

1.9. Completeness of EL

It was claimed in §1.7 that the proof theory provides an implicit systematic search procedure for a counter-example for a given sequent. Further it was claimed that should the procedure fail to find such a counter-example, then the sequent has a derivation. These claims will be justified here. They will be justified in a manner that should prepare the reader for the more difficult proof of the completeness and Cut-elimination theorem of ITT.

The meaning of *systematic attempt* is clarified by the description in Figure 1 in §1.9.1 of a procedure that given a sequent will find a *Cut-free* derivation for it, if one exists, and will otherwise define a *descending chain* for the sequent. Here a Cut-free derivation means one in which the Cut rule is not used; and a descending chain for a sequent is defined:

DEFINITION 23. *A Descending Chain for a Sequent.*
Let $\Gamma \vdash \Theta$ be a given sequent and let k be the number of formulas in $\Gamma \cup \Theta$. A descending chain for the sequent is a possibly infinite sequence of signed formulas $sF_1, \cdots, sF_n, \cdots$ without repetitions satisfying the following conditions:

1. *For each n, $k \leq n$, $sF_1, \cdots, sF_n$ is an open branch of a tree based on the sequent.*
2. *Let $\pm F_m$, $1 \leq m$ have the form of a premiss for some rule.*
   (a) *Let F_m be $[F \downarrow G]$. Then each of $-F$ and $-G$, respectively one of $+F$ or $+G$, follows $\pm[F \downarrow G]$ in the sequence.*
   (b) *Let F_m be $\exists P$. Then $+P(y)$, respectively each $-P(t)$, follows $\pm\exists P$ in the sequence, where y does not occur in a node of the sequence preceding $+P(y)$.*
   (c) *Let F_m be $(\lambda v.T)(t, s_1, \ldots, s_n)$. Then $\pm[t/x]T(s_1, \ldots, s_n)$ respectively follows $\pm(\lambda v.T)(t, s_1, \ldots, s_n)$ in the sequence.*

A descending chain is necessarily *consistent* in the sense that not both $+F$ and $-F$ are members for some F. It is also necessarily *saturated* in the sense that each possible application of a logical rule satisfying the restrictions of clauses (2a) to (2c) has a conclusion in the sequence.

A proof of the following theorem in §1.9.1 confirms that the procedure does what it is expected to do:

THEOREM 10.

1. *Let the procedure halt. Then the input sequent has either a Cut-free derivation or a finite descending chain.*
2. *Let the procedure not halt. Then an infinite descending chain for the sequent exists.*

One last step is needed to justify the claim that the proof theory provides an implicit systematic search procedure for a counter-example for a given sequent; namely that a sequent with a descending chain has a counter-example. This step is provided in §1.9.2 completing a proof for the following theorem, the ultimate goal of this section:

THEOREM 11. *Completeness of EL without Cut.*
A counter-example that is a model exists for each sequent that has no Cut-free derivation in EL.

An immediate corollary is:

COROLLARY 12. *If a sequent has a derivation in EL, then it has a Cut-free derivation.*

For consider a sequent with a derivation in which Cut is used. If the sequent has no Cut-free derivation then, by the completeness theorem, it has a counter-example. But that is impossible since by the consistency theorem the sequent is satisfied by every model.

1.9.1. A systematic search for a derivation. The definition of a descending chain is intended to describe the properties of a branch of a semantic tree that could be constructed in a failed systematic search for a Cut-free derivation for a given sequent. The description of a simplified *flow-chart* given in Figure 1 is meant to clarify what is meant by a "systematic search". The description makes use of the terminology introduced in §1.7.3, together with a method of naming subbranches of a tree first used in the example derivation of §1.7.2; each subbranch is labelled with a finite sequence of L's and R's called the subbranch's *signature*.

The tree to which the procedure is first applied consists of a single branch with nodes $\pm F$ for members F of the antecedent, respectively of the succedent. The root of this tree is labelled with the empty signature, nilsig, consisting of the empty sequence of L's and R's. Thus the first *current* branch to be processed is identified by the signature nilsig. If during the processing of the current branch identified by the signature sig an application of the $-\downarrow$ rule is made, the two subbranches resulting are identified by the signatures 'sigL' and 'sigR'; thus in particular if the current branch is identified by the signature nilsig, that is it is the initial tree, the two subbranches resulting are identified by the signatures 'L' and 'R'.

Since at least one of the sets Γ and Θ of formulas from a sequent $\Gamma \vdash \Theta$ is not empty, there is always a branch to choose as the first current branch to which the procedure can be applied. During the processing of the current branch, questions must be answered and decisions made based on the answers. The three most important of these questions are stated with their answers in the simplified flow-chart description of the procedure given in Figure 1. These questions are:

***A*.:** Is the current branch open?
***B*.:** Is there a node on the current branch that may serve as a premiss for a rule to extend the branch with its conclusion?
***C*.:** Is there an open branch to be made current?

Answering question ***A*** simply requires searching the current branch for a closing pair of nodes. Answering question ***B*** requires much more. By Lemma 7 of §1.7.3 the same signed formula need never occur twice on the same branch of a derivation, and the possible conclusions of an application of the $-\exists$ rule are determined by the occurrences of free variables and constants in nodes of the branch. This means that a node, other than one of the form $-\exists P$, need be used only once as a premiss, but that a node $-\exists P$ may be used any number of times but never with the same conclusion on the same branch. Thus

A. Is the current branch open?

Yes: ***B***. Is there a node on the current branch that may serve as a premiss for a rule to extend the branch with its conclusions?

Yes: Choose a node to ensure no repetitions of nodes and no omission of a possible premiss. Extend the branch with the conclusions and go to ***A***.

No: ***C***. Is there an open branch to be made current?

Yes: Choose the left-depth-first branch and make it current. Go to ***B***.

No: The process halts.

No: Go to ***C***.

FIGURE 1. Tree Search Procedure

information must be kept about nodes other than $-\exists P$ nodes on the current branch as to whether they have been used as premiss, and about the $-\exists P$ nodes as to whether a possible conclusion $-P(t)$ has already been drawn.

Answering question ***C*** requires determining whether a closing pair of nodes has been found on each of the possible branches of the current tree. Answering the question can be simplified if a specific order of search is followed for an open branch. For example a *left-depth-first* search proceeds as follows: An 'L' subbranch of a branch is extended until a closing pair of nodes is discovered, before any 'R' branch of the branch is extended. The result is that is the answer to question ***C*** is 'Yes', then the signature for the next current branch is known and the procedure can begin again with question ***A***.

Theorem 10 stated in the introduction to §1.9 summarizes what is expected of the procedure:

1. Let the procedure halt. Then the sequent either has a Cut-free derivation or it has a finite descending chain.
2. Let the procedure not halt. Then the sequent has an infinite descending chain.

PROOF OF THEOREM 10. If the answer to question ***C*** is 'No', then all possible branches of the tree have been explored. There are only two possibilities: It may be that all branches are closed, in which case the tree is a derivation; or it may be that there is an open branch and all possible rules with premiss on the branch have been used, in which case the open branch is a finite descending chain. If the process never halts then there is a branch on which there is no closing pair of nodes. Further eventually if the choices of premisses made in the 'Yes' of question ***C*** have been made properly an infinite sequence of signed formulas will be defined; although this will not be an infinite descending chain such a chain can be constructed by adding the conclusions of the $-\exists$ omitted because of clause (3) of Lemma 7 in §1.7.3. ⊣

The procedure could be the basis for a computer program for theorem proving, either fully automated or semi-automated. The most difficult and most important detail needing further clarification is the choice of a premiss and conclusion, particularly if a node $-\exists P$ appears on the current branch since then there are infinitely many conclusions possible. In a semi-automated version the user of the program is expected to assist in this choice. The program provides "bookkeeping" assistance as to unclosed branches and rules that may still be applied to extend them, but the truly creative decisions as to what rules and conclusions to use, in particular what eigen terms to use with the $-\exists$ rules, is left to the user of the program. In a fully automated program, such decisions are made by the program itself. There is an extensive literature on theorem proving; [47] provides an introduction to some of the latest techniques; [113] provides a description of a semi-automated theorem prover for a higher order logic.

1.9.2. A descending chain defines a counter-example. The last step in the proof of the completeness theorem is to construct a counter-example for a given sequent $\Gamma \vdash \Theta$ that has no Cut-free derivation. By Corollary 12 to Theorem 10, stated in §1.9 and proved in §1.9.1, such a sequent has a descending chain; its membership satisfies the conditions enumerated in the next lemma:

LEMMA 13. *Let Σ be the set of members of a descending chain for a sequent $\Gamma \vdash \Theta$. Then Σ satisfies the following conditions*:

1. $F \in \Gamma \Rightarrow +F \in \Sigma$;
   $F \in \Theta \Rightarrow -F \in \Sigma$.

2. *For no F is both $+F$ and $-F \in \Sigma$.*

3. $+[F{\downarrow}G] \in \Sigma \Rightarrow -F \in \Sigma$ and $-G \in \Sigma$;
   $-[F{\downarrow}G] \in \Sigma \Rightarrow +F \in \Sigma$ or $+G \in \Sigma$.

4. $+\exists P \in \Sigma \Rightarrow +P(t) \in \Sigma$, *for some* t:1;
   $-\exists P \in \Sigma \Rightarrow -P(t) \in \Sigma$, *for all* t:1.

5. *Let F be $([t/v]P)(t_1, \ldots, t_n)$ and let G be $(\lambda v.P)(t, t_1, \ldots, t_n)$. Then* $\pm G \in \Sigma \Rightarrow \pm F \in \Sigma$, *respectively.*

A proof of the lemma is left as an exercise. Sets of signed formulas satisfying (1) - (4) are sometimes called Hintikka sets since similar sets were first defined in [84].

The first step in constructing a counter-example for $\Gamma \vdash \Theta$ is to form the *closure* $\Sigma^\dagger$ of Σ, a set of signed formulas without free variables obtained from Σ by replacing each signed formula $\pm F$ in Σ by $\pm F^\dagger$, where the mapping $\dagger$ is defined:

DEFINITION 24. *The Closure of a Term and of a Descending Chain.*
Let $\dagger$ be the following one-to-one mapping of the constants $c_1, \ldots, c_i, \ldots$ and variables $v_1, \ldots, v_i, \ldots$ of type 1 *onto the constants of that type*:

$$c_i^\dagger \text{ is } c_{(2i-1)} \text{ and } v_i^\dagger \text{ is } c_{(2i)}.$$

The closure $T^\dagger$ of T is obtained by replacing each constant c_i by $c_i^\dagger$ and each free occurrence of a variable v_i by $c_i^\dagger$.
The closure $\Sigma^\dagger$ of Σ is obtained by replacing each member $\pm F$ of Σ by $\pm F^\dagger$.

Clearly the closure $\Sigma^\dagger$ of a descending chain Σ satisfies the appropriate closures of the conditions of Lemma 13.

The *possible values* function $\mathcal{E}(T)$ is defined next for each term T simultaneously with the domains $D\Sigma(\tau)$. The definition is by induction on the Definition 11 of types τ in §1.5.1. As its name suggests, $\mathcal{E}(T)$ is used to restrict a possible value $\Phi(T)$ of a term for a valuation Φ to membership in $\mathcal{E}(T)$.

DEFINITION 25. *The Domain Function $D\Sigma$ and Possible Values Function $\mathcal{E}$.*

1. *Let T:1. Then $\mathcal{E}(T)$ is $\{T^\dagger\}$.*
2. *Let T:[ ]. Then $\mathcal{E}(T)$ is the smallest subset of $\{+, -\}$ for which*
   $+T^\dagger \in \Sigma^\dagger \Rightarrow + \in \mathcal{E}(T)$.
   $-T^\dagger \in \Sigma^\dagger \Rightarrow - \in \mathcal{E}(T)$.
   $+T^\dagger \notin \Sigma^\dagger$, *and* $-\ T^\dagger \notin \Sigma^\dagger \Rightarrow +, - \in \mathcal{E}(T)$.

3.(a) *Let T:$[1^n]$, $1 \le n$. Then $\mathcal{E}(T)$ is the set of sets $\mathsf{S} \subseteq (D\Sigma(1) \times \cdots \times D\Sigma(1))$ for which for all $Q_1, \ldots, Q_n$:1*
   $+T^\dagger(Q_1^\dagger, \ldots, Q_n^\dagger) \in \Sigma^\dagger \Rightarrow \langle e_1, \ldots, e_n\rangle \in \mathsf{S}$,
   $-T^\dagger(Q_1^\dagger, \ldots, Q_n^\dagger) \in \Sigma^\dagger \Rightarrow \langle e_1, \ldots, e_n\rangle \notin \mathsf{S}$
   whenever $e_i \in \mathcal{E}(Q_i)$, $1 \le i \le n$.

3.(b) *Let T:[[ ]] Then $\mathcal{E}(T)$ is the set of sets $\mathsf{S} \subseteq D\Sigma([\,])$ for which for all Q:[ ]*
   $+T^\dagger(Q^\dagger) \in \Sigma^\dagger \Rightarrow e \in \mathsf{S}$,
   $-T^\dagger(Q^\dagger) \in \Sigma^\dagger \Rightarrow e \notin \mathsf{S}$,
   whenever $e \in \mathcal{E}(Q)$.

3.(c) *Let T:[[ ], [ ]]. Then $\mathcal{E}(T)$ is the set of sets $\mathsf{S} \subseteq (D\Sigma([\,]) \times D\Sigma([\,]))$ for which for all Q_1, Q_2:[ ]*
   $+T^\dagger(Q_1^\dagger, Q_2^\dagger) \in \Sigma^\dagger \Rightarrow \langle e_1, e_2\rangle \in \mathsf{S}$,
   $-T^\dagger(Q_1^\dagger, Q_2^\dagger) \in \Sigma^\dagger \Rightarrow \langle e_1, e_2\rangle \notin \mathsf{S}$,
   whenever $e_1 \in \mathcal{E}(Q_1)$ and $e_2 \in \mathcal{E}(Q_2)$.

3.(d) *Let* T:[[1]]. *Then* $\mathcal{E}(T)$ *is the set of sets* $\mathsf{S} \subseteq D\Sigma([1])$ *for which for all* Q:[1]:

$$+T^{\dagger}(Q^{\dagger}) \in \Sigma^{\dagger} \Rightarrow e \in \mathsf{S},$$
$$-T^{\dagger}(Q^{\dagger}) \in \Sigma^{\dagger} \Rightarrow e \notin \mathsf{S},$$

whenever $e \in \mathcal{E}(Q)$.

4. $D\Sigma(\tau)$ *is* $\cup\{\mathcal{E}(T) \| T{:}\tau\}$ *for each* τ.

LEMMA 14. *$\mathcal{E}(T)$ not empty.*
For no type τ or term $T{:}\tau$ is $\mathcal{E}(T)$ empty.

PROOF. Certainly $\mathcal{E}(T)$ is not empty if T:1 or T:[ ]. The remaining cases can be proved by induction on the Definition 11 of type in §1.5.1. ⊣

LEMMA 15. *A model $\Phi\Sigma$ with domain $D\Sigma$.*
Let $\Phi\Sigma$ be a function for which for each type τ and term $T{:}\tau$, $\Phi\Sigma(T) \in \mathcal{E}(T)$. Then

a) $D\Sigma$ *satisfies Definition* 15 *in* §1.5.2 *of* domain.
b) $\Phi\Sigma$ *satisfies Definition* 17 *in* §1.5.2 *of* valuation *to the domain* $D\Sigma$.
c) $\Phi\Sigma$ *satisfies Definition* 10 *in* §1.4.1 *of* model *with domain* $D\Sigma$.

PROOF. Consider the assertions of the lemma in turn.

a) That $D\Sigma(\tau)$ is not empty for any τ follows immediately from clause ***4*** of Definition 25 and Lemma 14.

b) The clauses of Definition 17 in §1.5.2 updated for $\Phi\Sigma$ and $D\Sigma$ are:

V.1.1: $\Phi\Sigma(cv) \in D\Sigma(\mathrm{t}[cv])$, for each constant or variable cv other than $\downarrow$ or $\exists$.

This follows immediately from the definitions of $\Phi\Sigma$ and $D\Sigma$.

V.1.2: • $\Phi\Sigma(\downarrow)$ is $\{\langle -, -\rangle\}$.
Let $\Phi\Sigma(\downarrow)$ be $\mathsf{S} \in \mathcal{E}(\downarrow)$ and let Q_1, Q_2:[ ]. Then

$$+\downarrow(Q_1^{\dagger}, Q_2^{\dagger}) \in \Sigma^{\dagger} \Rightarrow \langle e_1, e_2\rangle \in \mathsf{S}$$
$$-\downarrow(Q_1^{\dagger}, Q_2^{\dagger}) \in \Sigma^{\dagger} \Rightarrow \langle e_1, e_2\rangle \notin \mathsf{S}$$

whenever $e_1 \in \mathcal{E}(Q_1)$ and $e_2 \in \mathcal{E}(Q_2)$. But by clause (3) of Lemma 13 and clause ***2*** of Definition 25:

$$+\downarrow(Q_1^{\dagger}, Q_2^{\dagger}) \in \Sigma^{\dagger} \Rightarrow -Q_1^{\dagger}, -Q_2^{\dagger} \in \Sigma^{\dagger} \Rightarrow - \in \mathcal{E}(Q_1) \text{ and } - \in \mathcal{E}(Q_2),$$
$$-\downarrow(Q_1^{\dagger}, Q_2^{\dagger}) \in \Sigma^{\dagger} \Rightarrow +Q_1^{\dagger} \text{ or } +Q_2^{\dagger} \in \Sigma^{\dagger} \Rightarrow + \in \mathcal{E}(Q_1) \text{ or } + \in \mathcal{E}(Q_2).$$

Hence S is $\{\langle -, -\rangle\}$.
• $\Phi\Sigma(\exists)$ is a non-empty set of non-empty subsets of $D\Sigma([1])$.
Let $\Phi\Sigma(\exists)$ be $\mathsf{S} \in \mathcal{E}(\exists)$, and let $\oslash$ be the empty subset of $D\Sigma([1])$.

Then for all terms P:[1]

$$+\exists^{\dagger}(P^{\dagger}) \in \Sigma^{\dagger} \Rightarrow e \in \mathsf{S},$$
$$-\exists^{\dagger}(P^{\dagger}) \in \Sigma^{\dagger} \Rightarrow e \notin \mathsf{S},$$

whenever $e \in \mathcal{E}(P)$. But by clause (4) of Lemma 13 and clause ***3***.(***a***) of Definition 25

$$+\exists^{\dagger}(P^{\dagger}) \in \Sigma^{\dagger} \Rightarrow +P^{\dagger}(Q^{\dagger}) \in \Sigma^{\dagger} \text{ for some } Q{:}1 \Rightarrow \oslash \notin \mathcal{E}(P).$$
$$-\exists^{\dagger}(P^{\dagger}) \in \Sigma^{\dagger} \Rightarrow -P^{\dagger}(Q^{\dagger}) \in \Sigma^{\dagger} \text{ for all } Q{:}1 \Rightarrow \oslash \in \mathcal{E}(P).$$

Hence $\oslash \notin \mathsf{S}$.

V.2: Let $\Phi\Sigma(P)$ and $\Phi\Sigma(Q)$ be defined, P:$[\tau^{n+1}]$ and $Q{:}\tau$, where τ is $1 \Rightarrow 0 \le n$; τ is $[\,] \Rightarrow 0 \le n \le 1$; and τ is $[1] \Rightarrow n = 0$. Then when $n \ge 1$, $\Phi\Sigma((PQ))$ is the set of n-tuples $\langle e_1, \ldots, e_n\rangle$ for which

$$\langle \Phi(Q), e_1, \ldots, e_n\rangle \in \Phi(P);$$

when $n = 0$, $\Phi\Sigma((PQ))$ is $+$ if $\Phi\Sigma(Q) \in \Phi(P)$ and is $-$ otherwise.

Only the case τ is 1 is proved; the remaining two cases are left to the reader. The set $\Phi\Sigma(P)$ satisfies the following conditions for all $Q_0, Q_1, \ldots, Q_n$:1:

$$+P^{\dagger}(Q_0^{\dagger}, Q_1^{\dagger}, \ldots, Q_n^{\dagger}) \in \Sigma^{\dagger} \Rightarrow \langle e_0, e_1, \ldots, e_n\rangle \in \Phi\Sigma(P),$$
$$-P^{\dagger}(Q_0^{\dagger}, Q_1^{\dagger}, \ldots, Q_n^{\dagger}) \in \Sigma^{\dagger} \Rightarrow \langle e_0, e_1, \ldots, e_n\rangle \notin \Phi\Sigma(P),$$

whenever $e_i \in \mathcal{E}(e_i)$, $0 \le i \le n$. In particular, therefore for all $Q_1, \ldots, Q_n$:1:

$$+P^{\dagger}(Q^{\dagger}, Q_1^{\dagger}, \ldots, Q_n^{\dagger}) \in \Sigma^{\dagger} \Rightarrow \langle Q^{\dagger}, e_1, \ldots, e_n\rangle \in \Phi\Sigma(P),$$
$$-P^{\dagger}(Q^{\dagger}, Q_1^{\dagger}, \ldots, Q_n^{\dagger}) \in \Sigma^{\dagger} \Rightarrow \langle Q^{\dagger}, e_1, \ldots, e_n\rangle \notin \Phi\Sigma(P),$$

whenever $e_i \in \mathcal{E}(e_i), 1 \le i \le n$. Thus (V.2) follows in this case no matter the value of n since $\Phi\Sigma(Q)$ is $Q^{\dagger}$ and $P^{\dagger}(Q^{\dagger}, Q_1^{\dagger}, \ldots, Q_n^{\dagger})$ is $(P^{\dagger}Q^{\dagger})(Q_1^{\dagger}, \ldots, Q_n^{\dagger})$.

V.3: Let $\Phi\Sigma(P)$ be defined, P:$[1^n]$. Then when $n \ge 1$, $\Phi\Sigma(\lambda v.P)$ is the set of tuples $\langle \Phi^v(v), e_1, \ldots, e_n\rangle$ for which

$$\langle e_1, \ldots, e_n\rangle \in \Phi^v(P) \text{ for some } v\text{-variant } \Phi\Sigma^v \text{ of } \Phi\Sigma;$$

when $n = 0$, $\Phi\Sigma(\lambda v.P)$ is the set of values $\Phi\Sigma^v(v)$ for which $\Phi\Sigma^v(P)$ is $+$.

For all $Q_0, Q_1, \ldots, Q_n$:1, the following conditions are satisfied by the set $\Phi\Sigma((\lambda v.P))$:

$$+(\lambda v.P)^{\dagger}(Q_0^{\dagger}, Q_1^{\dagger}, \ldots, Q_n^{\dagger}) \in \Sigma^{\dagger} \Rightarrow \langle e_0, e_1, \ldots, e_n\rangle \in \Phi\Sigma((\lambda v.P)),$$
$$-(\lambda v.P)^{\dagger}(Q_0^{\dagger}, Q_1^{\dagger}, \ldots, Q_n^{\dagger}) \in \Sigma^{\dagger} \Rightarrow \langle e_0, e_1, \ldots, e_n\rangle \notin \Phi\Sigma((\lambda v.P)),$$

whenever $e_i \in \mathcal{E}(e_i), 0 \leq i \leq n$. In particular therefore

$+[Q_0^\dagger/v]P^\dagger(Q_1^\dagger, \ldots, Q_n^\dagger) \in \Sigma^\dagger \Rightarrow \langle Q_0^\dagger, e_1, \ldots, e_n\rangle \in \Phi\Sigma((\lambda v.P))$,

$-[Q_0^\dagger/v]P^\dagger(Q_1^\dagger, \ldots, Q_n^\dagger) \in \Sigma^\dagger \Rightarrow \langle Q_0^\dagger, e_1, \ldots, e_n\rangle \notin \Phi\Sigma((\lambda v.P))$,

whenever $e_i \in \mathcal{E}(e_i), 1 \leq i \leq n$. (V.3) follows when $\Phi\Sigma^v$ is the v-variant of $\Phi\Sigma$ for which $\Phi\Sigma^v(v)$ is $Q_0^\dagger$ by Lemma 5 of §1.5.4.

c) $\Phi\Sigma(T) \in D\Sigma(\tau)$, whenever $T{:}\tau$ by clause 4 of Definition 25 and the definition of $\Phi\Sigma$ in the statement of the lemma. ⊣

One last result is needed to complete the proof of Theorem 11, Completeness of EL without Cut, of §1.9 and its corollary. The theorem asserts that a counter-example that is a model exists for each sequent that has no Cut-free derivation in EL. Its corollary asserts that a derivable sequent of EL, has a Cut-free derivation. The needed result is:

For each formula F of EL, $\pm F^\dagger \in \Sigma^\dagger \Rightarrow \Phi\Sigma(F)$ is $\pm$, respectively.

But this is an immediate consequence of the definition of $\Phi\Sigma(F)$ and clause ***2*** of Definition 25.

EXERCISES §1.9.

1. Provide a proof for Lemma 13 of §1.9.2.
2. Complete the proof for case (V.2) of clause (***b***) of Lemma 15.
3. Theorem 11 of §1.9 is stronger than the completeness theorem for first order logic stated and proved by Henkin in [73]. The former states that a counter-example exists for each sequent that has no Cut-free derivation, while Henkin's completeness theorem states that a counter-example exists for each sequent that has no derivation. The redundancy of Cut is not a consequence of the Henkin completeness theorem. Indeed Henkin's completeness theorem can be proved by generalizing Definition 23 of a descending chain to one constructed by also using the Cut-rule. Lemma 13 can then be strengthened by replacing clause (2) by:

   2C) For each F, exactly one of $+F$ and $-F$ is a member of Σ.

   From such a Σ, a model that is a counter-example can be defined for the given sequent from which Σ is obtained. A corollary of this proof is the Lowenheim- Skolem theorem for first order logic [133], namely that a consistent set of first order formulas has a model to a domain for which each of $D([1^n])$, $n \geq 1$, is denumerable. This result follows since each of the sets $S[P]$, for a Σ satisfying (2C), has a single member. Complete the details.

CHAPTER 2

TYPE THEORY

2.1. The Type Theory TT

The types and the syntax of first order logic are mismatched in several ways. There are constants and variables of type 1 and of type [], but only type 1 variables can be quantified. There are constants of type $[1^n]$ for any $n \geq 1$, and a single constant $\exists$ of type [[1]], but there are no variables of these types. Further if the variables of these types are to be quantified, the constant $\exists$ will have to also be of any type $[[1^n]]$ and of type [[[1]]]. But no matter what type of predicate is admitted, it is possible to consider a higher type predicate of predicates of the given type. The logic Type Theory, abbreviated TT, is a logic with all such possible types.

DEFINITION 26. *The Types of TT.*
Types are defined inductively as follows:

1. 1 *is a type and* [] *is a* predicate *type and a type*;
2. *If* $\tau_1, \ldots, \tau_n$ *is a sequence of types,* $n \geq 1$, *then* $[\tau_1, \ldots, \tau_n]$ *is a* predicate *type and a type.*

As noted before, but for notation these are the types of the logic $\mathcal{F}^\omega$ of [9] and of the logic [128] without function types. They are a proper subset of the types of the logic known as the Simple Theory of Types of Church [28], which are defined as follows:

DEFINITION 27. *The Types of the Simple Theory of Types.*
Types are defined inductively as follows:

1. ι *and* o *are types*;
2. *If* α *and* β *are types then* $(\alpha\beta)$ *is a type.*

Here $(\alpha\beta)$ is the type of unary functions with argument type β and value of type α. The types ι and o correspond to the types 1 and [] of TT; that is, ι is the type of names of subjects and o the type of the truth values. But only some of the types $(\alpha\beta)$ correspond to types of TT. For example, there is no type of TT corresponding to the type $(\iota\iota)$ since this is the type of a function with an argument and value of type 1. A method for introducing such functions into the extension ITT of TT will be described in Chapter 5, but no types are admitted in TT or ITT corresponding to them.

Using Church's notation, the types of TT are ι and the value types, where these are defined:

1. ι is an argument type and o is an argument and value type;
2. If β is an argument type and α a value type, then $(\alpha\beta)$ is an argument and value type.

Thus ι can only be used as the type of an argument, never the type of a value; that is, ι can only be the type of a subject name and never the type of a predicate name.

The notation used for the types of TT is an application sugaring of the notation used for the corresponding types of Church's type theory. This becomes clearer when an alternative notation for the types of the latter theory is used. In the alternative notation, $(\alpha\beta)$ is written $(\beta \rightarrow \alpha)$, to indicate it is the type of a function from type β to type α. Then the type $[\tau_1, \ldots, \tau_n]$ of TT is written $(\tau_1^* \rightarrow (\tau_2^* \rightarrow \cdots (\tau_n^* \rightarrow o)$ as a type of the Simple Theory of Types, where τ_i^* is the type τ_i of TT expressed as a type of the Simple Theory of Types.

DEFINITION 28. *A Notation for Sequences.*
It is useful to have a shorthand notation for sequences, possibly empty, of types and of terms, and of type assignments for terms. A sequence $\tau_1, \ldots, \tau_n$ of types τ_i, where $0 \leq i \leq n$, is abbreviated to $\overline{\tau}$; when n is 0, $\overline{\tau}$ is said to be empty. Similarly a sequence $s_1, \ldots, s_n$ of terms s_i is abbreviated to $\overline{s}$. The assignment of the type τ_i to the term s_i for each i, $1 \leq i \leq n$, is abbreviated $\overline{s}{:}\overline{\tau}$. The notation is also employed within the type notation. Thus $[\overline{\tau}]$ abbreviates $[\tau_1, \ldots, \tau_n]$, where $0 \leq n$, so that $[\overline{\tau}]$ is $[\,]$ when $\overline{\tau}$ is empty. This notation will be used throughout the remainder of the book.

2.1.1. The terms of TT. There are denumerably many constants and variables for each type of TT. There are now three special constants: $\downarrow$ of type $[[\,],[\,]]$, $\exists$ of type $[[\tau]]$ and $=$ of type $[\tau, \tau]$ for each type τ. The latter two are examples of *polymorphic* notation; that is a notation in which a single expression may be assigned more than one type, with the type dependent upon the context in which the expression occurs. In the case of $\exists$ its type is determined when the type of the subject to which it is applied is known; similarly for $=$. The type of a constant or variable cv is denoted, as before, by $\mathrm{t}[cv]$. The usual infix notation for $=$ is introduced by defining $P = Q$ to be $((= P)Q)$.

An alternative to polytyping is to decorate with superscripts or subscripts the type of each symbol employed by the notation; this is done for example in [28] and in [9]. But the result is an unnecessarily complicated and untidy notation. Polymorphic typing is discussed in greater detail in §2.2.

The simplification and extension of the definition of type in Definition 26 results in a simplification and extension of the definition of term; that is of the strings that are assigned a type.

DEFINITION 29. *Term of TT.*

1. *A constant or variable cv is a term of type* $\mathrm{t}[cv]$.
2. P:$[\tau, \overline{\tau}\,]$ *and* Q:$\tau \Rightarrow (PQ)$:$[\,\overline{\tau}\,]$.
3. P:$[\,\overline{\tau}\,]$ *and v a variable of type* $\tau \Rightarrow ((\lambda v)P)$:$[\tau, \overline{\tau}\,]$.

The notation $\overline{\tau}$ used here is an example of the notation introduced in Notation 28, §2.1. For example $P[\tau, \overline{\tau}]$ abbreviates $P[\tau, \tau_1, \ldots, \tau_n]$, where n satisfies $0 \leq n$. The abbreviating notations $(\lambda v.P)$ and $\lambda v.P$ introduced in Notation 13 in §1.5.1 will continue to be used.

Definition 14, §1.5.1, of the depth of a subterm applies equally well to terms of TT.

It is important to note that clause (3) requires that P be a term of type $[\,\overline{\tau}\,]$. The expression $(\lambda v.v)$, when v:1 is not a term of TT, although it is a term if v:$[\,\overline{\tau}\,]$. However, in the latter case $(\lambda v.v)$:$[[\,\overline{\tau}\,], \overline{\tau}\,]$ and therefore $((\lambda v.v)(\lambda v.v))$ is not a term of TT.

Term, as defined here, is a term of the logic TT. A term of what is called the *pure lambda calculus*, described in [29], is obtained from this definition by removing all type restrictions on the definition, and not admitting constants in clause (1). Barendregt provides in [13] a thorough treatment of the lambda calculus.

Definition 19 in §1.5.3 of the substitution operator defines the operator for terms of TT, and Lemma 4 in §1.5.3 generalizes for TT.

As before a *formula* of TT is a term of type [ ]. Here, as in Definition 21 of §1.6, an alternative definition of formula is offered.

DEFINITION 30. *An Alternative Definition of Formula of TT.*

1. $cv(\,\overline{s}\,)$ *is a formula, when* cv:$[\,\overline{\tau}\,]$ *and* $\overline{s}$:$\overline{\tau}$.
2. *F and G are formulas* $\Rightarrow [F \downarrow G]$ *is a formula.*
3. $P(v)$ *a formula for some variable* $v \Rightarrow \exists P$ *is a formula.*
4. *P and Q of the same type* $\Rightarrow P = Q$ *is a formula.*
5. $([s/v]P)(\,\overline{s}\,)$ *a formula* $\Rightarrow (\lambda v.P)(s, \overline{s}\,)$ *is a formula.*

This definition is suggested by the definition of formula in [128]. To prove that the formulas defined in Definition 30 are exactly the terms of type [ ] of Definition 29 requires a study of lambda contractions that will be undertaken in §2.3.

EXERCISES §2.1.

1. Provide a proof for Lemma 4 of §1.5.3 for TT by induction on $ct[P]$.
2. Prove that a formula as defined in Definition 30 is a term of type [ ].

2.2. Polymorphic Typing of TT

Polymorphic typing, or briefly just *polytyping*, for TT means that a given term may have more than one type. Whitehead and Russell called it "typical ambiguity" and wrote the following in [149]:

> In spite of the contradictions which result from unnoticed typical ambiguity, it is not desirable to avoid words and symbols which have typical ambiguity. Such words and symbols embrace practically all the ideas with which mathematics and mathematical logic are concerned: the systematic ambiguity is the result of a systematic analogy. That is to say, in almost all the reasonings which constitute mathematics and mathematical logic, we are using ideas which may receive any one of an infinite number of different typical determinations, any one of which leaves the reasoning valid. Thus by employing typically ambiguous words and symbols, we are able to make one chain of reasoning applicable to any one of an infinite number of different cases, which would not be possible if we were to forego the use of typically ambiguous words and symbols.

The device of polytyping will be used to good advantage: No explicit typing is part of the notation for variables and constants; type declarations can be made to specify types when necessary, although often types can be inferred from context. It does introduce complications into the testing of a string of symbols to determine whether the string is or is not a term. But its advantages far exceed the complications as is argued in (A.10) of [124] and demonstrated in §2.5. Here a foundation for polymorphic typing is developed and applied to the logic TT in the form of an outline for an algorithm in Figure 3. It can be seen to be a formalization of Whitehead and Russell's typical ambiguity. It is not described in full detail but only to a level that should convince the reader that all the details can be provided. It is described in a manner that allows easy generalization to the more complicated context of ITT in Chapter 3.

The polytyping algorithm makes use of a *unification* algorithm described in Figure 2 and discussed in §2.2.3. The use of unification in the algorithm has been suggested by the algorithm described by Milner in [100] for polytyping computer programs; an earlier paper by Hindley [83] also describes an algorithm for combinatorial logic.

2.2.1. Typing strings of characters. By a *character* is meant here $\downarrow$, $\exists$, (λv) for any variable v, or, any constant or variable. By a *string* of characters, or just a string, is meant either a character or $(strastrb)$ where $stra$ and $strb$ are strings. By a *typeable* character is meant $\downarrow$, $\exists$, or any constant or variable.

A term is said to have *distinct variables* if it satisfies the following restriction: Each variable with a bound occurrence in the string is distinct from each variable with a free occurrence and from each variable that has an occurrence that is bound by a distinct λ. Every term has a bound variable variant that has distinct variables. There is therefore no loss in restricting attention to such terms. The polytyping algorithm to be described determines if a string can be assigned a type that is consistent with the term having distinct variables.

Each occurrence of a *defined* constant in a term is replaced with its definition. If the resulting string is replaced by one with distinct variables, then distinct occurrences of the same *defined* constant may be assigned distinct types.

By a *base for typing* of a string is a set of *type assignments* $ch{:}\tau$, one for each occurrence of a typeable character ch, in the string. A base for a typing of a string is *acceptable* if the typing of the string that results from the assignments results in a type that satisfies Definition 29 of term of TT.

2.2.2. Type expressions and their ranges. Since in general there is more than one acceptable base for a term, it is necessary to provide a notation for *type expressions* that define a *range* of types, as defined in Definition 32. An *assignment* $str{:}te$ of a type expression to a string str is to be understood as expressing that the string may be assigned any one of the types in the range of the type expression.

DEFINITION 31. *Type expressions.*

1. *Any type* τ *is a* type expression.
2. *Any finite sequence of lower case Greek letters* α, β,*and* γ, *and numerals beginning with a letter is a* type variable *and a* type expression.
3. *For each type variable* α, $\overrightarrow{\alpha}$ *is a* type sequence variable.
4. *Let* $\overline{te}$ *be a sequence* $te_1, \ldots, te_n$ *of type expressions, possibly empty, and let* $\overrightarrow{\alpha}$ *be any type sequence variable. Then each of* $[\overline{te}]$ *and* $[\overline{te}, \overrightarrow{\alpha}\,]$ *is a* type expression *and a* predicate *type expression.* $[\ldots]$ *abbreviates each of these type expressions.*
5. `Nil` *is a type expression.*

To assign the type sequence expression `Nil` to a string is to assert that the string cannot be typed.

The type sequence variable $\overrightarrow{\alpha}$ can only appear in type expressions like $[\overrightarrow{\alpha}\,]$ or $[\overline{te}, \overrightarrow{\alpha}\,]$. It is used when the arity of a predicate term is unknown. Thus all that can be inferred from the assignment $str{:}[\overrightarrow{\alpha}\,]$ concerning the arity n of str is $0 \leq n$. From the assignment $str{:}[\overline{te}, \overrightarrow{\alpha}\,]$ can be inferred that $m \leq n$, where m is the length of the sequence $\overline{te}$.

The set of all types of TT is denoted by TYP. Each type expression te has a range $\mathsf{RAN}(te)$ that is a subset of TYP defined as follows.

DEFINITION 32. *Range of type expressions.*

1. $\mathsf{RAN}(\tau)$ *is* $\{\tau\}$.
2. $\mathsf{RAN}(\alpha)$ *is* TYP, *for any type variable* α.
3. $\mathsf{RAN}([\overline{te}])$, *where* $\overline{te}$ *is any sequence of type expressions* $te_1, \ldots, te_n$, *is the set of all types* $[\overline{\tau}]$ *for which* $\tau_i \in \mathsf{RAN}(te_i), 0 \leq i \leq n$.

4. $\mathsf{RAN}([\overrightarrow{\alpha}])$ *is* $\bigcup\{\mathsf{RAN}([\alpha_1, \ldots, \alpha_n]) \| \alpha_1, \ldots, \alpha_n$ *distinct*; $n \geq 0\}$, *for any type sequence variable* $\overrightarrow{\alpha}$.
5. $\mathsf{RAN}([\overline{te}, \overrightarrow{\alpha}])$ *is the set of all types* $[\overline{\tau}, \overline{\sigma}]$ *for which* $[\overline{\tau}] \in \mathsf{RAN}([\overline{te}])$ *and* $[\overline{\sigma}] \in \mathsf{RAN}([\overrightarrow{\alpha}])$.
6. $\mathsf{RAN}(\mathtt{Nil})$ *is the empty set of types.*

Thus, for example, $\mathsf{RAN}([\alpha_1, \ldots, \alpha_n])$, when the type variables $\alpha_1, \ldots, \alpha_n$ are distinct, is the set of all types of arity n predicates. $\mathsf{RAN}([\overrightarrow{\alpha}])$ is therefore $\mathsf{TYP}-\{1\}$, the set of all predicate types. $\mathsf{RAN}([\alpha, \overrightarrow{\alpha}])$ is $\mathsf{TYP}-\{1, []\}$, the set of all types of predicates of arity n, $n \geq 1$. $\mathsf{RAN}([te, \overrightarrow{\alpha}])$ is the set of all types of predicates of arity n, $n \geq 1$, with first argument of type in $\mathsf{RAN}(te)$.

A *base range* for typing a string is similar to a base for typing except that each assignment *ch*:*tech* assigns a type expression *tech*, rather than a type, to each typeable character *ch* of the string. A base is a *member* of a base range for a string if for each typeable character *ch* of the string the type assigned to *ch* is in the range $\mathsf{RAN}(tech)$ of the type expression *tech* assigned to *ch* by the base range.

2.2.3. String assignments and unification. The polytyping algorithm described in Figure 3 requires a solution to the following problem: Given type expressions *tea* and *teb*, find a type expression *te* for which

(a) $$\mathsf{RAN}(te) \text{ is } \mathsf{RAN}(tea) \cap \mathsf{RAN}(teb).$$

Such a type expression *te* can be calculated for any *tea* and *teb* by using a slight generalization of *unification*, a technique first described under that name by J.A. Robinson in [123] and since widely used in theorem proving; see for example [47]. The application of the technique made here is similar to the applications made by Hindley in [82] and Milner in [100]. A slight generalization of Robinson's unification algorithm is necessary to accommodate type sequence variables.

Since there are no variable binding operators in type expressions, a substitution operator $[te/\alpha]$ can be defined for any type expression *te* and type variable α. It can be applied to any type expression to yield another type expression. Similarly if $\overline{te}$ is a possibly empty sequence of type expressions and $\overrightarrow{\alpha 1}$ and $\overrightarrow{\alpha}$ are type sequence variables, then a substitution operator $[\overline{te}, \overrightarrow{\alpha 1}/\overrightarrow{\alpha}]$ can be defined and applied to any type expression to yield another type expression. In particular $[/\overrightarrow{\alpha}]$ is a substitution of the empty sequence for the variable $\overrightarrow{\alpha}$.

Each of these substitutions takes the form $[nm/vr]$, where vr is either a type variable or a type sequence variable, and *nm* in the first case is a type expression and in the second case a sequence of type expressions, possibly the empty sequence. As described in §1.5.4 for substitutions of terms for variables, a sequence of such substitutions defines a *simultaneous* substitution operator. The identity substitution leaves a type expression unchanged.

DEFINITION 33. *Unifier.*
Let $\langle tea, teb\rangle$ *be a pair of type expressions. A* unifier *of the pair is a simultaneous substitution* Σ *for which* Σtea *is* Σteb. *A* unifier *of a sequence of pairs* $\langle tea_1, teb_1\rangle, \cdots, \langle tea_n, teb_n\rangle$ *is a simultaneous substitution* Σ *for which* Σtea_i *is* Σteb_i *for* $1 \leq i \leq n$.

LEMMA 16. *Let* Σ *be a unifier of tea and teb. Then*

$$\mathsf{RAN}(\Sigma tea) \subseteq \mathsf{RAN}(tea) \cap \mathsf{RAN}(teb).$$

PROOF. Clearly $\mathsf{RAN}(\Sigma tea) \subseteq \mathsf{RAN}(tea)$ and $\mathsf{RAN}(\Sigma teb) \subseteq \mathsf{RAN}(teb)$. ⊣

The *product* substitution $\Sigma 1 \circ \Sigma 2$ of two substitutions is defined as

$$(\Sigma 1 \circ \Sigma 2)te \text{ is } \Sigma 1(\Sigma 2te).$$

DEFINITION 34. *Most General Unifier (mgu).*
A most general unifier *of a sequence of pairs* $\langle tea_1, teb_1\rangle, \cdots, \langle tea_n, teb_n\rangle$ *of type expressions is a unifier* Σ *for which for each unifier* $\Sigma 1$ *of the sequence there is a substitution* $\Sigma 2$ *for which* $\Sigma 1$ *is* $\Sigma 2 \circ \Sigma$.

Figure 2 describes an algorithm that finds an mgu for any given sequence of pairs of type expressions listed on the list `LIST`, if one exists. Initially Σ is the identity substitution. The pairs are processed one at a time; a pair is said to be *done* when its processing has been completed. The processing of a pair $\langle tea, teb\rangle$ is described in a table in Figure 2. The headings on the table are Mem1, Mem2, $\Sigma 1$, Cond, and `Action`. The row in which the type expressions Σtea and Σteb are found under Mem1 and Mem2 determines how $\langle tea, teb\rangle$ is processed. $[\cdots]$ under Mem2 is to be understood as any predicate type. Under the heading `Action` appears one of A, B, and Halt. The first are to be understood as *go to* commands, and the third as the *Halt* command. Under the heading $\Sigma 1$ appears the substitution to be used in part B. Under the heading Cond appears the condition on the members of $\langle tea, teb\rangle$ that requires the algorithm to halt.

The correctness of the algorithm is proved in Lemma 17.

LEMMA 17. *Correctness of the Unification Algorithm.*
The algorithm of Figure 2 halts with any input. If it halts at the first Halt of A, then the current Σ *is an mgu for each of pairs on the initial* `LIST`; *if it halts elsewhere then at least one of the pairs has no unifier.*

PROOF. The following notation is used in the proof: For any substitution Σ, Σ`LIST` is the list obtained from any `LIST` by replacing each pair $\langle tea, teb\rangle$ by $\langle \Sigma tea, \Sigma teb\rangle$.

That the algorithm necessarily halts is proved first. Consider the set of distinct type variables and type sequence variables occurring in the type expressions of the pairs on `LIST`. Since there can be only a finite number on the initial `LIST`, and since with each execution of step B the number on Σ`LIST` decreases, the algorithm necessarily halts.

A. Halt if every pair on LIST is done. Else, choose the first not done pair $\langle tea, teb\rangle$ on LIST, mark it done. Process $\langle tea, teb\rangle$ using the row of the following table determined by Σtea and Σteb under the columns Mem1 and Mem2. Here Σ is the current substitution.

Mem1	Mem2	$\Sigma 1$	Cond	Action
α	teb	$[teb/\alpha]$		B
1	1			A
$[tea_1, \ldots, tea_m]$	$[teb_1, \ldots, teb_m]$			A
$[tea_1, \ldots, tea_m]$	$[teb_1, \ldots, teb_n, \overrightarrow{\beta}]$	$[tea_{n+1}, \ldots, tea_m/\overrightarrow{\beta}]$		B
$[tea_1, \ldots, tea_m, \overrightarrow{\alpha}]$	$[teb_1, \ldots, teb_n, \overrightarrow{\beta}]$	$[teb_{n+1}, \ldots, teb_m, \overrightarrow{\beta}/\overrightarrow{\alpha}]$		B
$[tea_1, \ldots, tea_m, \overrightarrow{\alpha}]$	$[teb_1, \ldots, teb_m, \overrightarrow{\alpha}]$			A
1	$[\cdots]$			Halt
$[tea_1, \ldots, tea_m]$	$[teb_1, \ldots, teb_n]$		$m \neq n$	Halt
$[tea_1, \ldots, tea_m]$	$[teb_1, \ldots, teb_n, \overrightarrow{\beta}]$		$m < n$	Halt
$[tea_1, \ldots, tea_m, \overrightarrow{\alpha}]$	$[teb_1, \ldots, teb_n, \overrightarrow{\alpha}]$		$m \neq n$	Halt

B. Halt if the variable of the substitution $\Sigma 1$ occurs in the term of the substitution. Else, $\Sigma = \Sigma 1 \circ \Sigma$, go to A.

FIGURE 2. Unifying Type Expressions

The remainder of the proof of the theorem is by induction on the number n of times the algorithm begins the Else part of A before halting.

Necessarily $1 \leq n$. The algorithm is said to *halts1* if it halts at the first Halt of A, and to *halts2* otherwise.

Let $\texttt{LIST}_i$, $1 \leq i \leq n$, be the list at the beginning of the i-th execution of the Else part of A. Thus $\texttt{LIST}_1$ is the initial list. Let Σ_i, $1 \leq i \leq n$, be the substitution current with $\texttt{LIST}_i$. Thus Σ_1 is the initial identity substitution and Σ_n the final. For each i, $1 \leq i \leq n-1$, there is a substitution $\Sigma 1_i$ for which Σ_{i+1} is $\Sigma 1_i \circ \Sigma_i$. $\Sigma 1_i$ may be the identity substitution.

Let $n = 1$. Then necessarily the algorithm halts2. In this case the correctness of the algorithm is apparent from the table in Figure 2 and the fact that a substitution $\Sigma 1$ failing the test in B has been obtained from a pair that cannot have a unifier.

Assume now that the lemma holds if the algorithm halts with $1 \leq n < k$. Consider the case $n = k$ and assume the algorithm halts1.

For each i, $2 \leq i \leq n$, let $\texttt{LIST}^*_i$ be the list obtained from $\Sigma 1\texttt{LIST}_i$ by removing its first pair. Consider the result of starting the algorithm with initial list $\texttt{LIST}^*_2$. The sequence of lists obtained in this case is $\texttt{LIST}^*_2, \cdots, \texttt{LIST}^*_n$. Let the current substitution when it halts be Σ^*_n. Then

$$\Sigma^*_n \text{ is } \Sigma 1_{n-1} \circ \cdots \circ \Sigma 1_2 \circ \Sigma_1 \text{ and } \Sigma_n \text{ is } \Sigma^*_n \circ \Sigma 1_1 \circ \Sigma_1.$$

By the induction assumption Σ^*_n is an mgu for the pairs on $\texttt{LIST}^*_2$.

Consider now the first pair $\langle tea, teb \rangle$ of $\mathtt{LIST}_1$; it is the pair processed during the first execution of the Else part of A. Since the initial substitution is the identity substitution, the possibilities for the two members of the pair and the substitution $\Sigma 1$ are given in the table of Figure 2. Clearly in each case Σ_n^* is an mgu for $\langle \Sigma 1 tea, \Sigma 1 teb \rangle$ since either $\Sigma 1 tea$ and $\Sigma 1 teb$ are identical or because each of their corresponding subtypes appear in pairs $\langle \Sigma 1 tea_i, \Sigma 1 teb_i \rangle$ on $\mathtt{LIST}_2^*$.

Consider now the case that the algorithm halts2. By the induction assumption at least one of the pairs on $\mathtt{LIST}_2^*$ has no unifier. Therefore at least one of the pairs on $\mathtt{LIST}_1$ has no unifier. ⊣

DEFINITION 35. $(tea \sqcap teb)$.
Let $\langle tea, teb \rangle$ be a pair of type expressions. Then $(tea \sqcap teb)$ is Σtea, if Σ is the most general unifier of the pair, and is otherwise the type expression `Nil`.

$(tea \sqcap teb)$ can, of course, equally well be defined to be Σteb if Σ is the most general unifier; it is a solution te to the identity (a) stated at the beginning of this subsection.

LEMMA 18. *Unification.*
Let tea and teb be type expressions. Then

$$\mathsf{RAN}(tea) \cap \mathsf{RAN}(teb) \subseteq \mathsf{RAN}((tea \sqcap teb)).$$

PROOF. A type in $\mathsf{RAN}(tea) \cap \mathsf{RAN}(teb)$ is defined by a unifier $\Sigma 1$. By Lemma 16 $\mathsf{RAN}(\Sigma 1 tea) \subseteq \mathsf{RAN}(tea) \cap \mathsf{RAN}(teb)$. By the definition of most general unifier there is a substitution $\Sigma 2$ for which $\Sigma 1$ is $\Sigma 2 \circ \Sigma$. Therefore $\mathsf{RAN}(\Sigma 1 tea)$ is $\mathsf{RAN}(\Sigma 2 \circ \Sigma tea) \subseteq \mathsf{RAN}(\Sigma tea) \subseteq \mathsf{RAN}((tea \sqcap teb))$. ⊣

2.2.4. Polytyping TT. Recall from §2.2.1 that an acceptable base for typing a string *str* is a type assignment $ch{:}\tau$, for each typeable character *ch* of *str*, that results in a type being assigned to *str* that satisfies Definition 29 of §2.1.1 for term of TT. Recall from §2.2.2 that a base range for typing a string is an assignment *ch*:*tech* of a type expression *tech* to each typeable character *ch* of the string. As defined in §2.2.2 a base is a member of a base range for a string if for each typeable character *ch* of the string, the type assigned to *ch* is in the range $\mathsf{RAN}(tech)$ of the type expression *tech* assigned to *ch* by the base range.

A *feasible* acceptable base for an assignment *str*:*te* is an acceptable base for *str* for which the type assigned to *str* is in $\mathsf{RAN}(te)$. The assignment is said to be *feasible* if there exists a feasible acceptable base for it. A *feasible acceptable* base range for *str*:*te* is one for which each base that is a member is a feasible acceptable base for *str*:*te*. A *feasible acceptable* base range for a finite list of assignments is one for which each base that is a member is a feasible acceptable base for each assignment on the list.

The goal of the algorithm described in Figure 3 is to determine if a list of assignments are simultaneously feasible and if so to determine all their feasible acceptable bases.

Initially the assignments appear as the only ones on a list LIST of assignments. The algorithm adds assignments for substrings *strc* of an initial string *str* to LIST. When the processing of an assignment on LIST has been completed it is marked *done*. The order of processing is such that no assignment for a substring of a string is done before an assignment of the string is done.

A substitution Σ is defined at each step of the algorithm; initially it is the identity substitution. For any list of assignments LIST, ΣLIST is the list with each *strc*:*tec* replaced by *strc*:Σ*tec*.

Should the algorithm halt at its first halt, then ΣLIST has a sublist that is a feasible acceptable base range for the input assignments. Further, it is a *most general* feasible acceptable base range in the sense that any feasible acceptable base for all the initial assignments is a member of the base range.

The not done assignments *strc*:*tec* on LIST are processed one at a time as described in the table of Figure 3. Two main cases are considered under the column headed '*strc* is'; when *strc* consists of a single character, and when *strc* has the form (*strastrb*). In the latter case two subcases arise, when *stra* has the form (λv) for any variable v, and when it has any *other* form. The corresponding row of the column headed 'ACTION' describes what has to be done to process the currently chosen not done assignment *strc*:*te*; here the operator $(tea \sqcap teb)$ defined in Definition 35 is used. Following the execution of the action, the processing of *strc*:*tec* is complete and it can be marked 'done' before control is returned to Begin.

THEOREM 19. *Correctness of Polytyping Algorithm for TT.*
For any initial LIST *the algorithm of Figure* 3 *halts. Let it halt at the first halt with final list* LISTf. *Then the base range defined as a sublist of* ΣLISTf *is the most general feasible acceptable range for the initial assignments on* LIST. *If it halts at any halt other than the first then the initial assignments on* LIST *are not feasible.*

PROOF. Since each execution of the algorithm for a string (*strastrb*) can at most add assignments for the substrings *stra* and *strb* to LIST, the algorithm must halt after finitely many executions.

The remainder of the proof of the theorem is by induction on the number n of distinct substrings in the strings of the initial assignments.

Let that number be 1, the smallest number possible since a string is a substring of itself. In this case the initial list LIST has a single member assigning a type expression to a character. The actions for the five possible cases are clearly the only ones possible. Consider for example $\exists$. The type of the constant must be a member of RAN([[α]]) so that ([[α]] $\sqcap \Sigma$*tec*) is the mgu, if one exists, of [[α]] and *tec*. Thus $\exists$:*tec* is feasible if and only if ([[α]] $\sqcap \Sigma$*tec*) $\neq$ Nil.

Assume now that the theorem holds whenever the strings in the initial assignments have fewer than n, $0 < n$, substrings. Consider an initial list

---

`Begin`
Halt if every assignment on `LIST` is done.
Else let *strc*:*tec* be the first not done.

<u>*strc* is</u>	<u>`ACTION`</u> [If not Halt, to be followed by: Return to `Begin`]
$\downarrow$	Halt if $([[\,],[\,]] \sqcap \Sigma tec)$=`Nil`.
	Else $\Sigma = ([[\,],[\,]] \sqcap \Sigma tec) \circ \Sigma$.
$\exists$	Halt if $([[\alpha]] \sqcap \Sigma tec)$=`Nil`.
	Else $\Sigma = ([[\alpha]] \sqcap \Sigma tec) \circ \Sigma$.
$=$	Halt if $([\alpha,\alpha] \sqcap \Sigma tec)$=`Nil`.
	Else $\Sigma = ([\alpha,\alpha] \sqcap \Sigma tec) \circ \Sigma$.
(λv)	Halt.
cv	Halt if $cv{:}teo$ on `LIST` and $(\Sigma teo \sqcap \Sigma tec)$=`Nil`.
	Else $\Sigma = (\Sigma teo \sqcap \Sigma tec) \circ \Sigma$.
$(strastrb)$	
<u>*stra* is</u>	
(λv)	If $((\lambda v)strb){:}teo$ on `LIST` and done,
	Halt if $(\Sigma teo \sqcap \Sigma tec)$=`Nil`;
	Else $\Sigma = (\Sigma teo \sqcap \Sigma tec) \circ \Sigma$.
	Else, Halt if $([\alpha, \overrightarrow{\alpha}] \sqcap \Sigma tec)$=`Nil`;
	Else $\Sigma = ([\alpha, \overrightarrow{\alpha}] \sqcap \Sigma tec) \circ \Sigma$;
	Add $v{:}\alpha$ and $strb{:}[\overrightarrow{\alpha}]$ to `LIST`.
not (λv)	If $(strastrb){:}teo$ on `LIST` and done,
	Halt if $(\Sigma teo \sqcap \Sigma tec)$=`Nil`;
	Else $\Sigma = (\Sigma teo \sqcap \Sigma tec) \circ \Sigma$.
	Else, Halt if $([\overrightarrow{\alpha}] \sqcap \Sigma tec)$=`Nil`;
	Else $\Sigma = ([\overrightarrow{\alpha}] \sqcap \Sigma tec) \circ \Sigma$;
	Add $stra{:}[\alpha, \overrightarrow{\alpha}]$ and $strb{:}\alpha$ to `LIST`.

FIGURE 3. Polytyping TT

`LIST` of assignments in which the strings of the assignments have n substrings.

• Consider first the case that the algorithm does not halt at the first halt.
Let *strc*:*tec* be the first not done assignment on `LIST` for the last execution of the algorithm before halting. Consider the possibilities for it and the resulting actions.

The first four cases when *strc* is $\downarrow$, $\exists$, $=$, and (λv) are simply repeats of the same cases when n=1. Because of the *distinct variable* assumption described in §2.2.1, no consideration has to be given for assignments $\exists{:}teo$ on `LIST`. However, the case cv must now consider the possibility that another assignment $cv{:}teo$ is on `LIST`. Clearly if $(\Sigma teo \sqcap \Sigma tec)$=`Nil` then the initial assignments are not feasible and halting, the assumed action, is the appropriate action.

Consider now the two subcases when *strc* is $(strastrb)$.

stra **is** $(\lambda \boldsymbol{v})$. The algorithm must necessarily halt for one of the two possible cases since otherwise $v{:}\alpha$ and $strb{:}[\overrightarrow{\alpha}]$ would be added to `LIST`, contradicting the assumption that $((\lambda v)strb){:}tec$ is the last not done assignment on `LIST` prior to halting.

If $((\lambda v)strb){:}teo$ is on `LIST` for some type expression *teo* and the assignment is done, then any new typing of $((\lambda v)strb)$ must be compatible with it. If
$(\Sigma teo \sqcap \Sigma tec)$=`Nil`, then the new proposed typing *tec* is not compatible with the typing *teo* and halting is the appropriate action.

If $((\lambda v)strb){:}teo$ is not on `LIST` for some type expression *teo* or it is but the assignment is not done, then it is only necessary to ensure that $((\lambda v)strb){:}tec$ is compatible with the most general typing possible for $((\lambda v)strb)$.

Any acceptable typing of $((\lambda v)strb)$ must be a member of $\mathsf{RAN}([\alpha, \overrightarrow{\alpha}])$. Thus if $([\alpha, \overrightarrow{\alpha}] \sqcap \Sigma tec)$=`Nil` then $((\lambda v)strb){:}tec$ is not feasible and halting is the correct action.

stra **is not** $(\lambda \boldsymbol{v})$. The argument in this case is similar to that of the previous case. The algorithm must halt, as assumed, at one of the two opportunities.

• Consider now the case that the algorithm halts at the first halt.
Let it halt after h executions of the actions of the table in Figure 3. Let $\texttt{LIST}_i$, $1 \leq i \leq h$, be the list of assignments at the beginning of the i-th execution of the algorithm. Thus $\texttt{LIST}_1$ is the initial list and $\texttt{LIST}_h$ the final. Let Σ_i, $1 \leq i \leq h$, be the substitution current with $\texttt{LIST}_i$. Thus Σ_1 is the initial identity substitution and Σ_h the final. For each i, $1 \leq i < h$, there is a substitution $\Sigma 1_i$ for which Σ_{i+1} is $\Sigma 1_i \circ \Sigma_i$. $\Sigma 1_i$ may be the identity substitution.

For each i, $2 \leq i \leq n$, let $\texttt{LIST}^*_i$ be the list obtained from $\Sigma_1\texttt{LIST}_i$ by removing its first assignment. Consider the result of starting the algorithm with initial list $\texttt{LIST}^*_2$. The sequence of lists obtained in this case is $\texttt{LIST}^*_2, \cdots, \texttt{LIST}^*_h$. Let the current substitution when it halts be Σ^*_h. Then

$$\Sigma^*_h \text{ is } \Sigma 1_{n-1} \circ \cdots \circ \Sigma 1_2 \circ \Sigma_1 \text{ and } \Sigma_h \text{ is } \Sigma^*_h \circ \Sigma 1_1 \circ \Sigma_1.$$

By the induction assumption the base range defined as a sublist of $\Sigma^*_h\texttt{LIST}^*_h$ is the most general feasible acceptable range for the assignments on $\texttt{LIST}^*_2$ which is $\Sigma_1\texttt{LIST}_2$ without its first member, say $strc{:}\Sigma_1 tec$. Since in this case the substitution Σ_1 is the identity substitution, it can be ignored.

Consider the possible forms of *strc* and the actions required. Since it is assumed that the algorithm halts at the first halt, no halt in the table of Figure 3 can be a possible action. The arguments for the cases that *strc* is a character present no difficulties. Consider, therefore, the two subcases when *strc* is $(strastrb)$.

stra **is** (λv). In this case $strb$:$[\overrightarrow{\alpha}]$ and v:α are added to $\mathtt{LIST}_1$ and therefore appear on $\mathtt{LIST}_2$. These assignments are the most general possible for these substrings of $((\lambda v)strb)$. Hence a typing for a list that includes $strb$:$\Sigma_1[\overrightarrow{\alpha}]$ and v:$\Sigma_1\alpha$ is feasible if and only if it is also feasible for a list that also includes $((\lambda v)strb)$:$\Sigma_1 tec$. In particular therefore a list that includes $strb$:$\Sigma_h^*\Sigma_1[\overrightarrow{\alpha}]$ and v:$\Sigma_h^*\Sigma_1\Sigma_1\alpha$ is feasible if and only if it is also feasible for a list that also includes $((\lambda v)strb)$:$\Sigma_h^*\Sigma_1 tec$. Since Σ_h is $\Sigma_h^*\Sigma_1$, the theorem holds for this case also.

stra **is not** (λv). The argument in this case is similar to the previous. Since only predicate types can be assigned to $(strastrb)$, the most general type expression that can be assigned is $[\overrightarrow{\alpha}]$, and the most general type expressions for *stra* and *strb* are respectively therefore $[\alpha, \overrightarrow{\alpha}]$ and α. Consequently a feasible typing for a list that includes $stra$:$\Sigma_1[\alpha, \overrightarrow{\alpha}]$ and :$\Sigma_1\alpha$ is feasible if and only if it is feasible for a list that includes $(strastrb)$:$\Sigma_1 tec$ also. Thus the theorem holds for this case also. ⊣

2.3. Lambda Contractions

Definition 20 of §1.5.4 defined β- and η-contractions. The definitions extend directly to contractions on the terms of TT. Here a fundamental theorem concerning these contractions will be proved.

DEFINITION 36. *Contractible Terms.*
A term CT is contractible if it has one of two forms:

(β): $(\lambda x.P)Q$; *or*
(η): $(\lambda x.Px)$, *where x has no free occurrence in P.*

The *contraction CT′* of CT is respectively $[Q/x]P$ and P. Note that the contraction of a term is of the same type as the term.

DEFINITION 37. *The Relation > Between Terms.*
Let P be a term with a contractible subterm CT, and let P' be either a bound variable variant of P, or obtained from P by replacing an occurrence of CT by CT' in P; then $P > P'$.

DEFINITION 38. *The Relation* $\gg$ *Between Terms.*
A contraction *sequence is a sequence $P_0, P_1, \ldots, P_k, P_{k+1}, \ldots$ of terms for which $P_k > P_{k+1}$, for $k \geq 0$. The relation $\gg$ holds between terms P and Q if they are members of a contraction sequence for which P is P_0 and Q is P_k, for some k, $k \geq 0$.*

The relation $\gg$ is reflexive and transitive; that is, $P \gg P$, and if $P \gg Q$ and $Q \gg R$ then $P \gg R$.

A fundamental theorem of the pure lambda calculus was first proved by Church and Rosser in [26], and has since been called the Church-Rosser theorem. A weak form of the theorem is given in [13]. Here it is proved as a lemma for TT.

LEMMA 20. *Weak Church-Rosser.*
Let $L > M$ and $L > N$. Then there exists an O for which $M \gg O$ and $N \gg O$.

PROOF. The conclusion is immediate if one or more of M or N is a bound variable variant of L; therefore assume otherwise. Let CTM be the occurrence of a contractible term in L that is replaced by its contraction CTM' in M. Let CTN and CTN' be similarly defined for L and N. There are two main cases to be considered:

• The occurrences CTM and CTN in L are disjoint in the sense that one of them does not occur within the other. Then O can be obtained from M by replacing the occurrence of CTN in M by CTN' and from N by replacing the occurrence of CTM in N by CTN'.

• The occurrence of CTN in L is within the occurrence of CTM. There are two subcases to consider:

(A) CTM has one of the forms (β) or (η) and CTN is in P.
(B) CTM is $(\lambda x.P)Q$ and CTN is in Q.

Consider (A) first. Let P' be obtained from P by replacing CTN by its contraction. Thus N is obtained from L by replacing P by P'; as a result CTM has been replaced by $(\lambda x.P')Q$ or $(\lambda x.P'x)$. Let O be obtained from N by replacing $(\lambda x.P')Q$ or $(\lambda x.P'x)$ by its contraction, respectively $[Q/x]P'$ or P'. M is obtained from L by replacing the occurrence of $(\lambda x.P)Q$ or $(\lambda x.Px)$ by respectively $[Q/x]P$ or P. Should CTM' be P then O can also be obtained from M by replacing CTN in CTM' by its contraction. Should CTM' be $[Q/x]P$ then O can also be obtained from M by replacing $[Q/x]CTN$ in CTM' by its contraction.

Consider now (B). Let Q' be obtained from Q by replacing the occurrence of CTN by CTN'. If there is more than one free occurrence of x in P, there is more than one occurrence of Q in M and therefore more than one occurrence of CTN. O is obtained from M by a sequence of contractions in which each occurrence of CTN is replaced by CTN'. In N, CTM is replaced by $(\lambda x.P)Q'$. Thus O can also be obtained from N by replacing the occurrence of $(\lambda x.P)Q'$ by its contraction $[Q'/x]P$. ⊣

A statement of the full Church-Rosser theorem of [13] is obtained from that of the lemma by replacing the assumptions $L > M$ and $L > N$ by $L \gg M$ and $L \gg N$.

A term is in *β-normal* form if it has no contractible subterm of the form (β), and is in *η-normal* form if it has no contractible subterm of the form (η). It is in *normal* form if it has no contractible subterm.

THEOREM 21. *Normal Form.*
Every term of TT has a normal form that is unique to within bound variable variants.

PROOF. The theorem will be proved in three stages. It will be proved first (I) that every term has a β-normal form, then (II) that every term with a β-normal form has an η-normal form, and finally (III) that the normal form is unique to within bound variable variants.

(I) By a contractible subterm CT of a term T is meant, in this first part of the proof, one of the form $(\lambda x.P)Q$, where $x, Q{:}\tau$, $P{:}[\overline{\tau}]$ and $(\lambda x.P){:}[\tau, \overline{\tau}]$. The proof of the first part has been adapted from one by Peter Andrews [8] for a version of Church's simple theory of types. It will be proved by induction on a lexicographic order defined on the pairs $\langle m[T], n[T]\rangle$ as follows:

$$\langle m', n'\rangle < \langle m, n\rangle \Leftrightarrow m' < m \text{ or } m' = m \text{ and } n' < n.$$

Here for a given term T, $m[T]$ and $n[T]$ are defined as follows:

• Definition of $m[T]$.
A measure $\#[\tau]$ of the complexity of a type τ is defined as follows: If τ is 1 or $[\,]$, then $\#[\tau]$ is 0; while if τ is $[\tau_1, \ldots, \tau_n]$, $n \geq 1$, then $\#[\tau]$ is $\#[\tau_1]+\ldots+\#[\tau_n]+n$.
Let CT be $(\lambda x.P)Q$, a contractible subterm of the term T. Define the type complexity $\#[CT]$ of CT to be $\#[\tau]$ where τ is the type of $(\lambda x.P)$. Define

$$m[T] \text{ is } max\{0, \#[CT] \| CT \text{ a contractible subterm of } T\}$$

so that if T is in normal form, then $m[T]$ is 0.

• Definition of $n[T]$.
A contractible subterm CT of T is *maximal* if $\#[CT]$ is $m[T]$. Define $n[T]$ to be the number of distinct occurrences of a maximal contractible subterm of T in T; thus if $n[T]$ is 0, then T is in normal form.

Let $n[T] \geq 1$ and let CT^* be the occurrence of a maximal contractible subterm of T which is furthest to the right in T, with the λ of CT determining its position in T. Let CT^* be $(\lambda x.P)Q$. Let T_1 be obtained from T by replacing the occurrence CT^* of $(\lambda x.P)Q$ in T by $[Q/x]P$. Then necessarily

$$\text{(a)} \qquad m[[Q/x]P] < \#[(\lambda x.P)Q] = m[T]$$

Since a contractible subterm (RS) of P or Q is a contractible subterm of $(\lambda x.P)Q$, it follows that $\#[(RS)] < \#[(\lambda x.P)Q]$.

Consider now a maximal contractible subterm (RS) of $[Q/x]P$ that is not a subterm of $(\lambda x.P)Q$. Necessarily either R or S is Q. If S is Q then (Rx) is a contractible subterm of P so that

$$\#[(Rx)] = \#[(RQ)] < \#[(\lambda x.P)Q].$$

Further, if R is Q then $\#[(QS)]$ is τ so that

$$\#[(\lambda x.P)Q] = \#[\tau] + \#[\tau_1, \dots, \tau_n] + 1 > \#[(QS)].$$

Thus (a) follows. Thus necessarily if T_1 results from T by replacing the right most contractible subterm CT^* of T by its contraction, then

$$\langle m[T_1], n[T_1]\rangle < \langle m[T], n[T]\rangle.$$

Thus every term has a β-normal form.

(II) Every term with a β-normal form has an η-normal form. Since each η-reduction of a term results in a term with one fewer occurrences of λ, it follows that every term of TT has a η-normal form and therefore a normal form.

(III) Two normal forms of the same term are bound variable variants of one another.

Let M and N be two normal forms for a term L. Thus there are contraction sequences

$$L, M_1, \dots, M_m \text{ and } L, N_1, \dots, N_n$$

for which M is M_m and N is N_n. Necessarily if one of m and n is 0, the other must be also since L is then in normal form. Thus it may be assumed that $1 \leq min\{m, n\}$.

By the weak Church-Rosser Lemma 20, there is a term O for which

$$L > M_1 \gg O \text{ and } L > N_1 \gg O.$$

If $m = 1$, then M_1 is O and therefore a bound variable variant of the normal form M_1 of L. Assume that whenever $m < k$ that O is a bound variable variant of the normal form M_m of L and consider the case $m = k$. But the sequence $M_1, \dots, M_m$ has one fewer members than the sequence $L, M_1, \dots, M_m$ so that by the induction assumption O is a normal form of M_1 that is a bound variable variant of M_m. Repeating the argument for $L, N_1, \dots, N_n$ results in the conclusion that O is a normal form of N_1 that is a bound variable variant of N_n. Thus the two normal forms M and N of L must be bound variable variants of each other. ⊣

By selecting one of the bound variable variants of a term in normal form as the representative normal form, each term of TT of predicate type can be assigned a unique normal form. By defining the normal form of a constant or variable cv:1 to be cv, every term of TT can be said to have a unique normal form. This justifies referring to *the* normal form of a given term of TT. Because the intension of a predicate is identified with its name, two predicate names with the same normal form are understood to have the same intension. The converse, however, is not necessarily true: Two predicate names with different normal forms may be assigned the same intension by definition.

Compare now the Definition 30 of §2.1.1 with the original definition of formula as a term of type []. That each formula of Definition 30 is a term of type [] can be proved by induction on the definition. The converse is proved next.

THEOREM 22. *Alternative Definition of Formulas.*
Each term of type [] *is a formula as defined in Definition* 30 *of* §2.1.1.

PROOF. Let T be a term of type []. T is either a constant or variable cv, or an application term, since it cannot be an abstraction term. By syntactic sugaring T may be assumed to be of the form $P(T_1, \ldots, T_n)$, where $n \geq 0$ and P is not an application term. From T a possibly infinite binary tree $\Im[T]$ of terms of type [] is constructed as follows:

1. T is the root of the tree.
2. Let $P(T_1, \ldots, T_n)$ be a leaf of the tree where $n \geq 0$ and P is not an application term. Then
   (a) Let P be $\downarrow$ so that n is 2. Then T_1 and T_2 are added as nodes below $P(T_1, \ldots, T_n)$.
   (b) Let P be $\exists$ so that n is 1. Therefore T_1 has type $[\tau]$ for some type τ. Let y be a variable of type τ that does not occur free in T_1. Then $(T_1 y)$, which is of type [], is added as a node below $P(T_1)$.
   (c) Let P be $=$ so that n is 2. Therefore T_1 and T_1 have the same type and $P(T_1, T_2)$ is an elementary formula.
   (d) Let P be an abstraction term $(\lambda v.Q)$. Thus $(\lambda v.Q)T_1$ is contractible. $([T_1/v]Q)(T_2, \ldots, T_n)$ is added as a node below $P(T_1, \ldots, T_n)$.

The only possible form for P other than those enumerated in clause (2) is a constant or variable cv. Thus each finite branch of $\Im[T]$ ends in a leaf of the form $cv(T_1, \ldots, T_n)$.

Necessarily each branch of $\Im[T]$ is finite. For consider a branch of the tree and a node of the branch that is not the root node. If the node has been added to the branch because of (2a), then T_1, respectively T_2, can be regarded as the contraction of the term $\downarrow(T_1, T_2)$ with the occurrence of $\downarrow$ understood to be respectively $(\lambda u, v.u)$ and $(\lambda u, v.v)$,where neither u or v has a free occurrence in T_1 or T_2. If the node has been added because of (2b), then $(T_1 y)$ can be regarded as the contraction of the term $\exists(T_1)$ with the occurrence of $\exists$ understood to be $(\lambda v.vy)$. Thus the branch can be understood to be the contraction sequence of a term and therefore by Theorem 21 necessarily finite.

The leaves of the finite tree $\Im[T]$ are formulas by Definition 30. Further, if the node or nodes immediately below a node are formulas by Definition 30, then so is the node. Thus by induction on the depth of the tree the root T of the tree is a formula by Definition 30. ⊣

2.4. The Proof Theory

In a reversal of the order of presentation in Chapter 1, the proof theory of TT will be presented before the semantics to provide some background to concepts of importance for the semantics.

A derivation in TT is similar in form to one in EL as defined in §1.7.2; it is a semantic tree constructed from applications of semantic rules in which each branch has a closing pair of nodes $+F$ and $-F$. The semantic rules are:

$$+\downarrow \quad \frac{+[F\downarrow G]}{-F} \quad \frac{+[F\downarrow G]}{-G} \qquad\qquad -\downarrow \quad \frac{-[F\downarrow G]}{+F \quad +G}$$

$$+\exists \quad \frac{+\exists P}{+P(y)} \qquad\qquad -\exists \quad \frac{-\exists P}{-P(t)}$$

y not free above conclusion

$$+= \quad \frac{+P=Q}{+\forall Z.[Z(P)\rightarrow Z(Q)]} \qquad\qquad -= \quad \frac{-P=Q}{-\forall Z.[Z(P)\rightarrow Z(Q)]}$$

Z not free in P or Q

$$+\lambda \quad \frac{+F}{+G} \qquad\qquad -\lambda \quad \frac{-F}{-G}$$

$$F > G$$

The rules for $\downarrow$, $\exists$, λ, and $=$ are the *logical* rules of TT. The remaining rule is:

$$Cut \quad \frac{\qquad}{+F \quad -F}$$

Compare the rules other than Cut with the steps in the construction of the binary tree $\Im[T]$ in the proof of Theorem 22 of §2.3. The T_1 and T_2 of (2a) are the F and G of the $\downarrow$ rules, the $(T_1 y)$ of (2a) is the $P(y)$ of the $+\exists$ rule, and $([T_1/v]Q)(T_1,\ldots,T_n)$ of (2c) is $[t/v]T(\overline{s})$ of the λ rules. Only the formula $P(t)$ in the conclusion of the $-\exists$ rule has no corresponding formula in the construction of the tree. That rule is the only one in which a choice must be made, namely the choice of the eigen term t in the conclusion of the rule.

As with the first order logic, it will be assumed that rules of deduction for all the usual logical connectives and for the universal quantifier $\forall$ have been provided by the reader. These will be used in the derivations given below.

EXERCISES §2.4.

1. Consider the following modified proof theory of TT: The $\pm\lambda$ rules are dropped and the definition of a closing pair enlarged to include any pair $+F$ and $-G$ for which $F \simeq G$ where $\simeq$ expresses that F and G have the same normal form. Prove that a sequent is derivable in the modified TT

if and only if it is derivable in TT. To prove the equivalence it is necessary to make use of the Cut rule in the modified TT.

2. Provide a definition for $\neg$ in terms of $\downarrow$ and derive rules $\pm\neg$. Assuming that $\neg$ is defined and that $\exists : [[\tau]]$ is primitive, provide a definition for $\forall : [[\tau]]$ and derive rules $\pm\forall$.

2.5. Intensional and Extensional Identity

An alternative to taking $=$ as a special constant of TT is to define it as follows

$$= \stackrel{\text{df}}{=} (\lambda x, y.\forall Z.[Z(x) \rightarrow Z(y)]).$$

In this definition the variables x and y may have any type τ so that the variable Z has the type $[\tau]$. Thus $=$ as defined here, as when it is taken to be a special constant, is a polymorphic notation; that is one like the constant $\exists$ which can have different types depending upon the context in which it is used. Once the types of the arguments of $=$ are known, the type of $=$ is known. For reasons that will become apparent, $=$ is called *intensional* identity.

Note that intensional identity is expressed in terms of the predicates that can be applied to its arguments. For example, if $c, d{:}1$, then $c = d$ means that the truth value resulting when a predicate of type [1] is applied to c is the same as the truth value when it is applied to d; there are no predicates that distinguish between c and d. It is Leibniz' identity of indiscernibles.

An *extensional* identity $=_e$ can also be defined:

$$=_e \stackrel{\text{df}}{=} (\lambda x, y.\forall \overline{u}.[x(\overline{u}) \leftrightarrow y(\overline{u})])$$

when $x, y{:}[\overline{\tau}]$ and $u_i{:}\tau_i$ for $0 \le i \le n$. The binary predicate $=_e$ defined here is of type $[[\overline{\tau}], [\overline{\tau}]]$. Thus like $=$, $=_e$ is a polymorphic notation; once the type $[\overline{\tau}]$ of the arguments of $=_e$ has been specified, the type of $=_e$ is determined. For example, if $\overline{\tau}$ is empty then

$$=_e (x, y) \stackrel{\text{df}}{=} x \leftrightarrow y.$$

The usual infix notation for $=_e$ is introduced by the definition scheme:

$$r =_e s \stackrel{\text{df}}{=\!=}_e (r, s),$$

where $r, s{:}[\overline{\tau}]$.

Note that extensional identity is expressed in terms of the subjects to which its arguments can be applied. The relationship between intensional and extensional identities is expressed in the following derivable sequent:

(IEId.1) $$\vdash \forall x, y.[x = y \rightarrow x =_e y].$$

Note that although $x{=}y$ is a term if $x, y{:}1$, $x{=}_e y$ is not; x and y must necessarily have the same predicate type.

An abbreviated derivation follows in which it is assumed that x, y:[1]. A derivation for any other case has a similar form. The derivation is annotated with the rules of deduction justifying each node; later these justifications will be left to the reader in all but the most difficult cases.

$-\forall x, y.[x = y \rightarrow x =_e y]$	
$-[x = y \rightarrow x =_e y]$	$-\forall$
$+x = y$	$- \rightarrow$
$-x =_e y$	$- \rightarrow$
$+\forall z.[z(x) \rightarrow z(y)]$	$+ =$
$+[P(x) \rightarrow P(y)]$	$+\forall$; $P \overset{\text{df}}{=} (\lambda u.\neg u =_e y)$

L R

$-P(x)$	$+ \rightarrow$
$-(\lambda u.\neg u =_e y)(x)$	df P
$-\neg x =_e y$	$-\lambda$
$+x =_e y$	$-\neg$

R

$+P(y)$	$+ \rightarrow$
$+(\lambda u.\neg u =_e y)(y)$	df P
$+\neg y =_e y$	$+\lambda$
$-y =_e y$	$+\neg$
$-\forall u.[y(u) \leftrightarrow y(u)]$	df of $=_e$
$-[y(u) \leftrightarrow y(u)]$	$-\forall$
$-[[y(u) \rightarrow y(u)] \wedge [y(u) \rightarrow y(u)]]$	df $\leftrightarrow$

RL RR

$-[y(u) \rightarrow y(u)]$	$-\wedge$
$+y(u)$	$- \rightarrow$
$-y(u)$	$- \rightarrow$

RR

$-[y(u) \rightarrow y(u)]$	$-\wedge$
$+y(u)$	$- \rightarrow$
$-y(u)$	$- \rightarrow$

The definition of the term P substituted in line 6 of the derivation is a *local* definition used only in the derivation; it is an abbreviation used in this case to make clearer the eigen term of the application of the $+\forall$ rule. A variation of this style of derivation is often useful; a substitution of an unknown term is made with the goal of discovering a definition for it that will close a branch.

The converse of (IEId.1), namely

(EIId) $$\vdash \forall x, y.[x =_e y \rightarrow x = y]$$

is not derivable. For consider the following open branch:

$-\forall x, y.[x =_e y \rightarrow x = y]$	
$-[x =_e y \rightarrow x = y]$	$-\forall$
$+x =_e y$	$-\rightarrow$
$-x = y$	$-\rightarrow$
$+\forall u.[x(u) \leftrightarrow y(u)]$	df $=_e$
$-\forall z.[z(x) \rightarrow z(y)]$	$+=$
$-[z(x) \rightarrow z(y)]$	$-\forall$
$+z(x)$	$-\rightarrow$
$-z(y)$	$-\rightarrow$

Any Cut-free derivation of (EIId) must include these initial nodes. As with first order logic, a sequent has a derivation in TT only if it has a Cut-free derivation. Although this will not be proved directly, the proof provided in Chapter 3 for ITT can be adapted for TT. Thus if there is a derivation of (IEId.2), then there is a derivation without Cut and with these initial nodes. But clearly no choice of an eigen term for an application of $+\forall$ with premiss $+\forall u.[x(u) \leftrightarrow y(u)]$ can result in a closing of the branch. This example is relevant to a discussion of the semantics for TT given in §2.6 below.

An *extensionality axiom* equivalent to the sequent (IEId.2) is often accepted in higher order logics and is essential in set theory; see, for example, [28] and [132]. But such an axiom is not acceptable for an applied logic trying to meet the needs of computer science. For in some such applications, the intension of a predicate is known only informally and its extension is provided by data entry. For example, consider again the predicates *Emp* and *Male* introduced in §1.5. A company database may be maintained by listing the extension of the *Emp* predicate with a value for each employee to indicate the employee's sex. The predicates *Male* and $(\lambda y.[Emp(y) \wedge Male(y)])$ can then be defined. The extension of the latter can then be retrieved from the database and printed; see, for example [60]. By an accident of hiring, however, the two predicates *Emp* and $(\lambda y.[Emp(y) \wedge Male(y)])$ may have the same extension; but clearly their intensions must be distinguished. For this reason an extensionality axiom concluding the intensional identity of predicates from their extensional identity is not assumed for TT or ITT.

A second example is provided in [108] where extensional and intensional occurrences of predicate variables are distinguished.

2.5.1. Extensional and intensional predicates. A predicate of predicates is *extensional* if its application to a subject that is a predicate depends only on the extension of the subject and not on its intension. A predicate *Ext* with extension the arity one extensional predicates of predicates can be defined using extensional identity:

$$Ext \stackrel{\text{df}}{=} (\lambda z.\forall x, y.[x =_e y \rightarrow [z(x) \rightarrow z(y)]]).$$

In this definition, z:[t[x]], where necessarily x, y:[$\overline{\tau}$].

EXAMPLE 1. The predicate $\exists$ of predicates is extensional; that is, the following sequent is derivable:

$$\vdash Ext(\exists).$$

A derivation follows:

$-Ext(\exists)$	
$-(\lambda z.\forall x, y.[x =_e y \rightarrow [z(x) \rightarrow z(y)]])(\exists)$	df of Ext
$-\forall x, y.[x =_e y \rightarrow [\exists(x) \rightarrow \exists(y)]]$	
$+x =_e y$	
$-[\exists(x) \rightarrow \exists(y)]$	
$+\exists(x)$	
$-\exists(y)$	
$+x(v)$	$+\exists$
$-y(v)$	$-\exists$
$+\forall u.[x(u) \leftrightarrow y(u)]$	df $=_e$
$+[x(v) \leftrightarrow y(v)]$	
$+[x(v) \rightarrow y(v)]$	

$-x(v)$ $\qquad$ $+y(v)$

EXAMPLE 2. Because $=_e$ is polymorphic the string $Ext(Ext)$ can be properly typed as a formula. However, it must be recognized that the first occurrence of Ext is of type [τ], where τ is the type of the second occurrence. Thus the two occurrences are not identical terms. In the derivation for the next sequent the two types are distinguished.

$$\vdash Ext(Ext)$$

A derivation follows:

$-Ext(Ext)$	
$-(\lambda z.\forall x, y.[x =_e y \rightarrow [z(x) \rightarrow z(y)]])(Ext)$	df of Ext
$-\forall x, y.[x =_e y \rightarrow [Ext(x) \rightarrow Ext(y)]]$	
$-\forall y.[x =_e y \rightarrow [Ext(x) \rightarrow Ext(y)]]$	
$-[x =_e y \rightarrow [Ext(x) \rightarrow Ext(y)]]$	
$+x =_e y$	
$-[Ext(x) \rightarrow Ext(y)]$	
$+Ext(x)$	
$-Ext(y)$	

$+(\lambda z.\forall u, v.[u =_e v \rightarrow [z(u) \rightarrow z(v)]])(x)$ df of Ext
$+\forall u, v.[u =_e v \rightarrow [x(u) \rightarrow x(v)]]$
$-(\lambda z.\forall u, v.[u =_e v \rightarrow [z(u) \rightarrow z(v)]])(y)$ df of Ext
$-\forall u, v.[u =_e v \rightarrow [y(u) \rightarrow y(v)]]$
$-[u =_e v \rightarrow [y(u) \rightarrow y(v)]]$
$+u =_e v$
$-[y(u) \rightarrow y(v)]$
$+y(u)$
$-y(v)$
$+[u = ev \rightarrow [x(u) \rightarrow x(v)]]$

L R
$-u =_e v$

R
$+[x(u) \rightarrow x(v)]$

RL RR
$-x(u)$
$-y(u)$

RR
$+x(v)$
$+y(v)$

2.5.2. Identity and string identity. Note that intensional identity $=$ is not to be understood as string identity. Consider, for example, the terms $(\lambda v.C(v))c$ and $C(c)$. The sequent

$$\vdash (\lambda v.C(v))c = C(c)$$

is derivable by one application of $-\lambda$. But as *strings* the two terms are clearly not identical. String identity is only recognizable in the metalanguage of TT and is not definable in the logic itself.

EXERCISES §2.5.

1. Provide a derivation for the following sequent:

$$\vdash \forall x, y.[x =_e y \rightarrow \forall z.[Ext(z) \rightarrow [z(x) \rightarrow z(y)]]]$$

2. Let $Z{:}[\overline{\tau}]$ be a variable without a free occurrence in $P{:}[[\overline{\tau}]]$. Prove that a sufficient condition for $\vdash Ext(P)$ to be derivable is that every elementary subformula in which Z has an occurrence takes the form $Z(\overline{s})$. Use this result to conclude that $\vdash Ext(Ext)$, example 1 of §2.5.1, is derivable.

3. Using the $\pm =$ rules, provide derivations for sequents expressing that $=$ is reflexive, symmetric, and transitive.

2.6. Semantics for TT

In §1.4 an interpretation was defined as a function that maps the subject names onto entities, and the predicate names of arity n onto n-tuples of entities. By assuming a fixed interpretation for each given subject name, that is for each constant of type 1, it is possible to regard the names as proxies for the entities they denote. Thus interpretations to a set of entities can be ignored and replaced by valuations to a domain $D(1)$ consisting of the constants of type 1. This simplification is used here and in later chapters.

An extensional semantics is provided for TT in §2.6.1; that is, all predicates of predicates of TT will be assumed to be extensional. It is a direct adaptation of the semantics provided for first order logic in Chapter 1, but because of the extended definition of type, Definition 26 in §2.1, the inelegant special cases of first order logic are absorbed into the general cases. Then an intensional semantics is provided in §2.6.2 in which predicates of predicates are not required to be extensional.

2.6.1. An extensional semantics for TT. An *extensional* valuation domain D for TT differs only in clause (D.3) from Definition 8 of §1.4. It is a function with domain the types, and range defined as follows:

DEFINITION 39. *Extensional Valuation Domain.*
An extensional *valuation domain for a valuation is a function D from the types of the logic to the following sets:*

$D(1)$ *is the set of constants of type* 1.

$D([\,])$ *is* $\{+, -\}$.

$D([\overline{\tau}])$ *is a non-empty subset of* $\wp(D(\tau_1) \times \cdots \times D(\tau_n))$, *the set of all subsets of the Cartesian product* $D(\tau_1) \times \cdots \times D(\tau_n)$.

The *standard* domain D is the one for which $D([\overline{\tau}])$ is $\wp(D(\tau_1) \times \cdots \times D(\tau_n))$ for all types $\tau_1, \ldots, \tau_n$, $n \geq 1$.

The definition of a valuation for first order logic completed with Definition 17 in §1.5.2 is extended here for TT. Because of the general form of the original definition, the definition for TT requires few changes. But the treatment of the special constant $=$ requires special care. The definition of the value $\Phi(T)$ of the extensional valuation function Φ for the argument T is by induction on $ct[T]$. Thus one might expect $\Phi(=)$ to be defined in (V.1.2) along with $\Phi(\downarrow)$ and $\Phi(\exists)$. But no suitable definition of $\Phi(=)$ can be given at that stage. As a consequence, $=$ must be treated in the definition as though it were defined:

$$=\stackrel{\text{df}}{=} (\lambda u, v.\forall Z.[Z(u) \rightarrow Z(v)])$$

and $ct[=]$ was $ct[(\lambda u, v.\forall Z.[Z(u) \rightarrow Z(v)])]$. Thus $\Phi(=)$ is not defined in (V.1.2) but is defined later in (V.4).

DEFINITION 40. *Extensional Valuation.*
An extensional *valuation is a function* Φ *defined for each term* T *of TT by induction on* $ct[T]$ *as follows*:

V.1.1. $\Phi(cv) \in D(\mathrm{t}[cv])$, *for each variable or constant* cv *other than* $\downarrow$, $\exists$, *or* $=$.

V.1.2. $\Phi(\downarrow)$ *is* $\{\langle -, -\rangle\}$, $\Phi(\exists)$ *is a non-empty set of non-empty subsets of* $D(\tau)\}$ *when* $\exists$:$[[\tau]]$, *and* $\Phi(=)$ *is* $\{\langle e, e\rangle \| e \in D(\tau)\}$ *when* $=$: $[\tau, \tau]$.

V.2. *Let* P:$[\tau, \overline{\tau}]$ *and* Q:τ. *Then when* $\overline{\tau}$ *is not empty*, $\Phi(PQ)$ *is the set of tuples* $\langle e_1, \ldots, e_n\rangle$ *for which* $\langle \Phi(Q), e_1, \ldots, e_n\rangle \in \Phi(P)$. *When* $\overline{\tau}$ *is empty*, $\Phi(PQ)$ *is* $+$ *if* $\Phi(Q) \in \Phi(P)$, *and is* $-$ *otherwise.*

V.3. *Let* P:$[\overline{\tau}]$ *and* v:τ. *Then when* $\overline{\tau}$ *is not empty*, $\Phi(\lambda v.P)$ *is the set of tuples* $\langle \Phi^v(v), e_1, \ldots, e_n\rangle$ *for which* $\langle e_1, \ldots, e_n\rangle \in \Phi^v(P)$ *for some* v*-variant* $\Phi^v(v)$ *of* Φ. *When* $\overline{\tau}$ *is empty*, $\Phi(\lambda v.P)$ *is the set* $\{\Phi^v(v) \| \Phi^v(P)$ is $+\}$.

It is assumed that $\Phi(\exists)$ in clause (V.1.2) is separately defined for each type τ. To be fully precise the type of $\exists$ should be another argument of Φ. However the need for such an additional argument will be left unsatisfied until the intensional semantics of ITT is defined in §3.3.

Thc definition of model in §1.4.1 remains valid for this extensional semantics for TT: a model with domain D is a valuation Φ for which $\Phi(P) \in D(\tau)$ for all P and τ for which P:τ. An extensional valuation that is a model is called an *extensional* model. Any valuation to the standard domain is an extensional model.

Extensional models to a non-standard domain are sometimes called *Henkin* models because they were first described in [74].

Consider again the sequent (IEId.2) of §2.5. It was claimed that it is underivable in TT. Therefore if the extensional semantics for TT were complete, there would be an extensional Henkin model Φ for which $\Phi(\forall x, y.[x =_e y \rightarrow x = y])$ is $-$ Consider the branch in §2.5 resulting from an attempt to construct a derivation for (IEId.2). It follows that the truth values indicated by the signed formulas must be the truth values for the formulas by Φ. In particular there is an x, y, z-variant $\Phi^{x,y,z}$ of Φ for which

$$\Phi^{x,y,z}(\forall u.[x(u) \leftrightarrow y(u)]) \text{ is } +, \Phi^{x,y,z}(z(x)) \text{ is } +, \text{ and } \Phi^{x,y,z}(z(y)) \text{ is } -.$$

But this is impossible because from the first assertion it follows that the extensions of $\Phi^{x,y,z}(x)$ and $\Phi^{x,y,z}(y))$ are identical, while the values of $\Phi^{x,y,z}(z(x))$ and $\Phi^{x,y,z}(z(y))$ depend only on these extensions. Indeed

$$\Phi(\forall x, y.[x =_e y \rightarrow x = y]) \text{ is } +.$$

It follows that extensional models are not the appropriate models for the proof theory of TT. Intensional models will be defined next.

2.6.2. An intensional semantics for TT. For the extensional valuations defined in §2.6.1, only the extensions of subject and predicate names are needed. For an intensional valuation Φ, both the intension and the extension of a name are needed: $\Phi(T)$ for a term T is

$$\langle \Phi_1(T), \Phi_2(T) \rangle,$$

where $\Phi_1(T)$ is the intension of T and $\Phi_2(T)$ its extension.

The terms of type 1 present a special case. A constant c:1 is the name of some entity. Its intension is taken to be its name c while its extension is the entity it names. Rather than introduce a special set of entities, the device used for first order logic will be used for the type theory; namely the constants c:1 are regarded as proxies for entities. The extension $\Phi_2(c)$ of c may be taken to be the constant c:1 itself. This is reflected in the first clause of the definition of DI given next. The definition of an *intensional* valuation domain $DI(\tau)$ for the type τ is by induction on the Definition 26 in §2.1 of the types of TT.

DEFINITION 41. *Intensional Valuation Domain DI for TT.*
N in each case is a term in normal form and without free variables.

$DI(1)$ *is* $\{\langle c, c\rangle \| c{:}1\}$.

$DI([\,])$ *is* $\{\langle N, \mathsf{s}\rangle \| N{:}[\,]$ *and* $\mathsf{s} \in \{+, -\}\}$.

$DI([\overline{\tau}])$ *is* $\{\langle N, S\rangle \| N{:}[\overline{\tau}], \mathsf{S} \in \mathsf{PD}\}$, *where* $\overline{\tau}$ *is* $\tau_1, \ldots, \tau_n$ $n \geq 1$, *and* PD *is a nonempty subset of the powerset* $\wp(DI(\tau_1) \times \cdots \times DI(\tau_n))$.

The *standard* intensional valuation domain for TT results when PD in the definition of $DI([\tau_1, \ldots, \tau_n])$ is $\wp(DI(\tau_1) \times \cdots \times DI(\tau_n))\}$.

The following definition of the value $\Phi(T)$ of an intensional valuation Φ is by induction on $ct[T]$. It modifies Definition 40 of §2.6.1 but treats $=$ in the same fashion.

DEFINITION 42. *An Intensional Valuation for TT to a Domain DI.*

V.1.1. *For a constant* $c{:}\tau$, *other than* $\downarrow$, $\exists$ *or* $=$, $\Phi_1(c)$ *is* c, *and* $\Phi_2(c)$ *satisfies* $\langle \Phi_1(c), \Phi_2(c)\rangle \in DI(\tau)$. *For a variable* $v{:}\tau$, $\Phi(v)$ *is any member of* $DI(\tau)$. *Thus* $\Phi_1(v)$ *is some* $N{:}\tau$ *in normal form and without free variables, and* $\Phi_2(v)$ *is such that* $\langle \Phi_1(v), \Phi_2(v)\rangle \in DI(\tau)$.

V.1.2. $\Phi_1(\downarrow)$ *is* $\downarrow$ *and* $\Phi_2(\downarrow)$ *is a set of pairs of pairs* $\langle\langle \Phi_1(M), -\rangle, \langle \Phi_1(N), -\rangle\rangle$, *for which* $M, N{:}[\,]$ *are in normal form and without free variables.* $\Phi_1(\exists)$ *is* $\exists$ *and* $\Phi_2(\exists)$, *when* $\exists{:}[[\tau]]$, *is a set of pairs* $\langle \Phi_1(P), \mathrm{S}\rangle$, *for which* $P{:}[\tau]$ *and* S *is a non-empty subset of* $DI(\tau)$. $\Phi_1(=)$ *is* $=$ *and* $\Phi_2(=)$ *is the set of pairs* $\{\langle e, e\rangle \| e \in D(\tau)\}$, *when* $= {:}[\tau, \tau]$.

Let $T{:}\tau$ and let $ct[T] > 0$. Let $x_1{:}\tau_1, \cdots x_n{:}\tau_n$ be all the variables with a free occurrence in T. $\Phi_1(x_i)$ is defined for each i, $1 \leq i \leq n$, and is a term without free variables. Define the term T^* to be

$$[\Phi_1(x_1)/x_1] \cdots [\Phi_1(x_n)/x_n]T$$

and $\Phi_1(T)$ to be the normal form of T^*. Thus for $ct[T] > 0$, to define $\Phi(T)$ it is sufficient to define $\Phi_2(T)$. In all cases $\Phi_2(T)$ is defined to be $\Phi_2(T^*)$ that is defined:

V.2. *Let $P{:}[\tau, \overline{\tau}]$ and $Q{:}\tau$. When $\overline{\tau}$ is not empty, $\Phi_2((PQ))$ is the set of tuples $\langle e_1, \ldots, e_n \rangle$ for which $e_i \in DI(\tau_i)$ and $\langle \Phi(Q), e_1, \ldots, e_n \rangle \in \Phi_2(P)$. When $\overline{\tau}$ is empty, $\Phi_2((PQ))$ is $+$ if $\Phi(Q) \in \Phi_2(P)$, and is $-$ otherwise.*

V.3. *Let $P{:}[\overline{\tau}]$ and $v{:}\tau$. When $\overline{\tau}$ is not empty, $\Phi_2((\lambda v.P))$ is the set of tuples $\langle \Phi^v(v), e_1, \ldots, e_n \rangle$ for which $e_i \in DI(\tau_i)$ and $\langle e_1, \ldots, e_n \rangle \in \Phi_2^v(P)$ for some v-variant Φ^v of Φ. When $\overline{\tau}$ is empty, $\Phi_2((\lambda v.P))$ is the set* $\{\Phi^v(v) \| \Phi_2^v(P)$ is $+$, Φ^v a v-variant of $\Phi\}$.

V.4. $\Phi(=)$ *is* $\Phi(\lambda u, v.\forall Z.[Z(u) \to Z(v)])$.

Intensional models are defined similarly to extensional ones: An *intensional* model with domain DI is a valuation Φ to DI for which $\Phi(T) \in DI(\tau)$ for all T and τ for which $T{:}\tau$.

Consider again the sequent (IEId.2) of §2.5. It is not valid for intensional models, since if Φ is intensional, the three displayed value assertions in §2.6.1 are compatible. Thus an intensional semantics, rather than an extensional, is appropriate for TT. Although this will not be confirmed directly with a proof of the completeness of TT for the intensional semantics, the proof given in Chapter 3 for ITT adapts directly to a proof for TT.

A predicate *Ext* with extension the arity one extensional predicates of predicates was defined in §2.5.1, and a derivation of the sequent of exercise (1) left as an exercise. The sequent is of interest because it suggests that instead of interpreting TT using extensional models and accepting this sequent as an axiom as is done in Church's simple theory of types, for example, intensional models may be used and quantifiers restricted to the *Ext* predicate when required.

CHAPTER 3

AN INTENSIONAL TYPE THEORY

3.1. The Terms of ITT

Church described in [30] a logic of sense and denotation based on the simple theory of types [28]; that is, using the terminology of Carnap, a logic of intension and extension. ITT differs from that logic in two ways: First ITT is based on TT and not on the simple theory of types; but of greater importance, ITT identifies the intension of a predicate with its name while the logic of Church treats intensions as separate entities with their own types and notation. The justification for this identification is the belief that in a given context a user discovers the intension of a predicate from its name.

The types of ITT, an intensional type theory, are the types of TT, and the constants and variables of ITT are those of TT. The terms of ITT are an extension of those of TT. Definition 29 of term of TT in §2.1.1 is extended for ITT by a fourth clause that introduces a *secondary typing* for some of the terms of ITT. Since secondary typing is the main feature of ITT that distinguishes it from TT, it will be motivated in §3.1.1 before being formally expressed in clause (4) of Definition 43 in §3.1.2.

3.1.1. Motivation for secondary typing. The purpose of secondary typing in ITT is to provide a simple but unambiguous way of distinguishing between an occurrence of a predicate name where it is being used and an occurrence where it is being mentioned. The necessity for recognizing this distinction has been stressed many times and a systematic use of quotes is traditionally employed for expressing it; see for example [24] or [120]. But the systematic use of quotes is awkward in a formal logic and subject to abuses as described by Church in footnote 136 of [31]. Secondary typing exploits the typing notation of TT to distinguish between a used predicate name and a mentioned predicate name and is not subject to the abuses cited by Church.

It is necessary to distinguish between a used predicate name and a mentioned predicate name in ITT because the logic is an attempt to provide a formalization for the nominalist view of Ayer expressed in this quote repeated from the introduction:

> ... the truths of logic and mathematics are analytic propositions ... we say that a proposition is analytic when its validity depends solely on the definitions of the symbols it contains.

As noted in the introduction, to pursue this thesis more carefully it is necessary to distinguish between the *intension* of a predicate and its *extension*. Since the intension of a predicate is identified by its name, and since to express membership in a predicate it is necessary to use the predicate name, a method of distinguishing between a used predicate name and a mentioned predicate name is essential. How the typing of TT is exploited for this purpose is described next.

Define Δ to be the set of all terms of TT in which no variable has a free occurrence. The set Δ has as its members the constants of each type τ including $\downarrow$ and $\exists$, and all application and abstraction terms in which no variable has a free occurrence. Each member of Δ of type 1 is a constant that may be the name for a subject for a predicate of type [1]. Each member of Δ of type other than 1 may be the name of a predicate the normal form of which is identified with the intension of the predicate. Such a predicate P may itself be the subject of an extensional predicate of the appropriate type. In this case only the extension $\Phi_2(P)$ that P is assigned by a given valuation Φ need be known; the fact that the intension $\Phi_1(P)$ is identified with the normal form of its name is irrelevant.

Consider, for example, the predicates $(\lambda x.[Emp(x) \wedge Male(x)])$ and Emp of §2.5 where it is assumed that $\Phi(Emp)$ and $\Phi(Male)$ have been defined for some valuation Φ. To determine the extension $\Phi_2((\lambda x.[Emp(x) \wedge Male(x)]))$ of the predicate $(\lambda x.[Emp(x) \wedge Male(x)])$ from the extensions $\Phi_2(Emp)$ and $\Phi_2(Male)$, it is unnecessary to know anything about

$$\Phi_1((\lambda x.[Emp(x) \wedge Male(x)])), \Phi_1(Emp) \text{ or } \Phi_1(Male).$$

Knowledge of the intensions $\Phi_1(Emp)$ and $\Phi_1(Male)$ is only required for those responsible for maintaining the extensions $\Phi_2(Emp)$ and $\Phi_2(Male)$. For a particular valuation Φ it is possible that

$$\Phi_2((\lambda x.[Emp(x) \wedge Male(x)])) =_e Emp) \text{ is } +,$$

although clearly

$$\Phi_2((\lambda x.[Emp(x) \wedge Male(x)])) = Emp) \text{ is } -.$$

In each case $=_e$ and $=$ are of type [[1], [1]].

Now consider the predicates $(\lambda x.Emp(x))$ and Emp. Since their names have the same normal form,

$$\Phi_2((\lambda x.Emp(x)) = Emp) \tag{1}$$

must be + for every valuation Φ. Again in this case $=$:[[1], [1]]. But this suggests that it is necessary to know $\Phi_2((\lambda x.Emp(x))$ and $\Phi_2(Emp)$ to determine that (1) is +, which is clearly not true; it is only necessary to know that the two terms $(\lambda x.Emp(x))$ and Emp have the same normal form. This can be expressed in ITT by the intensional identity $=$ of type $[1,1]$ provided the terms $(\lambda x.Emp(x))$ and Emp have not only the primary type $[1]$ but also the secondary type 1. Expanding the definition of $=$, two identical formulas expressing intensional identity for terms $(\lambda x.Emp(x))$ and Emp are obtained:

(2) $$\forall z.[z((\lambda x.Emp(x))) \to z(Emp)], \text{ where } z{:}[[1]].$$

(3) $$\forall z.[z((\lambda x.Emp(x))) \to z(Emp)], \text{ where } z{:}[1].$$

But note that they have very different meanings. (2) expresses that any predicate z of type $[[1]]$ when applied to the predicate $(\lambda x.Emp(x))$ of type $[1]$ yields the same truth value as when it is applied to the predicate Emp of type $[1]$. (3) on the other hand expresses that any predicate z of type $[1]$ when applied to the string $(\lambda x.Emp(x))$ of type 1 yields the same truth value as when it is applied to the string Emp of type 1. Given that the strings $(\lambda x.Emp(x))$ and Emp have the same normal form, and that names with the same normal form are identified, (3) is true in every valuation and the truth of (2) in any valuation should follow from it since the intension of a predicate is identified with the normal form of its name.

Consider now

(4) $$\forall z.[z((\lambda x.[Emp(x) \wedge Male(x)])) \to z(Emp)], \text{ where } z{:}[[1]].$$

(5) $$\forall z.[z((\lambda x.[Emp(x) \wedge Male(x)])) \to z(Emp)], \text{ where } z{:}[1].$$

From (4), by (IEId.1), follows

(6) $$\forall u.[(\lambda x.[Emp(x) \wedge Male(x)])(u) \leftrightarrow Emp(u)].$$

Since there is a valuation for which

$$\Phi_2((\lambda x.[Emp(x) \wedge Male(x)]) =_e Emp) \text{ is } -,$$

(6) must be false in the same valuation and therefore also (5). But since the truth value of (5) in a valuation does not depend upon

$$\Phi_2((\lambda x.[Emp(x) \wedge Male(x)]) \text{ and } \Phi_2(Emp),$$

(5) must be false in every valuation.

Another example may be helpful. Consider the predicate with name 'is a colour' in the elementary sentence

(7) Yellow is a colour.

The subject to which the predicate is applied is the predicate with name 'Yellow'. 'Yellow' is a predicate name of type $[1]$ and 'is a colour' a predicate name of type $[[1]]$. To interpret 'is a colour' as an extensional predicate means that the truth of (7) for a given valuation Φ depends only upon its extension $\Phi_2(Yellow)$ and not upon its intension $\Phi_1(Yellow)$, hardly a plausible interpretation. Identifying the intension of the predicate yellow with its name 'yellow' permits interpreting the predicate with name 'colour' as an intensional predicate; the meaning of the sentence in this interpretation is that of the sentence

(8) 'Yellow' is a colour word in English.

Higher order predication in this instance is given a *nominalist* interpretation: A predicate applied to a subject predicate is understood to be a predicate of a name of the subject predicate. This interpretation requires being able to distinguish between an occurrence of a predicate name that is being mentioned, and an occurrence that is being used. Assigning the predicate name 'yellow' the secondary type 1 and 'Colour' the primary type [1] permits (7) to be expressed as:

(9) Colour(yellow).

In a context in which the predicate name 'yellow' is of type [1], the name is being used; for example, "The pencil is yellow". In a context in which the predicate name has type 1 it is being mentioned; for example in the sentence (9).

The secondary typing of predicate names is accomplished by adding a fourth clause to the three of Definition 29 in §2.1.1. The clause assigns the secondary type 1 to terms in which at most a variable of type 1 has a free occurrence. This wider assignment of secondary typing is justified by a Herbrand interpretation.

Let now Δ^+ be Δ extended to include all terms in which at most a variable of type 1 has a free occurrence. Let P be a member of $(\Delta^+ - \Delta)$, and let $\overline{v}$ be a sequence of all the variables with a free occurrence in P. Thus if each member of the sequence $\overline{s}$ is a member of Δ, then $[\overline{s}/\overline{v}]P$ is a member of Δ also. In this way P can be interpreted as a function with arguments $\overline{s}$ and a value $[\overline{s}/\overline{v}]P$ in Δ. This interpretation is known as a *Herbrand* interpretation in recognition of its first presentation in [76].

It is important to note that an abuse of use and mention may result from an attempt to extend such a Herbrand interpretation to implicitly quoted terms with a free occurrence of a variable of type other than 1. The abandonment of this simple principle for a bit of ad hocery in the logic NaDSet described and used in the two papers [65, 66] was exploited in the proof of its inconsistency by Girard [67]. That a violation of this principle results in an inconsistent logic is demonstrated in §3.4.1.

3.1.2. Secondary typing. The secondary typing of predicate names is accomplished by adding a fourth clause to Definition 29 of §2.1.1:

DEFINITION 43. *Term of ITT.*

1. *A constant or variable cv is a term of type* t$[cv]$.
2. $P{:}[\tau,\overline{\tau}]$ *and* $Q{:}\tau \Rightarrow (PQ){:}[\overline{\tau}]$.
3. $P{:}[\overline{\tau}]$ *and v a variable of type* $\tau \Rightarrow (\lambda v.P){:}[\tau,\overline{\tau}]$.
4. *Let P be a term of type $[\overline{\tau}]$ in which at most variables of type* 1 *have a free occurrence. Then P has* primary *type $[\overline{\tau}]$ and* secondary *type* 1.

Here the abbreviating notation $(\lambda v.P)$ introduced in Notation 13 in §1.5.1 is used and will continue to be used.

It is important to note that clauses (1) and (4) are the only clauses in which terms of type 1 are defined. As a consequence any term P of type 1 that is an application or abstraction term has the type 1 because of clause (4). A term of type 1 that is not an application or abstraction term, that is a constant or variable of type 1, is said to have *primary* type 1 even though it does not have 1 as secondary type.

Since the intension of a predicate is identified with the normal form of its name, two terms P and Q of type $[\overline{\tau}]$ with the same normal forms have the same intension. Should they each have secondary type 1, then the identity of their intensions can be expressed by $P = Q$, where $=$:[1, 1].

Definition 14 in §1.5.1 of depth of a subterm extended to TT and extends again to terms of ITT. But although no term of type 1 of EL or TT has a proper subterm, such subterms do now appear because of clause (4). For example, let C be a constant of type [1] and x a variable of type 1. By (2) of Definition 43, $C(x)$ is a term of type [ ] with subterms C and x each of depth 1. By (4), $C(x)$ has secondary type 1 with the same subterms of the same depth. By (2) again $C(C(x))$ is a term of type [ ] with the first occurrence of C and $C(x)$ subterms of depth 1, and the second occurrence of C and x subterms of depth 2.

Lemma 4 of §1.5.3 generalized for TT. It generalizes again for ITT:

LEMMA 23. *Substitution.*
Let Q:t$[v]$, *where* v *is any variable. Then* $[Q/v]P$ *has the same type as* P.

PROOF. The only complication for ITT involves terms P of secondary type 1. But then if v has a free occurrence in P it is necessarily of type 1 so that Q is of type 1 by virtue of either clause (1) or clause (4). Then $[Q/v]P$ has type 1 by virtue of clause (4). ⊣

3.1.3. The normal form of a term of ITT. Definition 37 in §2.3 of the relation > between terms of TT can be taken unchanged as the definition of the same relationship between terms of ITT as can Definition 38 in §2.3 of the relation ≫. In TT the relation > can only hold between terms that are not of type 1. In ITT, on the other hand, the relation can also hold between terms with secondary type 1.

The proofs of the weak Church-Rosser Lemma 20 in §2.3 and the normal form Theorem 21 can be repeated for ITT. In particular it is meaningful to refer to *the* normal form of a term of ITT as a particular bound variable variant of a normal form of the term. As in TT, the normal form of a constant or variable cv:1 is defined to be cv; thus every term of ITT has a unique normal form.

3.2. Polytyping ITT

Polymorphic typing, or just polytyping, was discussed for TT in §2.2 and a polytyping algorithm was described in Figure 3 of §2.2.4. That one occurrence of a term of ITT may be assigned a predicate type τ and a second occurrence the secondary type 1 adds a complication to polytyping in ITT.

3.2.1. A polytyping algorithm for ITT. The goal here is to demonstrate that algorithms for polytyping in ITT can be constructed. The algorithm described in Figure 4 has been adapted from the one for TT described in Figure 3 in the following ways:

1. A tree `TREE` of assignments replaces the `LIST` of assignments for the algorithm of TT. Like the trees that are derivations, the root of the tree is above every other node of the tree. Each branch of `TREE` corresponds to a `LIST`.
2. As with the algorithm for TT, a *done* assignment is one for which the processing has been completed. A branch is *closed* if the assignments on it are not feasible. A branch is *open* if it is not closed and some assignment on it is not done. The order of the processing of assignments ensures that only not done assignments can occur below a not done assignment, and that no splitting of a branch occurs below a not done assignment. As a consequence each not done assignment determines a unique branch of `TREE`. The branch determined by the currently selected not done assignment is designated `Branch`.
3. A substitution is associated with each not closed branch.
4. Initially `TREE` consists of a single open branch of not done assignments. The first assignment on this branch is selected to begin the processing so that initially `Branch` is the only branch of `TREE`. The identity substitution is associated with this branch.
5. Apart from the absence of halts under `ACTIONS`, the only substantive change in Figure 4 for ITT from the corresponding Figure 3 for TT, is in the actions specified for the 'other' case. Secondary typing requires that two options be considered for the typing of $(strastrb)$, $stra$, and $strb$, the first being the only option for TT and the second being the option of $strb$ having a secondary type 1 indicated by the assignment of a new type expression $[\overrightarrow{\beta}] \cap 1$, discussed below, to $strb$. The variables $v_1, \ldots, v_k$ are all the variables with a free occurrence in $strb$. The current `Branch` may be extended and split as described in Figure 4 into $\texttt{Branch}_1$ and $\texttt{Branch}_2$.

The new type expression $[\overrightarrow{\beta}] \cap 1$ is an example of an *intersection* type expression that may take any form $te \cap 1$, where te is a *predicate* type expression. The types of TT and ITT are not extended to include an *intersection* type $\tau \cap 1$, but a type assignment $str{:}\tau \cap 1$ is to be understood as two assignments $str{:}\tau$

```
Begin
```
Halt if no branch of `TREE` is open.
Else, let *strc*:*tec* be the first not done assignment on the first open branch `Branch` of `TREE`. Σ is substitution for `Branch`.

<u>*strc* is</u>	<u>`ACTION`</u> [To be followed by: Return to `Begin`]
$\downarrow$	`Branch` closed if $([[\,],[\,]] \sqcap \Sigma tec)$=`Nil`. Else $\Sigma=([[\,],[\,]] \sqcap \Sigma tec) \circ \Sigma$.
$\exists$	`Branch` closed if $([[\alpha]] \sqcap \Sigma tec)$=`Nil`. Else $\Sigma=([[\alpha]] \sqcap \Sigma tec) \circ \Sigma$.
$=$	`Branch` closed if $([\alpha,\alpha] \sqcap \Sigma tec)$=`Nil`. Else $\Sigma=([\alpha,\alpha] \sqcap \Sigma tec) \circ \Sigma$.
(λv)	`Branch` closed.
cv	`Branch` closed if $cv{:}teo$ on `Branch` and $(\Sigma teo \sqcap \Sigma tec)$=`Nil`. Else $\Sigma=(\Sigma teo \sqcap \Sigma tec) \circ \Sigma$.
$(strastrb)$	
<u>*stra* is</u>	

(λv)	If $((\lambda v)strb){:}teo$ on `Branch` and done,
	    `Branch` closed if $(\Sigma teo \sqcap \Sigma tec)$=`Nil`;
	    Else $\Sigma=(\Sigma teo \sqcap \Sigma tec) \circ \Sigma$.
	Else, `Branch` closed if $([\alpha,\overrightarrow{\alpha}] \sqcap \Sigma tec)$=`Nil`;
	    Else $\Sigma=([\alpha,\overrightarrow{\alpha}] \sqcap \Sigma tec) \circ \Sigma$;
	    Add $v{:}\alpha$ and $strb{:}[\overrightarrow{\alpha}]$ to `Branch`.

not (λv)	If $(strastrb){:}teo$ on `Branch` and done,
	    `Branch` closed if $(\Sigma teo \sqcap \Sigma tec)$=`Nil`;
	    Else $\Sigma=(\Sigma teo \sqcap \Sigma tec) \circ \Sigma$.
	Else, `Branch` closed if $([\overrightarrow{\alpha}] \sqcap \Sigma tec)$=`Nil`;
	    Else, $\Sigma=([\overrightarrow{\alpha}] \sqcap \Sigma tec) \circ \Sigma$;
	    For i=1, 2, Σ_i is Σ;
	    $\texttt{Branch}_1$ is `Branch` with $stra{:}[\alpha,\overrightarrow{\alpha}]$ and $strb{:}\alpha$ added;
	    $\texttt{Branch}_2$ is `Branch` with $stra{:}[1,\overrightarrow{\alpha}]$, $strb{:}[\overrightarrow{\beta}] \cap 1$, and
	    $v_1{:}1,\ldots,v_k{:}1$ added.

FIGURE 4. Polytyping ITT

and *str*:1. Thus *str* can be assigned the secondary type 1 only if it has already been assigned a predicate type.

The definition of the range RAN(te) of a type expression *te* is not extended to include intersection type expressions. But an extension of Definition 35 of $(tea \sqcap teb)$ in §2.2.3 is provided in Definition 44.

DEFINITION 44. *Extended* $\sqcap$.
Let tea and teb be not intersection type expressions. Then $(tea \cap 1 \sqcap teb)$ *is* $(tea \sqcap teb \cap 1)$ *is* $(tea \cap 1 \sqcap teb \cap 1)$ *is* $(tea \sqcap teb)$.

EXAMPLE 1. This example provides a simple illustration of the effect of the new 'other' actions possible for ITT. The example contrasts the processing of constants with the processing of variables. For each proposed assignment two branches are created, and in each branch of each case the substitution $[/\overrightarrow{\alpha}]$ results from the test $([\overrightarrow{\alpha}]\sqcap\Sigma[\,])=$`Nil`. For each of the second branches created the substitution $[1/\overrightarrow{\beta}]$ results from the test $([\overrightarrow{\beta}]\sqcap\Sigma[1])=$`Nil`. As indicated in the figure by a wide line, only one branch of the four illustrated is not closed. It is the branch assigning C the primary type [1] and the secondary type 1. The corresponding branch for (XX) is closed by the conflict between the assignment $X{:}1$, needed because X is a variable with a free occurrence in X, and the assignment $X{:}[1]$.

$(CC){:}[\,]$		$(XX){:}[\,]$	
$C{:}[\alpha]$	$C{:}[1]$	$X{:}[\alpha]$	$X{:}[1]$
$C{:}\alpha$	$C{:}[1]\cap 1$	$X{:}\alpha$	$X{:}[1]\cap 1$
▬		▬	$X{:}1$
			▬

Assuming substitutions $[/\overrightarrow{\alpha}]$ and $[1,\overrightarrow{\beta}]$ have been applied.

EXAMPLE 2. This example illustrates how dual typing leads to a simple sequence of terms, all of type $[1]\cap 1$, generated by constants or constant terms 0 and S. The proposed assignment $S(S(0)){:}[1]$ is not feasible in TT. If 0 and S are defined: $0 \overset{\text{df}}{=} (\lambda u.\neg u = u)$ and $S \overset{\text{df}}{=} (\lambda u, v.u = v)$,then the terms of the sequence can be proved distinct; see example 3 and §4.4.1.

$$
\begin{array}{ll}
S(S(0)){:}[1] & \\
0{:}[1] & \\
S{:}[1,1,\overrightarrow{\alpha}] & ([1,\overrightarrow{\alpha}\sqcap[1]) \text{ is } [/\overrightarrow{\alpha}] \\
S(0){:}[\overrightarrow{\beta}]\cap 1 & \\
S{:}[1,\overrightarrow{\beta}] & ([1,\overrightarrow{\beta}]\sqcap[1,1]) \text{ is } [1/\overrightarrow{\beta}] \\
0{:}[\overrightarrow{\gamma}]\cap 1 & ([\overrightarrow{\gamma}]\cap 1\sqcap[1]) \text{ is } [1/\overrightarrow{\gamma}].
\end{array}
$$

EXAMPLE 3. Let also predicates RN and Lt be defined:

$$RN \overset{\text{df}}{=} \lambda Z, y.[y = 0 \vee \exists x.[Z(x) \wedge y = S(x)]].$$

$$Lt \overset{\text{df}}{=} \lambda wg, u.\forall Z.[\forall x.[wg(Z,x) \rightarrow Z(x)] \rightarrow Z(u)]).$$

Consider first the types that may be assigned to RN. There are two possibilities depending upon whether $S(x)$ is given the secondary type 1.

(a) Let $x{:}\tau$, where $\tau \neq 1$, so that $S(x)$ has type $[\tau]$ but not type 1.

Then necessarily $y{:}[\tau]$ and $Z{:}[\tau]$ in RN. The two occurrences of y have the same type so that $RN{:}[[\tau],[\tau]]$.

(b) Let $x{:}1$ and let $S(x)$ have the secondary type 1.

Then Z:[1] and y:1, so that RN:[[1], 1].

The type that RN receives in a given context depends upon the context. For consider now Lt. Let $x, u{:}\tau$, so that $Z{:}[\tau]$, $wg{:}[[\tau], \tau]$, and $wg(Z, x), Z(x){:}[\,]$. Hence $Lt{:}[[[\tau], \tau], \tau]$. Assume that $Lt(RN, x){:}[\,]$. Then necessarily RN:[[1], 1] and τ is 1. As an argument for the predicate Lt, RN can have only the type [[1], 1], and $Lt(RN)$ the type [1].

This example is returned to in §4.4 where Peano's axioms [115] for the natural numbers are shown to be derivable in ITT. It is demonstrated there that all the axioms can be derived under the typing (b) when a predicate N:[1] is defined to be $Lt(RN)$. This renders mute for ITT a common objection to type theories that polytyping permits the multiplication of ordinary mathematical concepts in infinitely many types. The sequence of terms defined by 0 and S are firmly anchored at type 1. This is the same reply to the objection that Rushby provides in (A.10) of [124], although in the type theory he discusses an axiom of infinity must be assumed to provide the anchoring.

THEOREM 24. *Correctness of Polytyping Algorithm for ITT.*
For any initial `TREE` *the algorithm of Figure* 4 *halts. If it halts with a* not closed *branch* `Branch` *with substitution* Σ, *then the base range defined as a sublist of* Σ`Branch` *is a feasible acceptable range for the initial assignments on* `TREE`. *If it halts without such a branch then the initial assignments are not feasible. Further if the initial assignments have a feasible acceptable typing then the typing is a member of the base range defined by an open branch.*

PROOF. Given the detailed proof provided for Theorem 19 in 2.2.4 of the correctness of the polytyping algorithm of TT described in Figure 3, it is sufficient to discuss the substantive change to the 'other' case described in item (5) of this subsection. The two tests and subsequent actions in the '`Branch` closed' lines in the ITT figure correspond to the tests and subsequent actions in the 'Halt' lines in the TT figure. The assignments added to `Branch` to create $\texttt{Branch}_1$ are those added to `LIST` in the TT figure. The assignments added to `Branch` to create $\texttt{Branch}_2$ are the most general possible to ensure that *strb* is assigned the secondary typing 1 so that (*strastrb*) is assigned $[\overrightarrow{\alpha}]$. ⊣

3.3. An Intensional Semantics for ITT

An intensional semantics for TT was defined in §2.6.2. Here a intensional semantics for ITT is defined in much the same way.

3.3.1. Domains and valuations. The definition of an intensional valuation domain DI for ITT differs from Definition 41 for TT in §2.6.2 only in the first clause; the other clauses are repeated here for convenience:

DEFINITION 45. *Intensional Valuation Domain for ITT.*
N in each of the following cases is a term in normal form and without free

variables:

$DI(1)$ *is* $\{\langle c, c\rangle \| c{:}1\} \cup \{\langle N, N\rangle \| N$ *of secondary type* $1\}$.

$DI([\,])$ *is* $\{\langle N, \mathsf{s}\rangle \| N{:}[\,]$ *and* $\mathsf{s} \in \{+, -\}\}$.

$DI([\overline{\tau}])$ *is* $\{\langle N, \mathsf{S}\rangle \| N{:}[\overline{\tau}], \mathsf{S} \in \mathsf{PD}\}$, *where* $\overline{\tau}$ *is* $\tau_1, \ldots, \tau_n$ $n \geq 1$,
$\mathsf{PD} \subseteq \wp(DI(\tau_1) \times \cdots \times DI(\tau_n))$, PD *not empty.*

The *standard* intensional valuation domain for ITT, as for TT, requires PD in the definition of $DI([\tau_1, \ldots, \tau_n])$ to be $\wp(DI(\tau_1) \times \cdots \times DI(\tau_n))\}$.

Because of secondary typing a valuation Φ for ITT is necessarily a function with two arguments, a type τ and a term $T{:}\tau$. The clauses (V.1) to (V.3) for $\Phi(\tau, T)$ below are obtained from the same clauses for $\Phi(T)$ for TT in Definition 42 of §2.6.2 by replacing $\Phi(T)$ with $\Phi(\tau, T)$ when τ is the primary type of T; the value of $\Phi(1, T)$ when T has secondary type 1 is defined in a new clause (V.4). Again in this definition $=$ must be given the special treatment described in §2.6.1.

DEFINITION 46. *An Intensional Valuation for ITT to a domain DI.*
An intensional *valuation Φ to a given domain DI is a function with arguments a type τ and a term T, $T{:}\tau$, with a value $\langle \Phi_1(\tau, T), \Phi_2(\tau, T)\rangle$ for which Φ_1 and Φ_2 satisfy the following conditions defined by induction on $ct[T]$:*

V.1.1. *For a constant $c{:}\tau$, other than $\downarrow$, $\exists$, or $=$, $\Phi_1(\tau, c)$ is c, and $\Phi_2(\tau, c)$ satisfies $\langle \Phi_1(\tau, c), \Phi_2(\tau, c)\rangle \in DI(\tau)$.*

For a variable $v{:}\tau$, $\Phi(\tau, v)$ is any member of $DI(\tau)$. Thus $\Phi_1(\tau, v)$ is some $N{:}\tau$ in normal form and without free variables, and $\Phi_2(\tau, v)$ is such that $\langle \Phi_1(\tau, v), \Phi_2(\tau, v)\rangle \in DI(\tau)$.

Let $x_1{:}\tau_1, \ldots, x_n{:}\tau_n$ be all the variables with a free occurrence in $T{:}\tau$. $\Phi_1(\tau_i, x_i)$ is defined for each $i, 1 \leq i \leq n$, and is a term without free variables. Let

$$T^* \text{ abbreviate } [\Phi_1(\tau_1, x_1)/x_1] \cdots [\Phi_1(\tau_n, x_n)/x_n]T$$

and define $\Phi_1(\tau, T)$ to be the normal form of T^*. Note that $\Phi_1(\tau, T)$ is the normal form of T^* when T is a constant or variable. The remaining cases defining conditions on Φ_2 for which $ct[T]$ is 0 are stated in (V.1.2). The cases for which $ct[T] > 0$ are stated in (V.2), (V.3), and (V.4).

V.1.2. *$\Phi_2([[\,], [\,]], \downarrow)$ is a set of pairs of pairs $\langle\langle \Phi_1([\,], F), -\rangle, \langle \Phi_1([\,], G), -\rangle\rangle$, for which $F, G{:}[\,]$. $\Phi_2([[\tau]], \exists)$ is a set of pairs $\langle \Phi_1([\tau], P), \mathsf{S}\rangle$, for which $P{:}[\tau]$, and S is a non-empty subset of $DI(\tau)$. $\Phi_2([\tau, \tau], =)$ is $\{\langle e, e\rangle \| e \in DI(\tau)\}$.*

V.2. *Let $P{:}[\tau, \tau_1, \cdots, \tau_n]$ and $Q{:}\tau$. $\Phi_2([\tau_1, \ldots, \tau_n], (PQ))$ is $\{\langle e_1, \ldots, e_n\rangle \| \langle \Phi(\tau, Q), e_1, \ldots, e_n\rangle \in \Phi_2([\tau, \tau_1, \ldots, \tau_n], P), e_i \in DI(\tau_i)\}$. $\Phi_2([\,], (PQ))$ is $+$ if $\Phi(\tau, Q) \in \Phi_2([\tau], P)$, and is $-$ otherwise.*

V.3. *Let* $P{:}[\tau_1, \ldots, \tau_n]$ *and* $v{:}\tau$. $\Phi_2([\tau, \tau_1, \ldots, \tau_n], (\lambda v.P))$ *is*

$$\{\langle \Phi^v(\tau, v), e_1, \ldots, e_n\rangle \| \langle e_1, \ldots, e_n\rangle \in \Phi_2^v([\tau_1, \ldots, \tau_n], P), \\ e_i \in DI(\tau_i), \Phi^v \text{ a } v\text{-variant of } \Phi\}.$$

$$\Phi_2([\tau], (\lambda v.P)) \text{ is } \{\Phi^v(\tau, v) \| \Phi_2^v([\,], P) \text{ is } +, \Phi^v \text{ a } v\text{-variant of } \Phi\}.$$

V.4. $\Phi_2(1, T)$ *is* $\Phi_1(1, T)$, *when* T *has secondary type* 1.

Lemma 5 of semantic substitution in §1.5.4 is updated here for ITT. The proof, although more complicated, is similar to that for EL.

LEMMA 25. *Semantic Substitution.*
Let Φ *be a valuation to a domain* DI. *Let* $x, R{:}\sigma$, *where* x *has no free occurrence in* R, *and let* $T{:}\tau$. *Let* Φ^x *be the* x*-variant of* Φ *for which* $\Phi^x(\sigma, x)$ *is* $\Phi(\sigma, R)$. *Then* $\Phi^x(\tau, T)$ *is* $\Phi(\tau, [R/x]T)$.

PROOF. It may be assumed that x has a free occurrence in T. No matter the form and type τ of T, $\Phi_1^x(\tau, T)$ is the normal form of

$$[\Phi_1^x(\sigma, x)/x][\Phi_1(\sigma_1, x_1)/x_1] \cdots [\Phi_1(\sigma_n, x_n)/x_n]T$$

where $x{:}\sigma, x_1{:}\sigma_1, \ldots, x_n{:}\sigma_n$ are all the variables with a free occurrence in T. Hence $\Phi_1^x(\tau, T)$ is the normal form of

$$[\Phi_1(\sigma, R)/x][\Phi_1(\sigma_1, x_1)/x_1] \cdots [\Phi_1(\sigma_n, x_n)/x_n]T,$$

which is the normal form of

$$[\Phi_1(\sigma_1, x_1)/x_1] \cdots [\Phi_1(\sigma_n, x_n)/x_n][\Phi_1(\sigma, R)/x]T$$

since none of the variables $x_1, \ldots, x_n$ have a free occurrence in $\Phi_1(\sigma, R)$. Therefore $\Phi_1^x(\tau, T)$ is $\Phi_1(\tau, [R/x]T)$ as required.

Let $ct[T]$ be 0 so that T is x. Then $[R/x]T$ is R and therefore $\Phi_2^x(\tau, T)$ is $\Phi_2(\tau, [R/x]T)$. Assume the conclusion of the lemma whenever $ct[T] < ct$. Let $ct[T]$ be ct, and consider the forms that T may take.

- T is (PQ). Let τ be $[\overline{\sigma}]$ so that $P{:}[\sigma, \overline{\sigma}]$ and $Q{:}\sigma$, for some σ.
  Let $\overline{\sigma}$ be not empty. $\Phi_2^x([\overline{\sigma}], (PQ))$ is by (V.2)

$$\{\langle \overline{e} \rangle \| \langle \Phi^x(\sigma, Q), \overline{e} \rangle \in \Phi_2^x([\sigma, \overline{\sigma}], P)\},$$

which is by the induction assumption

$$\{\langle \overline{e} \rangle \| \langle \Phi(\sigma, [R/x]Q), \overline{e} \rangle \in \Phi_2([\sigma, \overline{\sigma}], [R/x]P)\},$$

which is $\Phi_2([\overline{\sigma}], [R/x](PQ))$ as required.
  Let $\overline{\sigma}$ be empty. Then

$$\begin{aligned} \Phi_2^x([\,], (PQ)) \text{ is } + &\Leftrightarrow \Phi^x(\sigma, Q) \in \Phi_2^x([\sigma], P) \\ &\Leftrightarrow \Phi(\sigma, [R/x]Q) \in \Phi_2([\sigma], [R/x]P) \\ &\Leftrightarrow \Phi([\,], [R/x](PQ)) \text{ is } +. \end{aligned}$$

• T is $(\lambda y.P)$. Let $P{:}[\overline{\sigma}]$ and $y{:}\sigma$ so that τ is $[\sigma, \overline{\sigma}]$. It may be assumed that y has no free occurrence in R and is not x. Thus by (V.3) $[R/x](\lambda y.P)$ is $(\lambda y.[R/x]P)$.

Let $\overline{\sigma}$ be not empty. Then $\Phi_2^x([\sigma, \overline{\sigma}], (\lambda y.P))$ is

$$\{\langle \Phi^{x,y}(\sigma, y), \overline{e}\rangle \| \langle \overline{e}\rangle \in \Phi_2^{x,y}([\overline{\sigma}], P), \Phi^{x,y} \text{ a } y\text{-variant of } \Phi^x\},$$

which is

$$\{\langle \Phi^{y}(\sigma, y), \overline{e}\rangle \| \langle \overline{e}\rangle \in \Phi_2^{y}([\overline{\sigma}], [R/x]P), \Phi^{y} \text{ a } y\text{-variant of } \Phi\}$$

by the induction assumption and therefore is $\Phi_2([\sigma, \overline{\sigma}], [R/x](\lambda y.P))$ as required.

Let $\overline{\sigma}$ be empty. Then $\Phi_2^x([\sigma], \lambda y.P)$ is

$$\{\Phi^{x,y}(\sigma, y) \| \Phi_2^{x,y}([\,], P) \text{ is } +, \Phi^{x,y} \text{ a } y\text{-variant of } \Phi^x\},$$

which is

$$\{\Phi^{y}(\sigma, y) \| \Phi_2^{x}([\,], [R/x]P) \text{ is } +, \Phi^{y} \text{ a } y\text{-variant of } \Phi\}$$

by the induction assumption and therefore is $\Phi(\sigma, [R/x](\lambda y.P))$ as required.

• T is of secondary type 1 so that $x{:}1$.
Then by (V.5), $\Phi_2^x(1, T)$ is $\Phi_1^x(1, T)$ which is $\Phi_1(1, [R/x]T)$ and therefore is $\Phi_2(1, [R/x]T)$, as required. ⊣

LEMMA 26. *Let $T{>}T'$, where $T{:}\tau$. Then $\Phi(\tau, T)$ is $\Phi(\tau, T')$.*

PROOF. Let $T{>}T'$. Then either T' is a bound variable variant of T, or for some occurrence of a contractible subterm CT of T, T' is obtained from T by replacing the occurrence of CT by its contraction CT'. In either case T and T' necessarily have the same normal form so that $\Phi_1(\tau, T)$ is $\Phi_1(\tau, T')$. To prove the lemma it is therefore sufficient to prove

$$\Phi_2(\tau, T) \text{ is } \Phi_2(\tau, T').$$

This is also necessarily true if T' is a bound variable variant of T. The proof otherwise is by induction on the depth of CT in T.

Let the depth be 0. Then T is $(\lambda x.P)Q$ or $(\lambda x.Px)$, where in the latter case x has no free occurrence in P, and T' is $[Q/x]P$, respectively P. A proof for the case that T is $(\lambda x.P)Q$ follows; a proof for the case that T is $(\lambda x.Px)$ is similar.

Should τ be the secondary type 1, then the conclusion is immediate. Therefore let τ be $[\overline{\sigma}]$ and $x{:}\sigma$ so that $(\lambda x.P){:}[\sigma, \overline{\sigma}]$ and $Q{:}\sigma$.

Let $\overline{\sigma}$ be not empty. By (V.2), $\Phi_2(\tau, (\lambda x.P)Q)$ is

$$\{\langle \overline{e}\rangle \| \langle \Phi(\sigma, Q), \overline{e}\rangle \in \Phi_2([\sigma, \overline{\sigma}], (\lambda x.P))\}.$$

By (V.3), $\Phi_2([\sigma, \overline{\sigma}], (\lambda x.P))$ is

$$\{\langle \Phi^{x}(\sigma, x), \overline{e}\rangle \| \langle \overline{e}\rangle \in \Phi_2^{x}([\overline{\sigma}], P)\}.$$

Let Φ^x be the x-variant of Φ for which $\Phi^x(\sigma, x)$ is $\Phi(\sigma, Q)$. Then by Lemma 25,

$$\Phi_2^x([\overline{\sigma}], P) \text{ is } \Phi_2([\overline{\sigma}], [Q/x]P).$$

Thus $\Phi_2(\tau, T)$ is $\Phi_2(\tau, T')$ as required.

Let $\overline{\sigma}$ be empty. Then $\Phi_2([\,], T)$ is $+$ if $\Phi(\sigma, Q) \in \Phi_2([\sigma], (\lambda x.P))$ and is $-$ otherwise. By (V.3), $\Phi_2([\sigma], (\lambda x.P))$ is the set

$$\{\Phi^x(\sigma, x) \| \Phi_2^x([\,], P) \text{ is } +\}.$$

Let Φ^x be the x-variant of Φ for which $\Phi^x(\sigma, x)$ is $\Phi(\sigma, Q)$. Then by Lemma 25, $\Phi_2^x([\,], P)$ is $\Phi_2^x([\,], [Q/x]P)$ and therefore $\Phi_2(\tau, T)$ is $\Phi_2(\tau, T')$.

Assume the conclusion of the lemma when the depth of CT in T does not exceed d. When the depth is $d+1$, T is of the form RS, SR, or $(\lambda x.R)$ where CT is of depth d in R. From (V.2) and (V.3) it follows that if $\Phi_2(\sigma, R)$ is $\Phi_2(\sigma, R')$ then $\Phi_2(\tau, T)$ is $\Phi_2(\tau, T')$. ⊣

The following definition of intensional model for ITT modifies the definition for intensional model for TT given in §2.6.2 by an added condition.

DEFINITION 47. *Intensional Model of ITT.*
An intensional model for ITT with domain DI is a valuation Φ to DI for which

1. $\Phi(\tau, P) \in DI(\tau)$, *for each type τ and term $P{:}\tau$.*
2. *For each P and Q of primary type τ and secondary type* 1,
   $\Phi_2([\,], P{=}^1 Q)$ *is* $\Phi_2([\,], P{=}^\tau Q)$.

The added clause (2) does have a profound effect on the semantics of ITT, but nevertheless, the following theorem can still be proved:

THEOREM 27. *A valuation to the standard intensional domain is an intensional model.*

PROOF. As before, clause (1) is immediately satisfied since every possible value for $\Phi(\tau, P)$ is in $DI(\tau)$, for each type τ and term $P{:}\tau$. Consider, therefore, clause (2). By (V.4) of Definition 46

$$\Phi_2([\,], P{=}^\tau Q) \text{ is } \Phi_2([\,], \forall Z.[Z(P) \rightarrow Z(Q)])$$

for any τ for which $P, Q{:}\tau$. Although $\forall Z.[Z(P) \rightarrow Z(Q)]$ is defined in terms of $\exists$ and $\downarrow$, it can be shown that $\forall Z.[Z(P) \rightarrow Z(Q)]$ is $+$ if and only if

$$\Phi_2([\,], Z(P)) \text{ is } + \Rightarrow \Phi_2([\,], Z(Q)) \text{ is } +,$$

no matter the value of $\Phi_2([\tau], Z)$. Since by Definition 45 and the definition of the standard domain,

$$DI([\tau]) \text{ is } \{\langle N, \mathrm{S}\rangle \| N{:}[\tau], \mathrm{S} \subseteq \wp(DI(\tau)\},$$

it follows that for each $e \in DI(\tau)$ and each $N{:}[\tau]$, $\langle N, \{e\}\rangle \in DI([\tau])$.

Let P have normal form M so that $\langle M, \Phi_2(\tau, M)\rangle \in DI(\tau)$. Let $A{:}[\tau]$ be a constant for which $\langle A, \{\langle M, \Phi_2(\tau, M)\rangle\}\rangle \in DI([\tau])$. Let $\Phi([\tau], Z)$ be

$$\langle A, \{\langle M, \Phi_2(\tau, M)\rangle\}\rangle.$$

Thus $\Phi_2([\,], Z(P))$ is +, while $\Phi_2([\,], Z(Q))$ is + if and only if P and Q have the same normal form. Since this is true no matter the type τ, clause (2) holds. ⊣

3.3.2. Semantic basis for the proof theory. The following theorem updates Theorem 6 of §1.6 and provides the semantic basis for the proof theory of ITT. Clauses (5) and (6) of the theorem make use of the notations $P{=}^{\tau}Q$ and $P{=}^{1}Q$, where the type superscript on $=$ indicates the type of its arguments.

THEOREM 28. *Semantic Basis for the Proof Theory of ITT.*
Let Φ be a valuation that is a model. Then

1. $\Phi_2([\,], F)$ *is either* + *or* −, *but not both.*
2. $\Phi_2([\,], [F{\downarrow}G])$ *is* + $\Rightarrow \Phi_2([\,], F)$ *is* − *and* $\Phi_2([\,], G)$ *is* −;
   $\Phi_2([\,], [F{\downarrow}G])$ *is* − $\Rightarrow \Phi_2([\,], F)$ *is* + *or* $\Phi_2([\,], G)$ *is* +.
3. $\Phi_2([\,], \exists P)$ *is* + $\Rightarrow \Phi_2([\,], P(t))$ *is* + *for some* $t{:}\tau$;
   $\Phi_2([\,], \exists P)$ *is* − $\Rightarrow \Phi_2([\,], P(t))$ *is* − *for all* $t{:}\tau$;
   when $P{:}[\tau]$.
4. $\Phi_2([\,], F)$ *is* $\pm \Rightarrow \Phi_2([\,], G)$ *is* $\pm$, *respectively, when* $F > G$.
5. $\Phi_2([\,], P{=}^{\tau}Q)$ *is* $\pm \Rightarrow \Phi_2([\,], \forall Z.[Z(P) \rightarrow Z(Q)])$ *is* $\pm$, *respectively,*
   when $Z{:}[\tau]$ *has no free occurrence in* P *or* Q.
6. $\Phi_2([\,], P{=}^{1}Q)$ *is* $\pm \Rightarrow \Phi_2([\,], P{=}^{\tau}Q)$ *is* $\pm$, *respectively,*
   when P *and* Q *have primary type* τ *and secondary type* 1.

PROOF. 1. Since Φ is a model, $\Phi_2([\,], F)$ is $\pm$. That the value is unique follows from Φ being a function.
2. Follows immediately from (V.1.2).
3. Consider the first half.

$$\begin{aligned}\Phi_2([\,], \exists P) \text{ is } + &\Rightarrow \Phi([\tau], P) \in \Phi_2([[\tau]], \exists P)\\ &\Rightarrow \Phi([\tau], P) \text{ is some non-empty subset of } DI(\tau)\\ &\Rightarrow \Phi(\tau, t) \in \Phi_2([\tau], P), \text{ for some } t \text{ for which } \Phi(\tau, t) \in DI(\tau)\\ &\Rightarrow \Phi_2([\,], P(t)) \text{ is } +, \text{ for some } t{:}\tau.\end{aligned}$$

The second half can be similarly argued.
4. Is a special case of Lemma 26.
5. Follows immediately from (V.4) of Definition 46.
6. Follows immediately from the second clause of Definition 47. ⊣

3.3.3. Sequents and counter-examples. Here the definitions of §1.7 are updated for ITT. As defined there a sequent is an expression $\Gamma \vdash \Theta$ where Γ, the *antecedent* of the sequent, and Θ, the *succedent* of the sequent are finite, possibly empty, sets of formulas. A sequent $\Gamma \vdash \Theta$ is *satisfied* by a valuation Φ, if $\Phi_2([\,], F)$ is − for some F in the antecedent or is + for some F in the succedent. A valuation Φ is a *counter-example* for a sequent if $\Phi_2([\,], F)$ is + for every F in the antecedent and − for every F in the succedent. A sequent is *valid* if it is satisfied by every valuation that is a model.

As with first order logic, the proof theory of ITT described next provides a systematic search procedure for a counter-example for a given sequent $\Gamma \vdash \Theta$.

Should the procedure fail to find such a valuation, and if it does fail it will fail in a finite number of steps, then that $\Gamma \vdash \Theta$ is valid follows from the completeness theorem to follow. The steps resulting in a failure are recorded as a *derivation*. Thus a derivation for a sequent is constructed under the assumption that a counter-example Φ exists for the sequent. *Signed* formulas are introduced to abbreviate assertions about the truth value assigned to a formula F by Φ. Thus $+F$ is to be understood as an abbreviation for $\Phi_2([\,], F)$ is $+$ and $-F$ for $\Phi_2([\,], F)$ is $-$, for some conjectured counter-example Φ.

3.4. Proof Theory for ITT

The proof theory of ITT extends the proof theory of TT described in §2.4 with the addition of rules *Int* of intensionality. However the rules of TT are repeated here for ease of reference.

$$+\downarrow \quad \frac{+[F\downarrow G]}{-F} \quad \frac{+[F\downarrow G]}{-G} \qquad\qquad -\downarrow \quad \frac{-[F\downarrow G]}{+F \quad +G}$$

$$+\exists \quad \frac{+\exists P}{+P(y)} \qquad\qquad -\exists \quad \frac{-\exists P}{-P(t)}$$

y not free above conclusion

$$+\lambda \quad \frac{+F}{+G} \qquad\qquad -\lambda \quad \frac{-F}{-G}$$

$$F > G$$

$$+= \quad \frac{+P = Q}{+\forall Z.[Z(P) \to Z(Q)]} \qquad\qquad -= \quad \frac{-P = Q}{-\forall Z.[Z(P) \to Z(Q)]}$$

$$+Int \quad \frac{+P{=^1}Q}{+P{=^\tau}Q} \qquad\qquad -Int \quad \frac{-P{=^1}Q}{-P{=^\tau}Q}$$

P and Q have primary type τ and secondary type 1.

$$Cut \quad \frac{}{+F \quad -F}$$

It follows from Theorem 27 of §3.3.1 that there are models for ITT. Thus a semantic proof of the following theorem can be given in much the same manner as the proof of Theorem 9 in §1.8.

THEOREM 29. *Consistency of ITT.*
A derivable sequent of ITT is valid.

A proof of the theorem is left as an exercise.

3.4.1. An essential restriction. A predicate term, it is stressed in §3.1.1 and stated in clause (4) of Definition 43 in §3.1.2, can have a secondary type 1 only if no variable of type other than 1 has a free occurrence in it. Here an example is given that demonstrates that ITT is inconsistent if this restriction is ignored.

A variant of Russell's set is defined:

$$R \stackrel{\text{df}}{=} (\lambda u.\exists X.[\neg X(u) \wedge u = X]).$$

Here u:1 and X:[1]. In violation of the restriction on clause (4), X is given the secondary type 1 so that the occurrence of $=$ is of type [1, 1] and R is a term of type [1]. Since R has no free variable, it also has secondary type 1 so that $R(R)$:[ ]. Then both the sequents

$$\vdash R(R) \text{ and } R(R) \vdash$$

are derivable using only the $+=$ rule and neither of the $\pm Int$ rules. A derivation of the second of these sequents follows:

$$\begin{array}{ll}
+R(R) & \\
+\exists X.[\neg X(R) \wedge R =^1 X] & \\
+[\neg X(R) \wedge R =^1 X] & \\
+\neg X(R) & \\
+R =^1 X & \\
-X(R) & \\
+\forall Z.[Z(R) \rightarrow Z(X)] & += \\
+[(\lambda U.U(R))(R) \rightarrow (\lambda U.U(R))(X)] & \\
+[R(R) \rightarrow X(R)] & \\
\hline
-R(R) \qquad\qquad +X(R) &
\end{array}$$

A derivation of the sequent $\vdash R(R)$ is left as an exercise.

3.4.2. Derivable rules for identity. Here derivable rules are provided which permit the shortening of derivations involving the intensional and extensional identities.

$$+= \quad \frac{+s = t}{-P(s) \qquad +P(t)} \qquad\qquad -= \quad \frac{-s = t}{\begin{array}{c}+Z(s)\\ -Z(t)\end{array}}$$

Here if $s, t{:}\tau$, then $P, Z{:}[\tau]$, with Z new to the branch. It is important to note that the $-=$ rule differs from every other previously described rule in as much as its two conclusions do not split a branch but must be simultaneously added to the branch as conclusions.

Similar rules can be derived for extensional identity:

$$+=_e \quad \frac{+P=_eQ}{\begin{array}{cc} +P(\bar{s}) & -P(\bar{s}) \\ +Q(\bar{s}) & -Q(\bar{s}) \end{array}} \qquad -= \quad \frac{-P=_eQ}{\begin{array}{cc} +P(\bar{x}) & -P(\bar{x}) \\ -Q(\bar{x}) & +Q(\bar{x}) \end{array}}$$

Here $\overline{x}$ is a sequence of distinct variables of types appropriate for P and Q that are new to the branch, and $\overline{s}$ is a sequence of terms of the same types as $\overline{x}$.

3.4.3. Relationship between the identities. A derivation in TT, which is also a derivation in ITT, was provided in §2.5 for the following sequent:

(IEId.1.) $$\vdash \forall x, y.[x = y \to x =_e y]$$

where necessarily x and y are of some type $[\overline{\sigma}]$. The sequent expresses that predicates with the same intensions have the same extensions. The sequent is clearly not derivable if $x, y{:}1$ since then $x =_e y$ is not a formula. In ITT, on the other hand, each instance of the sequent scheme

(IEId.2.) $$P =^1 Q \vdash P =_e Q$$

is derivable when P and Q have primary type τ and secondary type 1 and $=:[1, 1]$. Here is an abbreviated derivation that requires an application of $+Int$:

$+P =^1 Q$
$-P =_e Q$
$+P =^\tau Q$ $\quad +Int$
$+\forall Z.[Z(P) \to Z(Q)]$ $\quad +=$

$-(\lambda v.P =_e v)(P)$ $\qquad +(\lambda v.P =_e v)(Q)$
$-P =_e P$ $\qquad +P =_e Q$

It is important to understand the distinction between the sequent (IEId.1) and the sequent scheme (IEId.2). The former refers to a single sequent of ITT, while the latter refers to a denumerable collection of sequents. Each sequent of (IEId.2) has a formula $P = Q$ as antecedent and a formula $P =_e Q$ as succedent. In the formula $P = Q$ each of P and Q has the secondary type 1, while in the formula $P =_e Q$ each has the primary type τ. The sequent scheme (IEId.2) will be used directly or indirectly in many derivations to follow.

3.4.4. Properties of derivations. The following lemma updates Lemma 7 in §1.7.3 for ITT.

LEMMA 30. *Properties of Derivations in ITT.*
If a sequent $\Gamma \vdash \Theta$ *has a derivation, then it has a derivation with the following properties*:

1. *The same signed formula does not occur twice as a node of the same branch.*

2. *Each branch of the derivation has exactly one closing pair.*
3. *Each variable with a free occurrence in the eigen term t of an application of $-\exists$ has a free occurrence in a node above the conclusion of the rule. With the possible exception of a single constant for each type, each constant occurring in the eigen term t of an application of $-\exists$ has an occurrence in a node above the conclusion of the rule.*
4. *Let the sequent $\Gamma \vdash \Theta$ have a derivation. Then there is a derivation for a sequent $\Gamma' \vdash \Theta'$, where $\Gamma' \subseteq \Gamma$ and $\Theta' \subseteq \Theta$, for which each node of the derivation has a descendent that is a closing node.*

A proof of the lemma is left to the reader. The properties are useful when attempting to find a derivation since they narrow the choices available.

EXERCISES §3.4.

1. Provide a proof for Theorem 29, the consistency of ITT.
2. Provide derivations for the $\pm=$ and $\pm=_e$ rules.
3. In §1.7.5 a method was described that transforms derivations with signed formulas as nodes into derivations with sequents as nodes. Update the method for ITT.
4. Provide a derivation for the sequent $\vdash R(R)$ under the same assumption made for the derivation of $R(R) \vdash$.
5. Provide a proof for Lemma 30.

3.5. Completeness of ITT

The proof of the completeness theorem for ITT is similar in form to that given for the first order logic EL in Theorem 11 of §1.9. A review of that proof will provide a reader with an overview of the definitions, lemmas, and corollaries of this section. For example, the following is an update of Definition 23 of §1.9. Note that clause 2(b) reflects clause (3) of Lemma 30.

DEFINITION 48. *Descending Chain for a Sequent of ITT.*
Let $\Gamma \vdash \Theta$ be a given sequent and let k be the number of formulas in $\Gamma \cup \Theta$. A descending chain for the sequent is a possibly infinite sequence of signed formulas $sF_1, \cdots, sF_n, \cdots$ without repetitions satisfying the following conditions:

1. *For each n, $k \leq n$, $sF_1, \cdots, sF_n$ is an open branch of a tree based on the sequent.*
2. *Let $\pm F_m$, $1 \leq m$ be a possible premiss for some rule.*
   (a) *Let F_m be $[F \downarrow G]$. Then each of $-F$ and $-G$, respectively one of $+F$ and $+G$, follows $\pm[F \downarrow G]$ in the sequence.*
   (b) *Let F_m be $\exists P$. Then $+P(y)$, respectively each $-P(t)$, follows $\pm\exists P$ in the sequence, where y has no free occurrence in a signed formula preceding $+P(y)$, and where each variable with a free occurrence in t has a free occurrence in a signed formula preceding $-P(t)$. Further,*

with the possible exception of a single constant for each type, each constant occurring in t has an occurrence in a signed formula preceding $-P(t)$.

(c) *Let* sF_m *be* $\pm F$. *Then* $\pm G$, *respectively, follows* $\pm F$ *in the sequence when* $F{>}G$.

(d) *Let* sF_m *be* $\pm P{=}^{\tau}Q$. *Then* $\pm\forall Z.[Z(P) \rightarrow Z(Q)]$, *respectively, follows* $\pm P{=}^{\tau}Q$ *in the sequence when P and Q have either primary or secondary type* τ *and* $Z{:}[\tau]$.

(e) *Let* sF_m *be* $\pm P{=}^{1}Q$. *Then* $\pm P{=}^{\tau}Q$, *respectively, follows* $\pm P{=}^{1}Q$ *in the sequence when P and Q have primary type* τ *and secondary type* 1.

From clause (1) of the definition it follows that a descending chain is consistent in the sense that not both $+F$ and $-F$ are members for some F. From clause (2) it follows that a descending sequent is saturated in the sense that each possible application of a logical rule satisfying the restrictions of clauses 2(a) to 2(e) is an actual application.

In §1.9.1 a search algorithm for a descending chain for a sequent of EL was sketched with the properties that are confirmed in Theorem 10 of §1.9. An updating of the algorithm for ITT and a proof of the following theorem for the updated algorithm is left to the reader.

THEOREM 31. *Properties of Search Algorithm for ITT.*

1. *Let the procedure halt. Then the input sequent has either a Cut-free derivation or a finite descending chain.*
2. *Let the procedure not halt. Then an infinite descending chain for the sequent exists.*

The following corollary is immediate:

COROLLARY 32. *A sequent without a Cut-free derivation in ITT has a descending chain.*

The justification for the claim that the proof theory of EL provides an implicit systematic search procedure for a counter-example for a given sequent was completed with the proof of Theorem 11 of §1.9. For ITT the claim is justified by the proof of the following theorem given in §3.5.1.

THEOREM 33. *Completeness of ITT without Cut.*
A counter-example that is a model exists for each sequent that has no Cut-free derivation in ITT.

As with EL, an immediate corollary for ITT is:

COROLLARY 34. *If a sequent has a derivation in ITT, then it has a Cut-free derivation.*

For consider a sequent with a derivation in which Cut is used. If the sequent has no Cut-free derivation then, by the completeness theorem, it has a counter-example. But that is impossible since by the consistency theorem for ITT, Theorem 29 in §3.4, the sequent is satisfied by every model.

3.5.1. A counter-example that is an intensional model. The following definition is essentially a repeat of Definition 24 in §1.9.2:

DEFINITION 49. *The Closure of a Term and a Descending Chain.*
Let $\dagger$ *be the following one-to-one mapping of the constants* $c_1, \ldots, c_i, \ldots$ *and variables* $v_1, \ldots, v_i, \ldots$ *of each type onto the constants of that type*:

$$c_i^\dagger \text{ is } c_{(2i-1)} \text{ and } v_i^\dagger \text{ is } c_{(2i)}.$$

The closure $T^\dagger$ *of a term* T *is obtained by replacing each constant* c_i *by* $c_i^\dagger$ *and each free occurrence of a variable* v_i *in* T *by* $c_i^\dagger$. *The closure* $\Sigma^\dagger$ *of a descending chain* Σ *is obtained by replacing each signed formula* $\pm F$ *of* Σ *by* $\pm F^\dagger$.

The next lemma restates Lemma 13 of §1.9.2 for ITT with the conditions of that lemma appropriately closed.

LEMMA 35. *Let* $\Sigma^\dagger$ *be the closure of a descending chain* Σ *for a sequent* $\Gamma \vdash \Theta$. *Then* $\Sigma^\dagger$ *satisfies the following conditions*:

1. $F \in \Gamma \Rightarrow +F^\dagger \in \Sigma^\dagger$;
   $F \in \Theta \Rightarrow -F^\dagger \in \Sigma^\dagger$.
2. *For no* F *is both* $+F^\dagger$ *and* $-F^\dagger \in \Sigma^\dagger$.
3. $+[F{\downarrow}G]^\dagger \in \Sigma^\dagger \Rightarrow -F^\dagger \in \Sigma^\dagger$ *and* $-G^\dagger \in \Sigma^\dagger$;
   $-[F{\downarrow}G]^\dagger \in \Sigma^\dagger \Rightarrow +F^\dagger \in \Sigma^\dagger$ *or* $+G^\dagger \in \Sigma^\dagger$.
4. *For each type* τ:
   $+\exists P^\dagger \in \Sigma^\dagger \Rightarrow +(P(Q))^\dagger \in \Sigma^\dagger$, *for some* $Q{:}\tau$;
   $-\exists P^\dagger \in \Sigma^\dagger \Rightarrow -(P(Q))^\dagger \in \Sigma^\dagger$, *for all* $Q{:}\tau$.
5. $\pm F^\dagger \in \Sigma^\dagger \Rightarrow \pm G^\dagger \in \Sigma^\dagger$, *respectively, whenever* $F > G$.
6. $\pm P^\dagger =^\tau Q^\dagger \in \Sigma^\dagger \Rightarrow \pm\forall Z.[Z(P^\dagger) \rightarrow Z(Q^\dagger)] \in \Sigma^\dagger$, *respectively, when* P *and* Q *have primary or secondary type* τ *and* $Z{:}[\tau]$.
7. $\pm P^\dagger =^1 Q^\dagger \in \Sigma^\dagger \Rightarrow \pm P^\dagger =^\tau Q^\dagger \in \Sigma^\dagger$, *respectively, when* P *and* Q *have primary type* τ *and secondary type* 1.

A proof of the lemma is left as an exercise.

Let $\Gamma \vdash \Theta$ be a sequent of ITT without a Cut-free derivation and therefore, by Corollary 32 of §3.5, with a descending chain Σ. The following is an extension of Definition 25 of §1.9.2 for ITT. An intensional valuation domain $D\Sigma$, as defined in Definition 45 of §3.3.1, is defined simultaneously with a possible values function $\mathcal{E}$ obtained from $\Sigma^\dagger$, the closure of Σ. In effect intensional valuations $\Phi\Sigma$, as defined in Definition 46 of §3.3.1, are defined simultaneously with $D\Sigma$.

DEFINITION 50. *The Possible Values Function* $\mathcal{E}$ *and Valuation Domain* $D\Sigma$.
Let $\Sigma^\dagger$ *be the closure of* Σ. *The domain* $D\Sigma(\tau)$ *is defined to be* $\cup\{\mathcal{E}(\tau, T) \| T{:}\tau\}$, *for each* τ, *where* $\mathcal{E}(\tau, T)$ *is defined for each term* $T{:}\tau$ *with normal form* N, *by induction on* τ:

$\mathcal{E}(1, T)$ *is* $\{\langle N^\dagger, N^\dagger\rangle\}$,
    where T *is of primary or secondary type* 1.

$\mathcal{E}([\,], T)$ *is the smallest subset of* $\{\langle N^\dagger, +\rangle, \langle N^\dagger, -\rangle\}$ *for which*
$+N^\dagger \in \Sigma^\dagger \Rightarrow \langle N^\dagger, +\rangle \in \mathcal{E}([\,], T)$,
$-N^\dagger \in \Sigma^\dagger \Rightarrow \langle N^\dagger, -\rangle \in \mathcal{E}([\,], T)$, *and*
$+N^\dagger \notin \Sigma^\dagger$ *and* $-N^\dagger \notin \Sigma^\dagger \Rightarrow \langle N^\dagger, +\rangle, \langle N^\dagger, -\rangle \in \mathcal{E}([\,], T)$.

$\mathcal{E}([\tau_1, \ldots, \tau_n], T)$ *is the set of pairs* $\langle N^\dagger, \mathsf{S}\rangle$ *for which for all* $Q_i{:}\tau_i$, *with normal form* M_i, *and all* $e_i \in \mathcal{E}(\tau_i, Q_i)$,
$+(N(M_1, \ldots, M_n))^\dagger \in \Sigma^\dagger \Rightarrow \langle e_1, \ldots, e_n\rangle \in \mathsf{S}$, *and*
$-(N(M_1, \ldots, M_n))^\dagger \in \Sigma^\dagger \Rightarrow \langle e_1, \ldots, e_n\rangle \notin \mathsf{S}$.

Note that $\mathcal{E}(1, T)$, when T is $c{:}1$, is $\{\langle c^\dagger, c^\dagger\rangle\}$, since $c^\dagger$ is, by definition, its own normal form.

In the remainder of this section, D 45, D 46 and D 47 refer to the definitions of those numbers in §3.3.1, D 49 and D 50 to the definitions, and L 35 and L 36 to the lemmas of those numbers in §3.5.1; and (V.1.1) - (V.5) to the clauses of D 46.

A proof of the following lemma is left to the reader:

LEMMA 36. *The domain* $D\Sigma$ *is an intensional valuation domain for ITT as defined in D* 45.

The following definition introduces an overline notation for finite sequences of expressions. The expression $\Phi\Sigma_2([\overline{\tau}], T)$ is to be understood to represent $\Phi\Sigma_2([\tau_1, \cdots, \tau_n], T)$, where $1 \leq n$. The expression $T(\overline{Q})$ is to be understood to be $T(Q_1, \cdots, Q_n)$. The expression $\langle\overline{\Phi\Sigma(\tau, Q)}\rangle$ is to be understood to be

$$\langle\Phi\Sigma(\tau_1, Q_1), \cdots, \Phi\Sigma(\tau_n, Q_n)\rangle.$$

DEFINITION 51. *The mapping* $\Phi\Sigma$ *to the domain* $D\Sigma$.
The mapping $\Phi\Sigma$ *to* $D\Sigma$, *where* $\Phi\Sigma(\tau, T)$ *is* $\langle\Phi\Sigma_1(\tau, T), \Phi\Sigma_2(\tau, T)\rangle$ *for each* τ *and each* $T{:}\tau$, *is defined as follows*: $\Phi\Sigma_1(\tau, T)$ *is the normal form of* $T^\dagger$, *while* $\Phi\Sigma_2(\tau, T)$ *is defined by induction on the type of* T:

$\Phi\Sigma_2(1, T)$ *is* $\Phi\Sigma_1(1, T)$, *when* T *has primary or secondary type* 1.

$\Phi\Sigma_2([\,], T)$ *is* $+$ *if* $\Phi\Sigma_1([\,], T) \in \Sigma^\dagger$, *and is otherwise* $-$.

$\Phi\Sigma_2([\overline{\tau}], T)$ *is* $\{\langle\overline{\Phi\Sigma(\tau, Q)}\rangle \| +\Phi\Sigma_1([\,], T(\overline{Q}) \in \Sigma^\dagger\}$.

As before $\Phi\Sigma_2([\tau, \tau], =)$ *is* $\Phi\Sigma_2([\tau, \tau], (\lambda u, v.\forall Z.[Z(u) \rightarrow Z(v)]))$.

LEMMA 37. *The mapping* $\Phi\Sigma$ *is an intensional valuation to the domain* $D\Sigma$.

PROOF. It is required to prove that $\Phi\Sigma$ and $D\Sigma$ satisfy the conditions imposed on Φ and DI in D 46. Consider each clause of D 46 in turn, the definition of $\Phi\Sigma(\tau, T)$ of that clause given in D 51, and the conditions that $\Phi\Sigma(\tau, T)$ is required to satisfy by the given clause.

It is easy to confirm that the definition of $\Phi\Sigma_1(\tau, T)$ conforms to the requirements of D 46, so it is only necessary to prove that $\Phi\Sigma_2(\tau, T)$ conforms. Consider the clauses of D 46 in turn

(V.1.1) When T is a constant $c{:}\tau$, other than $\downarrow$, $\exists$, or $=$, or a variable $v{:}\tau$

$$\langle \Phi\Sigma_1(\tau, T), \Phi\Sigma_2(\tau, T)\rangle \in D\Sigma(\tau).$$

(V.1.2) $\Phi\Sigma_2([[\,],[\,]], \downarrow)$ as defined in D 51 is

$$\{\langle\langle \Phi\Sigma_1([\,], F), -\rangle, \langle \Phi\Sigma_1([\,], G), -\rangle\rangle \| +\Phi\Sigma_1([\,], \downarrow(F, G)) \in \Sigma^\dagger\},$$

meeting the requirements of D 46. $\Phi\Sigma_2([[\tau]], \exists)$ as defined in D 51 is

$$\{\Phi\Sigma([\tau], P) \| +\Phi\Sigma_1([\,], \exists(P)) \in \Sigma^\dagger\}.$$

Thus it is sufficient to prove that $\Phi\Sigma_2([\tau], P)$ is not empty; this follows directly from (4) of L 35.

(V.2) $\Phi\Sigma_2([\overline{\tau}], (PQ))$ as defined in D 51 is

$$\{\langle\overline{\Phi\Sigma([\tau], Q)}\rangle \| +\Phi\Sigma_1([\,], (PQ)(\overline{Q})) \in \Sigma^\dagger\}.$$

Thus it is sufficient to prove that

$$+\Phi\Sigma_1([\,], (PQ)(\overline{Q})) \in \Sigma^\dagger \Leftrightarrow \langle \Phi\Sigma(\tau, Q), \overline{\Phi\Sigma(\tau, Q)}\rangle \in \Phi\Sigma_2([\tau, \overline{\tau}], P).$$

But $(PQ)(\overline{Q})$ is $P(Q, \overline{Q})$ so that the equivalence follows from D 51. The argument for $\Phi\Sigma_2([\,], (PQ))$ is similar

(V.3) $\Phi\Sigma_2([\tau, \overline{\tau}], (\lambda v.P))$ as defined in D 51 is

$$\{\langle \Phi\Sigma(\tau, Q), \overline{\Phi\Sigma([\tau], Q)}\rangle \| +\Phi\Sigma_1([\,], (\lambda v.P)(Q, \overline{Q})) \in \Sigma^\dagger\}.$$

Thus it is sufficient to prove: When $\Phi\Sigma^v$ is a v-variant of $\Phi\Sigma$ for which $\Phi\Sigma^v(\tau, v)$ is $\Phi\Sigma(\tau, Q)$, that

$$+\Phi\Sigma_1([\,], (\lambda v.P)(Q, \overline{Q})) \in \Sigma^\dagger \Leftrightarrow \langle\overline{\Phi\Sigma(\tau, Q)}\rangle \in \Phi\Sigma_2^v([\overline{\tau}], P).$$

But $\Phi\Sigma_1([\,], (\lambda v.P)(Q, \overline{Q}))$ is $\Phi\Sigma_1^v([\,], P(\overline{Q}))$ so that the equivalence follows from D 51. The argument for $\Phi\Sigma_2([\tau], (\lambda v.P))$ is similar.

That $\Phi\Sigma$ satisfies also (V.4) and (V.5) follows immediately from D 51. ⊣

THEOREM 38. *A valuation $\Phi\Sigma$ is a model with domain $D\Sigma$.*

PROOF. It is required to prove:

1. $\Phi\Sigma(\tau, T) \in D\Sigma(\tau)$, for each type τ and term $T{:}\tau$;
2. For each P and Q of primary type τ and secondary type 1, $\Phi\Sigma_2([\,], P{=}^1 Q)$ is $\Phi\Sigma_2([\,], P{=}^\tau Q)$.

By D 50, the domain $D\Sigma(\tau)$ is defined to be $\cup\{\mathcal{E}(\tau, T) \| T{:}\tau\}$, for each τ. But clearly by D 50 and D 51, $\Phi\Sigma(\tau, T) \in \mathcal{E}(\tau, T)$.

Consider clause (2). Since $P = Q$ is a form of $= (P, Q)$, $\Phi\Sigma_2([\,], P =^1 Q)$ is a special case of the third clause of D 51. Thus

$$\Phi\Sigma_2([\,], P =^1 Q) \text{ is } + \Leftrightarrow +\Phi\Sigma_1([\,], P =^1 Q) \in \Sigma^\dagger.$$

But

$$\Phi\Sigma_1([\,], P =^1 Q) \text{ is } =^1 (\Phi\Sigma_1(1, P), \Phi\Sigma_1(1, Q)).$$

Thus by (7) of L 35

$$+\Phi\Sigma_1([\,], P =^1 Q) \in \Sigma^\dagger \Leftrightarrow +\Phi\Sigma_1([\,], P =^\tau Q) \in \Sigma^\dagger$$

and clause (2) follows immediately. ⊣

COROLLARY 39. *$\Phi\Sigma$ is a counter-example for the sequent $\Gamma \vdash \Theta$.*

PROOF. For $\Phi\Sigma$ to be a counter-example for $\Gamma \vdash \Theta$ it is sufficient to prove that for each $F \in \Sigma$ that $\Phi\Sigma_2([\,], F)$ is $+$, and that for each $G \in \Theta$ that $\Phi\Sigma_2([\,], G)$ is $-$. But this follows directly from clause (1) of L 35 and Definition 51. ⊣

3.5.2. Denumerable models. The valuations Φ to the domain $D\Sigma$ in general have non-denumerable domains $D\Sigma(\tau)$ when τ is neither 1 nor [] since there will in general be non-denumerably many members of each set $\mathcal{E}(\tau, T)$. The exercise of §1.9.2 asked for the construction of denumerable models for first order logic, essentially to establish a result first proved by Skolem in [133]. The method suggested for first order logic adapts to ITT as well. The set Σ is extended by including *Cut* as a rule in the generation of a descending chain for a closed sequent. This is related to the manner in which Henkin proved the completeness of first order logic in [73] and of the simple theory of types in [74]. With *Cut* so used, clause (2) of Lemma 35 can be replaced by:

(2C) For each F exactly one of $\pm F$ is in Σ.

As a consequence, each set $\mathcal{E}(\tau, T$ has a single member.

EXERCISES §3.5.

1. Prove Lemma 35 and Lemma 36.
2. Although in ITT the rules of intensionality are said to replace the axiom of extensionality in a usual theory of types, it is plausible that an axiom of extensionality can be consistently added to ITT. Is this true?

3.6. SITT

The logic presented in Principia Mathematica is known as the *ramified* theory of types. It had not only the type restrictions of TT but also level restrictions. The types and levels remove from the logic the ability to form impredicative definitions in the sense of Poincaré (see Feferman [45]). In this strict sense the logic TT, and therefore also ITT, is impredicative. The extension SITT to ITT briefly described here permits the definition of predicates that cannot be defined in ITT, and that are related to the paradoxical predicates that gave rise to the ramified theory of types in the first place. For this reason, SITT is called a *strongly* impredicative type theory.

The extension of ITT is accomplished as follows. The role of a variable in the definition of term of ITT is purely that of a placeholder. Nevertheless in the proof theory the role of a variable in the λ rules is distinct from its role as an eigen variable in the conclusion of the $+\exists$ rule. These two roles can be served in formulas by syntactically distinct variables, abstraction variables for the λ rules and quantification variables for the $+\exists$ rule. The quantification variables are interpreted in SITT as they are in ITT but the abstraction variables receive a strict interpretation as placeholders for predicate names with both a primary and the secondary type 1. The resulting logic is a consistent extension of ITT in which such strongly impredicative sets such as Russell's can be defined and reasoned about. The paper of Jamie Andrews [6] achieves related goals in a weakly typed logic. SITT is perhaps at best of historical interest because of its treatment of Russell's set and no serious attempt has been made to find a useful application for it.

3.6.1. Types and Terms of SITT. The types of SITT extend the types of ITT with *intersection* types. Predicate names that are assigned both a primary type τ and the secondary type 1 in ITT are assigned a third type, the intersection type $\tau \cap 1$ as well.

DEFINITION 52. *The Types of SITT.*

1. *1 is a type that is not a* predicate type. *[] is a predicate type and a type.*
2. *If $\overline{\sigma}$ is a sequence of types, then $[\overline{\sigma}]$ is a* predicate type *and a type.*
3. *If τ is a predicate type then $\tau \cap 1$ is an* intersection type, *and a type, but not a predicate type.*

Intersection types are so called because they identify a subtype of the type 1 through use of a higher order type. Thus, for example, $[1] \cap 1$ is a subtype of the type 1 defined by the type $[1]$. This will be made explicit when domains $D(\tau)$ are defined in §3.6.3 for each type τ of SITT. Note, however, that $[[1] \cap 1]$ is a predicate type but not an intersection type, so that $[[1] \cap 1] \cap 1$ is an intersection type.

The intersection types defined here are also intersection types in the sense of [121]: in the notation used there, $\tau \cap 1$ would be written $\tau \& 1$. However [121] does not require the second argument of & to be the equivalent of the type 1.

For the type 1 and for each predicate type there are denumerably many variables called *q-variables*, short for quantification variables, and denumerably many constants. For each predicate type there are denumerably many variables called *a-variables*, short for abstraction variables. For any variable or constant cv, $\mathrm{t}[cv]$ is its given type; note that it is never an intersection type. Thus a possible value for $\mathrm{t}[v]$ is $[[1] \cap 1]$ but not $[1] \cap 1$. The type 1 is not a possible value of $\mathrm{t}[v]$ for an a-variable v.

As with TT and ITT there is a constant $\downarrow$ of type $[[\,], [\,]]$, and a constant $\exists$ of any type $[[\tau]]$, where τ is any type that is not an intersection type; thus $\mathrm{t}[\downarrow]$

is $[[\,],[\,]]$ and t$[\exists]$ is $[[\tau]]$, for some τ appropriate for the context in which an occurrence of $\exists$ appears.

DEFINITION 53. *The Terms of SITT.*

1. *A constant or variable cv is a term of type* t$[cv]$.
2. $Q{:}\tau$ *and* $P{:}[\tau,\overline{\sigma}\,] \Rightarrow (PQ){:}[\overline{\sigma}\,]$.
3. $P{:}[\overline{\sigma}\,]$ *and* $x{:}\tau \Rightarrow (\lambda x.P){:}[\tau,\overline{\sigma}\,]$, *for a quantification variable* x.
   $P{:}[\overline{\sigma}\,]$ *and* $v{:}\tau \Rightarrow (\lambda v.P){:}[\tau \cap 1,\overline{\sigma}\,]$, *for an abstraction variable* v.
4. $P{:}[\overline{\sigma}\,] \Rightarrow P{:}1$, *provided no q-variable of type other than* 1 *has a free occurrence in* P.
5. $P{:}[\overline{\sigma}\,]$ *and* $P{:}1 \Rightarrow P{:}[\overline{\sigma}\,] \cap 1$.

A term of ITT is necessarily a term of SITT of the same type. For the first four clauses of definition 53 are the clauses of Definition 43 in §3.1.2 of term of ITT if all references to abstraction variables are excluded.

The type 1 assigned to a term P by clause (4) is as before said to be the *secondary* type of P. Thus every abstraction variable has the secondary type 1, since it has a predicate type as its primary type and no q-variable has a free occurrence in it. Note that

$$P{:}[\overline{\sigma}\,] \cap 1 \Leftrightarrow P{:}[\overline{\sigma}\,] \text{ and } P{:}1$$

since clause (5) is the only way in which a term can be assigned an intersection type.

In the definition of term the constant $\exists$ is not treated differently than any other constant of its type. But that it is of type $[[\tau]]$, where τ is not an intersection type has the following effect: Let $\exists(\lambda x.F)$ be a formula; that is $\exists(\lambda x.F){:}[\,]$. Then necessarily $(\lambda x.F){:}[\tau]$ for some type τ that is not an intersection type. Hence by clause (3), x is a quantification variable.

Definition 19 in §1.5.4 of the substitution operator $[Q/v]$ extends to SITT as it did for TT and ITT, except now it is required that $Q{:}\tau$ if v is a quantification variable, and $Q{:}\tau \cap 1$ if v is an abstraction variable. The substitution operator does not diminish the types of the term to which it is applied, but can increase it.

3.6.2. Lambda contractions. The relation $>$ of immediate lambda contraction is defined for terms of SITT as it is defined in Definition 37 of §2.3 for the terms of TT and ITT. But it is essential that the difference between a type τ that is not an intersection type and the intersection type $\tau \cap 1$ be recognized in the definition of a contraction.

A term CT is β -contractible if it has the form $(\lambda v.P)Q$ and its contraction is $[Q/v]P$. But by clauses (2) and (3) of Definition 53, for $(\lambda v.P)Q$ to be a term it is necessary that Q have the type τ if v is a q-variable and the type $\tau \cap 1$ if v is an a-variable. Hence if v is an a-variable, $(\lambda v.P)Q$ is not a term if $Q{:}\tau$ and Q does not have the secondary type 1. In particular, if Q is a q-variable, then $(\lambda v.P)Q$ is not a term. On the other hand if v is a q-variable, and u is an a-variable for which u:t[v], then $(\lambda v.P)u$ is a term even though u also has type t[v] $\cap$ 1.

The proof of theorem 21 in §2.3 and its corollary can be repeated for SITT. Thus every term of SITT can be understood to have a normal form that is unique to within bound variable variants.

3.6.3. Semantics of SITT. The semantics for SITT is an extension of the semantics for ITT. Definition 45 in §3.3.1 for the domains for intentional valuations of ITT is extended by a definition for the domains $DI(\tau \cap 1)$ for intersection types. Only clause (V.4) of Definition 46 §3.3.1 of intensional valuation for ITT is changed: its replacement provides valuations for terms with an intersection type as type; valuations for terms with a type that is not an intersection type are defined as before.

DEFINITION 54. *Domains for SITT.*
The clause to be added to Definition 45 *of* §3.3.1 *is the following*:

> $DI(\tau \cap 1)$ *is the set of pairs* $\langle N, N \rangle$ *for which* $N{:}\tau \cap 1$, *where* N *is in normal form and without free variables.*

DEFINITION 55. *Intensional Valuation for SITT.*
Definition 46 *in* §3.3.1 *of a valuation for ITT is unchanged but for clause* (V.4) *that is replaced by*

V.4. $\Phi_2(1, T)$ *is* $\Phi_2(\tau \cap 1, T)$ *is* $\Phi_1(\tau \cap 1, T)$, *when* $T{:}\tau$ *has secondary type* 1.

Lemma 25 of semantic substitution for ITT, after updating for intersection types, can be proved for SITT. Similarly Lemma 26 for ITT in §3.3.1 can be updated for SITT. The definition of an intensional model for SITT is unchanged from Definition 47 for ITT in §3.3.1. Theorem 28 in §3.3.2 provides a semantic basis for the proof theory of ITT. It can be repeated for SITT.

3.6.4. Proof Theory for SITT. The semantic rules for the proof theory of SITT are identical in appearance with those for the proof theory of ITT in §3.4. But it is important to note that the relation $=$ of the two Int rules for SITT differs from the same relation for the rules of ITT. However for terms P and Q of SITT that make no use of abstraction variables $P = Q$ holds in SITT if and only if it holds in ITT. Thus every sequent derivable in ITT is derivable in SITT.

Theorem 29 asserting the consistency of ITT can be proved for SITT.

THEOREM 40. *Consistency of SITT.*
A derivable sequent of SITT is valid.

Similarly Theorem 33 asserting the completeness of ITT without Cut can be repeated for SITT.

3.6.5. Russell's set. The Russell set R is defined to be the extension of the predicate $(\lambda u.\neg u(u))$, where u:[1] is an abstraction variable so that u:[1] $\cap$ 1. Since u has secondary type 1, $u(u)$:[] when the second occurrence of u is assigned the secondary type. Therefore $\neg u(u)$:[ ], and $(\lambda u.\neg u(u))$:[[1] ∩ 1] as required by clause (3) of Definition 53. Arguments for R must therefore have the type [1] ∩ 1. Examples of such arguments are therefore the empty predicate ⊘ defined to be $(\lambda x.\neg x = x)$ and the universal predicate V defined to be $(\lambda x.x = x)$, where x is a q-variable of type 1; but R is not an example since R:[[1]∩1] or R:[[1]∩1]∩1. Therefore the expression R(R) is not a term of SITT since no type can be assigned to it.

Each of the two sequents

$$\vdash R(\oslash) \text{ and } \vdash \neg R(\mathsf{V})$$

is derivable. Note, however, that although $\vdash R(\oslash)$ is derivable, $\vdash \exists x.R(x)$ is not since for no q-variable x is $R(x)$ a formula. Thus although R is a term of SITT and can like other terms appear in derivable sequents, it cannot be treated like a term of ITT when quantification is involved.

EXERCISES §3.6.

1. Repeat the proofs of the consistency and completeness theorems of ITT for SITT.
2. Provide derivations for $\vdash R(\oslash)$ and $\vdash \neg R(\mathsf{V})$.

CHAPTER 4

RECURSIONS

4.1. Introduction

The purpose of this chapter is to provide some fundamental results upon which the many applications of recursive definitions can be based. It is the simplicity of these foundations that distinguishes ITT from other higher order logics.

The recursive definitions most commonly used in mathematics and computer science are least predicate definitions and more rarely greatest predicate definitions. In §4.2 a general method is described for providing these definitions that makes use of terms called *recursion generators*. A recursion generator with *parameters* of type $\overline{p}$ is any term of type $[\overline{p}, [\overline{\tau}], \overline{\tau}]$ without any free occurrences of variables; here $\overline{p}$ may be empty but $\overline{\tau}$ may not. It is used to define recursively a predicate of type $[\overline{p}, \overline{\tau}]$ with parameters of type $\overline{p}$. A recursion generator without parameters is therefore of type $[[\overline{\tau}], \overline{\tau}]$.

From a recursion generator RG of type $[[\overline{\tau}], \overline{\tau}]$, a map c in the sense of [111] or an operator F in the sense of [103] can be defined to be

$$(\lambda X, x.RG(X, x)).$$

For if $X{:}[\overline{\tau}]$, then

$$(\lambda X, x.RG(X, x))(X){:}[\overline{\tau}].$$

Conversely, given an operator F that is a function of argument and value of type $[\overline{\tau}]$, a recursion generator can be defined as

$$(\lambda X, x.F(X)(x)),$$

where $F(X)(x)$ is to be understood as asserting that x is a member of the value $F(X)$ of F, which is of type $[\overline{\tau}]$, for the argument X which is also of type $[\overline{\tau}]$. Recursion generators, rather than maps or operators, have been chosen as the fundamental concept for recursion theory because ITT is a logic of *predicates*, not *functions*.

Recursive predicates are defined from recursion generators using two operators Lt and Gt that are terms of type $[[[\overline{\tau}], \overline{\tau}], \overline{\tau}]$ defined in §4.2. The terms $Lt(RG)$ and $Gt(RG)$ are read as the least, respectively the greatest, predicate defined by RG. Induction rules can be derived for $Lt(RG)$ and $Gt(RG)$

without any assumptions on the recursion generator RG other than its type. However, few useful properties of $Lt(RG)$ and $Gt(RG)$ beyond these rules can be derived without additional assumptions on RG.

Monotonic recursion generators RG are defined in §4.3, and a study of the properties of the predicates $Lt(RG)$ and $Gt(RG)$ for such RG is undertaken. It is shown that these predicates are respectively the least and greatest predicates satisfying an extensional identity. This is the reverse of the fixed point method of defining recursive predicates described in [117]; there recursive predicates are defined as fixed points of intensional identities. The formulas $Lt(RG)(\overline{x})$ and $Gt(RG)(\overline{x})$, resulting when the predicates $Lt(RG)$ and $Gt(RG)$ are supplied with appropriate arguments, are related to the convergence formulas of [111].

Recursion generators, when used in the logic ITT with secondary types, provide a simple but powerful method of defining and reasoning about both well-founded and non-well-founded recursive predicates. An example is offered in §4.4: The zero 0, successor S, recursion generator RN and term N defined in Example 3 of §3.2 are used to derive all of Peano's axioms. Thus an axiom of infinity is not needed in ITT.

$\exists$-continuous and $\forall$-continuous recursion generators are defined in §4.5. Decidable sufficient conditions for a recursion generator to be monotonic and for it to be $\exists$-continuous and $\forall$-continuous are provided in §4.6.

Horn clauses, as they are now known, were first identified as useful formulas of first order logic in [86]. They have since been suggested many times as a means for defining recursive predicates; see, for example, [56], [34], [89], [57], and [12]. This use is the basis for the programming language Prolog, described in [138], for example, and the more recent programming language Gödel described in [80]. A first order semantics is provided for them in [94] and [109]. A higher order version of Horn clause definitions based on [28] is described in [108]. In §4.7 a higher order version of Horn clauses called *Horn sequents* is defined in terms of recursion generators. It is proved that a Horn sequent defined by a recursion generator RG and the converse of the sequent are satisfied by either $Lt(RG)$ or $Gt(RG)$. These results are generalized for simultaneous recursions in §4.7.2. Numerous examples and exercises involving Horn sequents are offered.

An iteration predicate It is defined and applied in §4.8. $It(wg, ws)(x)$ is the predicate that results from iterating the recursion generator wg x times beginning with ws. The following sequents are derivable:

$$[Mon(wg) \wedge Con\exists(wg)] \vdash Lt(wg) =_e (\lambda v.[\exists x{:}N].IT(wg, \oslash)(x, v),$$
$$[Mon(wg) \wedge Con\forall(wg)] \vdash Gt(wg) =_e (\lambda v.[\forall x{:}N].IT(wg, \mathsf{V})(x, v),$$

where $Mon(wg)$ expresses that the recursion generator $wg{:}[[\tau], \tau]$ is monotonic and $Con\exists(wg)$ and $Con\forall(wg)$ that it is $\exists$- and $\forall$-continuous; further $\oslash{:}[\tau]$ is the empty predicate and V:[τ] the universal. In addition the sequent

$$[Mon(wg) \wedge Con\exists(wg)] \\ \vdash [\forall x{:}N].\forall v.[CIt(wg,\oslash)(x,v) \rightarrow CIt(wg,\oslash)(S(x),v)]$$

is derivable where CIt is *cumulative* iteration, defined in terms of It. It follows that for each natural number n

$$CIt(wg,\oslash)(S^n(0)) \subseteq CIt(wg,\oslash)(S^{n+1}(0)) \subseteq Lt(wg),$$

where $\subseteq$ is defined in the obvious way. Thus a series of *approximations* can be defined for a predicate $Lt(wg)$ for which wg is monotonic and $\exists$-continuous. Although this suggests a method of *computing* $Lt(wg)$ for particular arguments, *computations* must be distinguished from *definitions*. A discussion of this distinction is postponed to Chapter 8.

The chapter concludes with the definition and study in §4.9 of *potentially* infinite predicates; that is, predicates with a finite but unbounded extension. Such predicates are fundamental for intuitionist/constructive logic and computer science.

4.2. Least and Greatest Predicate Operators

The basis for least and greatest predicate definitions of predicates of type [τ] are the predicates Lt and Gt defined as follows:

$$Lt \stackrel{\text{df}}{=} (\lambda wg, u.\forall Z.[\forall x.[wg(Z,x) \rightarrow Z(x)] \rightarrow Z(u)]),$$

$$Gt \stackrel{\text{df}}{=} (\lambda wg, u.\exists Z.[\forall x.[Z(x) \rightarrow wg(Z,x)] \wedge Z(u)]).$$

Here $u, x{:}\tau$, $Z{:}[\tau]$, and $wg{:}[[\tau],\tau]$; thus the variable wg has the type of a recursion generator without parameters. These definitions can be generalized in the obvious way for a recursion generator of any type. Given a particular recursion generator RG with parameters of type $\overline{p}$,

$$\lambda\overline{w}.Lt(RG(\overline{w})) \text{ and } \lambda\overline{w}.Gt(RG(\overline{w}))$$

are two recursive predicates defined by RG when $\overline{w}{:}\overline{p}$, where $\overline{w}$ is a sequence of distinct variables none of which have a free occurrence in RG.

A recursion generator is so called because of the following two derivable sequents that express a general form of induction for the unary predicates $Lt(wg)$ and $Gt(wg)$:

(Lt.1) $\vdash \forall wg, Y.[\forall x.[wg(Y,x) \rightarrow Y(x)] \rightarrow \forall x.[Lt(wg)(x) \rightarrow Y(x)]]$
(Gt.1) $\vdash \forall wg, Y.[\forall x.[Y(x) \rightarrow wg(Y,x)] \rightarrow \forall x.[Y(x) \rightarrow Gt(wg)(x)]]$,

where in each case $wg{:}[[\tau],\tau]$. Derivations are given below.

Two useful rules of deduction can be derived from (Lt.1) and (Gt.1):

LtInd.
$$\frac{-\forall x.[Lt(RG(\overline{r}))(x) \rightarrow P(x)]}{-\forall x.[RG(\overline{r})(P,x) \rightarrow P(x)]}$$

GtInd.
$$\frac{-\forall x.[P(x) \rightarrow Gt(RG(\overline{r}))(x)]}{-\forall x.[P(x) \rightarrow RG(\overline{r})(P,x)]}$$

where $RG{:}[\overline{p},[\tau],\tau]$, $\overline{r}{:}\overline{p}$, and where x has no free occurrence in P or in any member of $\overline{r}$.

Similar + rules can be derived but are not particularly useful. The LtInd and GtInd rules are the only induction rules needed; special cases of them, such as mathematical induction for Peano arithmetic, can be derived from them. The GtInd rule is related to what in [103] is called *co-induction*. The two rules are related to what Park calls in [111] *fixpoint induction*. It is noteworthy that their derivations require no assumption on RG other than it has the type of a recursion generator.

The abbreviated derivations below have a format that will be followed in the remainder of the monograph. The column to the right of the signed formulas that are nodes of the tree is intended to help the reader understand the derivation. Some nodes are followed by the letters a, b, c, ... in parenthesis. For example, the 3rd node is followed by '(a)' and the 7th by '(b)'. Other nodes such as the 8th and the 11th are followed by the first the symbol '⊣', the reverse of the symbol '⊢' that separates antecedent and succedent in a sequent, and then a brief reference to help the reader justify the node. Thus '⊣ df Lt' indicates that the 8th node has been obtained from a previous node, in this case the 6th, by replacing the defined predicate Lt with its definition. The reference '⊣ (a)' on the last node indicates that the node has been obtained from node (a) while the reference '⊣ (b)' opposite the closing bar indicates that node (b) is a closing node. Although there are no examples in these derivations, both a reference and a justification can appear together on the same line; thus '(c)⊣ df Lt', for example, might appear following a node.

To avoid as much branching of the derivation tree as possible, the following derived rules will be used without further justification:

$$+[F \rightarrow G], +F \Rightarrow +G \text{ and } +[F \rightarrow G], -G \Rightarrow -F.$$

Derivation of (Lt.1)

$-\forall wg,Y.[\forall x.[wg(Y,x) \rightarrow Y(x)] \rightarrow \forall x.[Lt(wg)(x) \rightarrow Y(x)]]$
$-[\forall x.[wg(Y,x) \rightarrow Y(x)] \rightarrow \forall x.[Lt(wg)(x) \rightarrow Y(x)]]$
$+[\forall x.[wg(Y,x) \rightarrow Y(x)]$ (a)
$-\forall x.[Lt(wg)(x) \rightarrow Y(x)]$
$-[Lt(wg)(x) \rightarrow Y(x)]$
$+Lt(wg)(x)$

$-Y(x)$ (b)
$+(\lambda wg, u.\forall Z.[\forall x.[wg(Z,x) \rightarrow Z(x)] \rightarrow Z(u)])(wg)(x)$ $\dashv$df Lt
$+\forall Z.[\forall x.[wg(Z,x) \rightarrow Z(x)] \rightarrow Z(x)]$
$+[\forall x.[wg(Y,x) \rightarrow Y(x)] \rightarrow Y(x)]$
$+Y(x)$ $\dashv$(a)
$\dashv$(b)

Derivation of (Gt.1)

$-\forall wg, Y.[\forall x.[Y(x) \rightarrow wg(Y,x)] \rightarrow \forall x.[Y(x) \rightarrow Gt(wg)(x)]]$
$-[\forall x.[Y(x) \rightarrow wg(Y,x)] \rightarrow \forall x.[Y(x) \rightarrow Gt(wg)(x)]]$
$+\forall x.[Y(x) \rightarrow wg(Y,x)]$ (a)
$-\forall x.[Y(x) \rightarrow Gt(wg)(x)]$
$-[Y(x) \rightarrow Gt(wg)(x)]$
$+Y(x)$ (b)
$-Gt(wg)(x)$
$-(\lambda wg, u.\exists Z.[\forall x.[Z(x) \rightarrow wg(Z,x)] \wedge Z(u)])(wg)(x)$ $\dashv$df Gt
$-\exists Z.[\forall x.[Z(x) \rightarrow wg(Z,x)] \wedge Z(x)]$
$-[\forall x.[Y(x) \rightarrow wg(Y,x)] \wedge Y(x)]$
$\dashv$(a)(b),$-\wedge$

EXERCISES §4.2.

1. Provide derivations for the rules LtInd and GtInd.

4.3. Monotonic Recursion Generators

Let RG be a recursion generator without parameters. Few useful properties of $Lt(RG)$ and $Gt(RG)$, beyond the instantiations of (Lt.1) and (Gt.1) of §4.2, can be derived without assuming that RG has properties other than its typing. The first of the additional properties to be studied is *monotonicity*

$$Mon \stackrel{\text{df}}{=} (\lambda wg.\forall U, V.[\forall x.[U(x) \rightarrow V(x)] \rightarrow \forall y.[wg(U,y) \rightarrow wg(V,y)]]).$$

In this definition it is assumed that wg:[[τ], τ]. The definition can be generalized in the obvious way for wg:[[$\overline{\tau}$], $\overline{\tau}$], where $\overline{\tau}$ is not empty.

RG is said to be *monotonic* if the sequent

$$\vdash Mon(RG)$$

is derivable.

Although the definition of Mon is expressed in terms of the derivability of a sequent, and is therefore not necessarily decidable, a strong decidable sufficient condition for this property to hold of a recursion generator is defined in §4.6.

4.3.1. Properties of monotonic recursion generators. The following sequents are derivable:

Lt.2)	$\vdash [\forall wg{:}Mon].\forall x.[wg(Lt(wg), x) \rightarrow Lt(wg)(x)]$
Lt.3)	$\vdash [\forall wg{:}Mon].\forall x.[Lt(wg)(x) \rightarrow wg(Lt(wg), x)]$
Gt.2)	$\vdash [\forall wg{:}Mon].\forall x.[Gt(wg)(x) \rightarrow wg(Gt(wg), x)]$
Gt.3)	$\vdash [\forall wg{:}Mon].\forall x.[wg(Gt(wg), x) \rightarrow Gt(wg)(x)]$
LtGt)	$\vdash [\forall wg{:}Mon].\forall x.[Lt(wg)(x) \rightarrow Gt(wg)(x)]$,

Note that $Lt(wg)(x)$ can be replaced by $Lt(wg, x)$ using the syntactic sugaring notation of §1.2.

The sequents (Lt.2) and (Lt.3), respectively (Gt.2) and (Gt.3), can be understood to express that $Lt(RG)$, respectively $Gt(RG)$, is a fixed point of an extensional identity when RG is monotonic. The following definition of the predicate $FixPt$ makes use of the extensional identity:

$$FixPt \stackrel{\text{df}}{=} \lambda wg, X.wg(X) =_e X.$$

The derivable sequents expressing the fixed point results are

FixLt)	$\vdash [\forall wg{:}Mon].FixPt(wg, Lt(wg))$
FixGt)	$\vdash [\forall wg{:}Mon].FixPt(wg, Gt(wg))$.

Thus for predicates $Lt(RG)$ and $Gt(RG)$ for which RG is monotonic, the following sequents are derivable:

$$\vdash RG(Lt(RG)) =_e Lt(RG)$$

$$\vdash RG(Gt(RG)) =_e Gt(RG).$$

The sequent $(LtGt)$ expresses that $Lt(RG)$ is the *lesser* and $Gt(RG)$ the *greater* of the two fixed point of the extensional identity $RG(X) =_e X$.

It is important to note that the derivability of these fixed point equations requires only the assumption that RG is monotonic; the predicates $Lt(RG)$ and $Gt(RG)$ are not defined as the fixed points of the equation. This is in contrast to the fixed point method of defining recursive predicates described in [117] where recursive predicates are defined as the fixed points of an intensional identity.

Derivation of (Lt.2)

$-[\forall wg{:}Mon].\forall x.[wg(Lt(wg), x) \rightarrow Lt(wg)(x)]$	
$-\forall wg.[Mon(wg) \rightarrow \forall x.[wg(Lt(wg), x) \rightarrow Lt(wg)(x)]]$	
$+Mon(wg)$	
$-\forall x.[wg(Lt(wg), x) \rightarrow Lt(wg)(x)]$	
$-[wg(Lt(wg), x) \rightarrow Lt(wg)(x)]$	
$+wg(Lt(wg), x)$	(a)
$-Lt(wg)(x)$	
$-\forall Z.[\forall x.[wg(Z, x) \rightarrow Z(x)] \rightarrow Z(x)]$	$\dashv$df Lt
$+\forall x.[wg(Z, x) \rightarrow Z(x)]$	(b)

$-Z(x)$
$-wg(Z, x)$ (c)
$+[\forall x.[wg(Z, x) \rightarrow Z(x)] \rightarrow \forall x.[Lt(wg)(x) \rightarrow Z(x)]]$ $\dashv$(Lt.1)
$+\forall x.[Lt(wg)(x) \rightarrow Z(x)]$ (d) $\dashv$(b)
$+\forall X, Y.[\forall x.[X(x) \rightarrow Y(x)] \rightarrow \forall x.[wg(X, x) \rightarrow wg(Y, x)]]$ $\dashv Mon(wg)$
$+[\forall x.[Lt(wg)(x) \rightarrow Z(x)] \rightarrow \forall x.[wg(Lt(wg), x) \rightarrow wg(Z, x)]]$ $\dashv +\forall$
$+\forall x.[wg(Lt(wg), x) \rightarrow wg(Z, x)]$ $\dashv$(d)
$+[wg(Lt(wg), x) \rightarrow wg(Z, x)]$
$+wg(Z, x)$ $\dashv$(a)
——— $\dashv$(c)

Derivation of (Gt.2)

$-[\forall wg{:}Mon].\forall x.[Gt(wg)(x) \rightarrow wg(Gt(wg), x)]$
$-\forall wg.[Mon(wg) \rightarrow \forall x.[Gt(wg)(x) \rightarrow wg(Gt(wg), x)]$
$+Mon(wg)$
$-\forall x.[Gt(wg)(x) \rightarrow wg(Gt(wg), x)]$
$+Gt(wg)(x)$
$-wg(Gt(wg), x)$ (a)
$+\exists Z.[\forall x.[Z(x) \rightarrow wg(Z, x)] \wedge Z(x)]$ $\dashv$df Gt
$+[\forall x.[Z(x) \rightarrow wg(Z, x)] \wedge Z(x)]$
$+\forall x.[Z(x) \rightarrow wg(Z, x)]$ (b)
$+Z(x)$
$+wg(Z, x)$ (c)
$+\forall wg, Y.[\forall x.[Y(x) \rightarrow wg(Y, x)] \rightarrow \forall x.[Y(x) \rightarrow Gt(wg)(x)]]$ $\dashv$(Gt.1)
$+[\forall x.[Z(x) \rightarrow wg(Z, x)] \rightarrow \forall x.[Z(x) \rightarrow Gt(wg)(x)]]$ $\dashv +\forall$
$+\forall x.[Z(x) \rightarrow Gt(wg)(x)]$ (d)$\dashv$(b)
$+\forall X, Y.[\forall x.[X(x) \rightarrow Y(x)] \rightarrow \forall x.[wg(X, x) \rightarrow wg(Y, x)]]$ $\dashv Mon(wg)$
$+[\forall x.[Z(x) \rightarrow Gt(wg)(x)] \rightarrow \forall x.[wg(Z, x) \rightarrow wg(Gt(wg), x)]]$ $\dashv +\forall$
$+\forall x.[wg(Z, x) \rightarrow wg(Gt(wg), x)]$ $\dashv$(d)
$+[wg(Z, x) \rightarrow wg(Gt(wg), x)]$
$+wg(Gt(wg), x)$ $\dashv$(c)
——— $\dashv$(a)

EXERCISES §4.3.

1. Provide derivations for (Lt.3), (Gt.3), and (LtGt).
2. Provide a derivation for the following sequents

   ERGLt) $\vdash [\forall wg1, wg2{:}Mon].[wg1 =_e wg2 \rightarrow Lt(wg1) =_e Lt(wg2)]$
   ERGGt) $\vdash [\forall wg1, wg2{:}Mon].[wg1 =_e wg2 \rightarrow Gt(wg1) =_e Gt(wg2)]$.

3. The term RN defined in Example 3 of §3.2 is a recursion generator of type [[1], 1] with = :[1, 1]. Prove that it is monotonic; that is, provide a derivation for

$$\vdash Mon(RN).$$

4.4. Zero, Successor, and Natural Numbers

To illustrate the recursion generator method of defining recursive predicates, the definitions in Exercise 2 of §3.2 of zero 0, successor S, the recursion generator RN, and the predicate N are repeated here and Peano's axioms are derived. In addition a zero and the binary successor of ordered pair are defined.

Although all of Peano's axioms are derived for 0, S, and N, the sequence

$$0, S(0), S(S(0)), \ldots$$

of terms is not the sequence of natural numbers 0, 1, 2, ... , according to the Frege-Russell definition of these terms in [126], but is rather a sequence of terms from which *counter sets* in the sense of [120] can be defined; but any such sequence is all that is needed for counting; see Exercise (8) of §4.4.

4.4.1. A zero and successor. Consider the following definitions repeated from Example 2 of §3.2.1:

$$0 \stackrel{\text{df}}{=} (\lambda u.\neg u = u) \quad \text{and} \quad S \stackrel{\text{df}}{=} (\lambda u, v.u = v).$$

Here 0:[1] but 0 also has the secondary type 1, since no variable has a free occurrence in it. Further, since S:[1, 1] it follows that $S(0)$:[1] and $S(0)$:1; similarly

$$S(0), S(S(0)), \ldots, S(S(\cdots S(0))\cdots))$$

are all of type 1 as is $S(x)$ when x:1. Thus because of secondary typing, S may be used as the name of a function with argument and value of type 1.

The successor S has been defined in a simpler manner than in most set theories. For example, in both [132] and [120], the successor function is defined, using the notation of ITT, to be $(\lambda u, v.[u = v \vee v(u)])$. But this expression cannot be typed in ITT and is therefore not a term. But, as will be seen, the more complicated definition of S is unnecessary to produce a provably infinite sequence of distinguishable terms.

The secondary typing of $S(x)$ is exploited in derivations of the following two sequents that are related to two axioms of Peano for the natural numbers [115]:

S.1) $\vdash \forall x, y.[S(x) = S(y) \rightarrow x = y]$

S.2) $\vdash \forall x.\neg S(x) = 0,$

These two sequents ensure that the terms in the sequence 0, $S(0)$, $S(S(0))$, ... are distinct. In TT this is accomplished by adding an axiom of infinity complicating recursive definitions.

Derivations of (S.1) and (S.2) follow. They illustrates the use of the Cut rule for the reuse of derivations, as described in §1.7.4, by making use of the derivation of (IEId.1) provided in §2.5. The derivations are abbreviated and use the format described in §4.2.

Derivation of (S.1)

$-\forall x, y.[S(x){=}S(y) \to x{=}y]$
$-[S(x){=}S(y) \to x{=}y]$
$+S(x){=}^{1}S(y)$ (a)
$-x{=}y$ (b)
$+S(x){=}^{[1]}S(y)$ $\dashv$(a),+Int
$+S(x){=}_{e}S(y)$ $\dashv$IEId.1
$+\forall z.[S(x)(z) \leftrightarrow S(y)(z)]$ $\dashv$df $=_e$
$+[S(x)(y) \leftrightarrow S(y)(y)]$
$+[S(y)(y) \to S(x)(y)]$ $\dashv$df $\leftrightarrow$

$-S(y)(y)$	$+S(x)(y)$	$\dashv + \to$
$-y{=}y$	$+x{=}y$	
———	———	$\dashv$(b)

Derivation of (S.2)

$-\forall x.\neg S(x){=}0$
$-\neg S(x){=}0$
$+S(x){=}^{1}0$
$+S(x){=}^{[1]}0$ $\dashv$+Int
$+S(x){=}_{e}0$ $\dashv$IEId.1
$+\forall z.[S(x)(z) \leftrightarrow 0(z)]$ $\dashv$df $=_e$
$+[S(x)(x) \leftrightarrow 0(x)]$
$+[S(x)(x) \to 0(x)]$

$-S(x)(x)$	$+0(x)$
$-x{=}x$	$+\neg x{=}x$
———	$-x{=}x$
	———

4.4.2. Nil and ordered pair. An arity 2 successor can be defined in terms of ordered pair; the following definition is taken from [29].

$$\langle\,\rangle \stackrel{\text{df}}{=} (\lambda u, v, w.w(u, v)).$$

If $u{:}\sigma$ and $v{:}\tau$, then $w{:}[\sigma, \tau]$ and $\langle\,\rangle{:}[\sigma, \tau, [\sigma, \tau]]$, and $\langle u, v\rangle{:}[[\sigma, \tau]]$. The usual infix notation for ordered pairs will be used. Ordered n-tuples, $n \geq 2$, can be defined in the expected fashion from ordered pair. Derivable sequents are

OP.1) $\vdash \forall x1, y1, x2, y2.[\langle x1, y1\rangle{=}\langle x2, y2\rangle \to x1{=}x2 \wedge y1{=}y2]$
OP.2) $\vdash \forall x, y.\neg\langle x, y\rangle{=}Nil$,

where Nil is defined $Nil \stackrel{\text{df}}{=} (\lambda u.\neg u{=}u)$.

The type assumed by the polymorphic $=$ depends upon the types of the variables: If $x1, x2{:}\sigma$ and $y1, y2{:}\tau$, then $\langle x1, y1\rangle, \langle x2, y2\rangle{:}[[\sigma, \tau]]$ and therefore $=$ $:[\,[[\sigma, \tau]], [[\sigma, \tau]]]$. But if $x1, x2, y1, y2{:}1$, then each of $\langle x1, y1\rangle$ and $\langle x2, y2\rangle$ can have the secondary type 1 so that $=$ $:[1, 1]$.

Nil inherits its type from =. If = has type $[[[\sigma,\tau]],[[\sigma,\tau]]]$, then *Nil* has type $[[\sigma,\tau]]$ and u type $[\sigma,\tau]$. If = has type [1,1] then u:1 and *Nil* is 0.

Note the similarity of (S.1) and (S.2) and (OP.1) and (OP.2) when = has type [1,1]. The importance of these sequents will become evident in §4.4.4 and §4.7.

Derivation of (OP.1)

$-\forall x1, y1, x2, y2.[\langle x1, y1\rangle=\langle x2, y2\rangle \rightarrow x1=x2 \wedge y1=y2]$
$+\langle x1, y1\rangle=\langle x2, y2\rangle$
$+\langle x1, y1\rangle=_e\langle x2, y2\rangle$ ⊣IEId.1 or IEId.2
$-x1=x2 \wedge y1=y2$
$+\forall w.[\langle x1, y1\rangle(w) \leftrightarrow \langle x2, y2\rangle(w)]$ ⊣df $=_e$

L *R*
$-x1=x2$
$+\forall w.[\langle x1, y1\rangle(w) \rightarrow \langle x2, y2\rangle(w)]$
$+[\langle x1, y1\rangle(T1) \rightarrow \langle x2, y2\rangle(T1)]$ ⊣$T1 \overset{df}{=} (\lambda u, v.x1=u)$

LL *LR*
$-\langle x1, y1\rangle(T1)$
$-(\lambda w.w(x1, y1))(T1)$ ⊣df $\langle\rangle$
$-S1(x1, y1)$
$-x1=x1$

LR
$+\langle x2, y2\rangle(T1)$
$+(\lambda w.w(x2, y2))(T1)$ ⊣df $\langle\rangle$
$+T1(x2, y2)$
$+x1 = x2$

R
$-y1=y2$
$\cdots$ use $T2 \overset{df}{=} (\lambda u, v.x2 = u)$ instead of $T1$.

Derivation of (OP.2)

$-\forall x, y.\neg\langle x, y\rangle=Nil$
$-\neg\langle x, y\rangle=Nil$
$+\langle x, y\rangle=Nil$
$+\langle x, y\rangle=_e Nil$ ⊣IEId.1 or IEId.2
$+\forall w.[\langle x, y\rangle(w) \rightarrow Nil(w)]$ ⊣df $=_e$
$+[\langle x, y\rangle(T) \rightarrow Nil(T)]$ $T \overset{df}{=} (\lambda u, v.x=u)$

L R

$-\langle x, y\rangle(T)$
$-(\lambda u, v, w.w(u, v))(x, y, T)$ $\quad\dashv$df $\langle\ \rangle$
$-T(x, y)$
$-x{=}x$

R
$+Nil(T)$
$+(\lambda w.\neg w{=}w(T)$ $\quad\dashv$df Nil
$+\neg T{=}T$
$-T{=}T$

4.4.3. Projections of ordered pair. The definitions from [29] of projection functions that are intended to select respectively the first and second members of a pair are:

$$Hd^* \stackrel{\text{df}}{=} (\lambda w.w(\lambda u, v.u)) \quad\text{and}\quad Tl^* \stackrel{\text{df}}{=} (\lambda w.w(\lambda u, v.v)).$$

But in ITT these definitions of ordered pair are ill formed if $u, v{:}1$, and don't have the desired properties otherwise. For if $u{:}1$ then $(\lambda u, v.u)$ is not a term of TT or ITT since u does not have type $[\overline{\tau}]$ as required by clause (3) of Definition 43 of §3.1.2. If on the other hand $u, v{:}[1]$ then $(\lambda u, v.u){:}[[1],[1],1]$, $w{:}[[[1],[1],1]]$, and $Hd^*{:}[[[[1],[1],1]]]$. But then for $Hd^*(\langle x, y\rangle)$ to be a term it is necessary that $\langle x, y\rangle{:}[[[1],[1],1]]$, while if $x, y{:}[1]$ then $\langle x, y\rangle{:}[[[1],[1]]]$.

The following definitions of predicates Hd and Tl do however have the desired properties:

$$Hd \stackrel{\text{df}}{=} (\lambda z, x.\exists v.z{=}\langle x, v\rangle) \quad\text{and}\quad Tl \stackrel{\text{df}}{=} (\lambda z, y.\exists u.z{=}\langle u, y\rangle).$$

The following sequents are derivable; the provision of derivations is left as an exercise.

Hd.1) $\quad\vdash \forall x, y.Hd(\langle x, y\rangle, x)$
Tl.1) $\quad\vdash \forall x, y.Tl(\langle x, y\rangle, y)$
Hd.2) $\quad\vdash \forall z, x, y.[[Hd(z, x) \wedge Hd(z, y)] \rightarrow x{=}y]$
Tl.2) $\quad\vdash \forall z, x, y.[[Tl(z, x) \wedge Tl(z, y)] \rightarrow x{=}y]$
Hd.3) $\quad\vdash [\forall z{:}Pr].\exists!x.Hd(z, x)$
Tl.3) $\quad\vdash [\forall z{:}Pr].\exists!y.Tl(z, y)$.

Here Pr is defined:

$$Pr \stackrel{\text{df}}{=} \lambda z.\exists u, v.z{=}\langle u, v\rangle,$$

and $\exists!$ is the unique existential quantifier defined as

$$\exists!v.P(v) \stackrel{\text{df}}{=} \exists v.[P(v) \wedge \forall u.[P(u) \rightarrow u{=}v]].$$

As will be seen in Chapter 5, these sequents provide a justification for the use of a functional notation for Hd and Tl.

4.4.4. A recursion generator for the S-sequence. A recursive definition of a predicate N that includes all the members of the sequence 0, $S(0)$, $S(S(0))$, ... in its extension provides a simple example of recursive definitions using recursion generators. Other such sequences can be defined in terms of other definitions of a zero and a successor; for example, Nil and ordered pair may be used. As noted in the introduction to this section, such a sequence is not the sequence of natural numbers 0, 1, 2, ... , according to the Frege-Russell definition of these terms in [126], but is rather a sequence of terms from which counter sets in the sense of [120] can be defined; but any such sequence is all that is needed for counting; see (8) of the exercises for §4.4.

A recursion generator of type [[1], 1] is defined when = :[1, 1]:

$$RN \stackrel{\text{df}}{=} \lambda Z, x.[x{=}0 \vee \exists u.[Z(u) \wedge x{=}S(u)]].$$

As noted in example 3 of §3.2, it is possible for RN to have a type other than [[1], 1], but when used in the context of the predicate Lt, this is the only type RN can have. The arity 1 predicate N is defined in terms of RN and Lt:

$$N \stackrel{\text{df}}{=} Lt(RN).$$

One of Peano's axioms is mathematical induction, a rule which is a direct application of the general induction rule LtInd of §4.2:

(NInd.)
$$\frac{-\forall x.[N(x) \to Q(x)]}{-Q(0) \qquad -\forall x.[Q(x) \to Q(S(x))]}$$

As was noted, two other of Peano's axioms are expressed by (S.1) and (S.2) of §4.4.1. The remaining two of the five axioms of Peano are

N.1) $\qquad \vdash N(0)$

N.2) $\qquad \vdash \forall x.[N(x) \to N(S(x))]$.

These sequents are immediate consequences of the derivable sequent (Lt.2) when RN replaces wg. Providing derivations for these sequents is an exercise of §4.4.

Derivation of (NInd). A derivation of the rule NInd follows immediately from the following derivation of the sequent

$$Q(0), \forall x.[Q(x) \to Q(S(x))] \vdash \forall x.[N(x) \to Q(x)]$$

$+Q(0)$	(a)
$+\forall x.[Q(x) \to Q(S(x))]$	(b)
$-\forall x.[N(x) \to Q(x)]$	
$-\forall x.[Lt(RN)(x) \to Q(x)]$	$\dashv$df N
$-\forall x.[RN(Q, x) \to Q(x)]$	$\dashv$LtInd
$+RN(Q, x)$	
$-Q(x)$	(c)

$+(\lambda Z, y.[y{=}0 \vee \exists x.[Z(x) \wedge y{=}S(x)]])(Q, x)$ ⊣df RN
$+[x{=}0 \vee \exists u.[Q(u) \wedge x{=}S(u)]]$

L R

$+x{=}0$
$-Q(0)$ ⊣(c),=
⊣(a)

R
$+\exists u.[Q(u) \wedge x{=}S(u)]$
$+[Q(u) \wedge x{=}S(u)]$
$+Q(u)$ (d)
$+x{=}S(u)$ (e)
$+Q(S(u))$ ⊣(a), (d)
$+Q(x)$ ⊣(e), =
⊣(c)

4.4.5. Understanding $Gt(RN)$. The predicate N, that is $Lt(RN)$, applies to each member of the sequence 0, $S(0)$, $S(S(0))$, ... Because of (LtGt) of §4.3.1, so does the predicate $Gt(RN)$. Consider now the predicate GnN defined

$$GnN \stackrel{\text{df}}{=} (\lambda x.[Gt(RN)(x) \wedge \neg Lt(RN)(x)]).$$

Is it perhaps necessarily empty? That is, is the sequent

a) $\vdash \forall x.[Gt(RN)(x) \rightarrow Lt(RN)(x)]$

derivable? That (a) is not derivable is demonstrated by the following unclosable branch of a semantic tree:

$-\forall x.[Gt(RN)(x) \rightarrow Lt(RN)(x)]$
$-[Gt(RN)(x_0) \rightarrow Lt(RN)(x_0)]$
$+Gt(RN)(x_0)$
$-Lt(RN)(x_0)$
$+RN(Gt(RN), x_0)$ ⊣(Gt.2)
$-RN(Lt(RN), x_0)$ ⊣(Lt.2)
$+[x_0 = 0 \vee \exists u.[Gt(RN)(u)) \wedge x_0 = S(u)]]$
$-[x_0 = 0 \vee \exists u.[Lt(RN)(u)) \wedge x_0 = S(u)]]$
$+\exists u.[Gt(RN)(u)) \wedge x_0 = S(u)]$
$-\exists u.[Lt(RN)(u)) \wedge x_0 = S(u)]$
$+[Gt(RN)(x_1)) \wedge x_0 = S(x_1)]$
$+x_0 = S(x_1)$
$+Gt(RN)(x_1)$
$-Lt(RN)(x_1)$
...
$+x_1 = S(x_2)$

$+Gt(RN)(x_2)$
$-Lt(RN)(x_2)$
$+x_0 = S(S(x_2))$
$\ldots$

The branch cannot be closed because the sequence $x_0, x_1, \ldots, x_k, \ldots$ of distinct variables of type 1 can be continued indefinitely with nodes

$$+x_k = S(x_{k+1}), \quad +Gt(RN)(x_{k+1}), \quad \text{and} \quad -Lt(RN)(x_{k+1})$$

added to the branch for any $k \geq 0$. Thus for any $n \geq 1$,

$$+x_0 = S^n(x_n)$$

can be added to the branch; that is, for any n there is always an x_n of which x_0 is the $n'th$ successor. But it must be stressed that each member of GnN is a term and therefore necessarily a *finite* string of symbols.

By the completeness theorem of ITT there is a model of ITT in which each member of GnN is not *well-founded* in the sense that each member of the sequence $S(0)$, $S(S(0))$, $\cdots$ is well-founded; that is each is the $n'th$ successor of 0 for some fixed n. For this reason $Lt(RN)$ is said to be *well-founded* and $Gt(RN)$ to be *non-well-founded.*

Models in which predicates similar to GnN are not empty form the basis for what is called *non-standard* arithmetic and the *non-standard* analysis of Abraham Robinson [122].

Later predicates $Gt(RG)$ are examined for recursion generators RG other than RN. In all these cases $Gt(RG)$ is non-well-founded in the same sense as $Gt(RN)$.

EXERCISES §4.4.

1. Provide derivations for (Hd.1), (Tl.1), (Hd.2), (Tl.2), (Hd.3), and (Tl.3) of §4.4.3.
2. Provide derivations for (N.1) and (N.2) of §4.4.4.
3. An extensional version of ordered pair can be defined as follows

$$\langle\,\rangle_e \stackrel{\text{df}}{=} \lambda u, v, w.[Ext(w) \wedge w(u, v)],$$

where Ext is an appropriately modified version of the predicate defined in §2.5. Derivable sequents corresponding to the sequents (OP.1) and (OP.2) are

OPE.1) $\vdash \forall x1, y1, x2, y2.[\langle x1, y1\rangle_e =_e \langle x2, y2\rangle_e \rightarrow x1 =_e x2 \wedge y1 =_e y2]$
OPE.2) $\vdash \forall x, y.\neg\langle x, y\rangle_e =_e Nil$

The converse of (OP.1) follows immediately from properties of $=$. The converse of (OPE.1) is

OPE.3) $\vdash \forall x1, y1, x2, y2.[x1 =_e x2 \wedge y1 =_e y2 \rightarrow \langle x1, y1\rangle_e =_e \langle x2, y2\rangle_e]$.

Provide derivations for (OPE.1), (OPE.2), and (OPE.3).

4. Let $\leq$ be admitted as a constant of ITT of type $[1, 1]$. Let $R\leq$ be the recursion generator without parameters defined as follows:

$$R\leq \stackrel{\mathrm{df}}{=} \lambda Z, x, y.[\,[x = 0 \wedge N(y)] \vee \exists u, v.[Z(u, v) \wedge x = S(u) \wedge y = S(v)]\,].$$

Let $\leq$ and $<$ be defined by

$$\leq \stackrel{\mathrm{df}}{=} Lt(R\leq) \text{ and } < \stackrel{\mathrm{df}}{=} \lambda x, y. \leq (S(x), y).$$

Provide derivations for the following sequents:

O.1) $\vdash \forall y.[N(y) \leftrightarrow 0 \leq y]$
O.2) $\vdash \forall x, y.[x \leq y \leftrightarrow S(x) \leq S(y)]$
O.3) $\vdash \forall x, y.[x \leq y \rightarrow N(x) \wedge N(y)]$
O.4) $\vdash \forall x, y.[x \leq y \rightarrow x \leq S(y)]$
O.5) $\vdash [\forall x, y{:}N].[x \leq y \vee y < x]$
O.6) $\exists x.x < 0 \vdash$

Use LtInd with $\leq$ whenever possible.

5. It is sometimes useful to use what Kleene calls in [88] *course-of-values* induction in place of NInd of §4.4.4. In ITT it takes the following form:

$$\text{(CVInd.)} \quad \frac{-\forall x.[N(x) \rightarrow Q(x)]}{-Q(0) \quad - \forall x.[\forall u.[u \leq x \rightarrow Q(u)] \rightarrow Q(S(x))]}$$

Provide a derivation for CVInd.

6. The predicate *Least* is defined as

$$Least \stackrel{\mathrm{df}}{=} \lambda X, y.[N(y) \wedge X(y) \wedge \forall x.[x < y \rightarrow \neg X(x)]].$$

A y for which $Least(X, y)$ holds is necessarily unique; that is the following sequent is derivable:

Least.1) $\forall X.[\forall y, y'{:}N].[Least(X, y) \wedge Least(X, y') \rightarrow y = y'].$

A sufficient condition for $Least(X, y)$ to hold for some y is that N and X have a member in common; that is, the following sequent is derivable:

Least.2) $\forall X.[[\exists x{:}N].X(x) \rightarrow \exists y.Least(X, y)].$

Provide derivations for these two sequents. For the sequent (Least.2) export $[\exists x{:}N]$ to become an initial $[\forall x{:}N]$ and use course-of-values induction.

7. Let up and down *chains* be defined:

$$UCh \stackrel{\mathrm{df}}{=} (\lambda X.[\forall x{:}N].\forall y.[X(x, y) \rightarrow X(S(x), y)])$$
$$DCh \stackrel{\mathrm{df}}{=} (\lambda X.[\forall x{:}N].\forall y.[X(S(x), y) \rightarrow X(x, y)]).$$

These definitions are related in the following way:

UDCh) $\vdash \forall X.[UCh(X) \leftrightarrow DCh(\lambda x, y.\neg X(x, y))]$.

Chains get their names because of the following two sequents:

UCh) $UCh(X) \vdash [\forall x1, x2{:}N].[x1 \leq x2 \rightarrow \forall y.[X(x1, y) \rightarrow X(x2, y)]$.
DCh) $DCh(X) \vdash [\forall x1, x2{:}N].[x1 \leq x2 \rightarrow \forall y.[X(x2, y) \rightarrow X(x1, y)]$.

The membership of $\lambda X(S^n(0), y)$ *increases* with increasing n for an up chain and *decreases* for a down chain. Provide derivations for these three sequents.

8. In the introduction to this section it was noted that although all of Peano's axioms are derived for 0, S, and N, the sequence 0, $S(0)$, $S(S(0))$, $\cdots$ of terms is not the sequence of natural numbers 0, 1, 2, ... , according to the Frege-Russell definition of these terms in [126], but is rather a sequence of terms from which counter sets in the sense of [120] can be defined. Nevertheless any such sequence is all that is needed for counting. For example, the natural number 3, according to the Frege-Russell definition, is the predicate of predicates with extension with exactly three members. A predicate with extension with exactly three members is one with extension for which there exists a one-to-one correspondence with the extension of the predicate

$$3Cnt \overset{\text{df}}{=} \lambda u.[u = 0 \vee u = S(0) \vee u = S(S(0))].$$

Thus the *predicate* 3 is defined as

$$3 \overset{\text{df}}{=} \lambda X.\exists Z.11Cor(X, 3Cnt, Z),$$

where the predicate 11Cor, one-to-one correspondence, is defined

$$11Cor \overset{\text{df}}{=} \lambda X, Y, Z.[\forall x, y.[Z(x, y) \rightarrow X(x) \wedge Y(y)] \wedge [\forall x{:}X].[\exists! y{:}Y].Z(x, y)] \wedge [\forall y{:}Y].[\exists! x{:}X].Z(x, y)].$$

Provide definitions for the natural number zero and for the successor predicate for the natural numbers. Define a recursion generator for the predicate *NNum* of natural numbers and derive all of Peano's axioms.

9. The addition predicate $\oplus$ for members of N is defined as $\oplus \overset{\text{df}}{=} Lt(R\oplus)$, where

$$R\oplus \overset{\text{df}}{=} \lambda Z, x, y, z.[[x = 0 \wedge N(y) \wedge z = y] \vee \exists u, v.[x = S(u) \wedge z = S(v) \wedge Z(u, y, v)]].$$

Prove that $\oplus$ is functional by first providing a derivation for

$\oplus$.1) $\vdash [\forall x, y{:}N].[\exists z{:}N]. + (x, y, z)$, and then for

$\oplus$.2) $\vdash [\forall x, y{:}N].[\exists! z{:}N]. + (x, y, z)$.

4.5. Continuous Recursion Generators

A recursion generator RG without parameters is said to be $\exists$-*continuous*, respectively $\forall$-*continuous* if the sequent

$$\vdash Con\exists(RG), \text{ respectively } \vdash Con\forall(RG)$$

is derivable where

$Con\exists \stackrel{\text{df}}{=} (\lambda wg.[\forall X{:}UCh].$
$\quad \forall y.[wg((\lambda v.[\exists x{:}N].X(x,v)), y) \to [\exists x{:}N].wg((\lambda v.X(x,v)), y)])$

$Con\forall \stackrel{\text{df}}{=} (\lambda wg.[\forall X{:}DCh].$
$\quad \forall y.[[\forall x{:}N].wg((\lambda v.X(x,v)), y) \to wg((\lambda v.[\forall x{:}N].X(x,v)), y)]).$

and UCh and DCh are defined in (7) of the exercises of §4.4. In these definitions it is assumed that $wg{:}[[\tau],\tau]$ and that $X{:}[1,\tau]$; they can be generalized in an obvious way for $wg{:}[[\overline{\tau}],\overline{\tau}]$, where $\overline{\tau}$ is not empty.

The relationship between $Con\exists$ and $Con\forall$ is expressed in the following derivable sequent

ECCA) $\quad \forall wg.[Mon(wg) \to [Con\exists(wg) \to Con\forall(CC(wg))]],$

where

$$CC \stackrel{\text{df}}{=} \lambda wg, U, v.\neg wg((\lambda v.\neg U(v)), v)).$$

A derivation of the sequent is left as an exercise.

The following sequents are also derivable:

RG$\exists$) $\quad \vdash [\forall wg{:}Mon].\forall X, y.[[\exists x{:}N].wg((\lambda v.X(x,v)), y) \to$
$\qquad\qquad wg((\lambda v.[\exists x{:}N].X(x,v)), y)]$

RG$\forall$) $\quad \vdash [\forall wg{:}Mon].\forall X, y.[wg((\lambda v.[\forall x{:}N].X(x,v)), y) \to$
$\qquad\qquad [\forall x{:}N].wg((\lambda v.X(x,v)), y)].$

Note their relationship to the definition of $Con\exists$ and $Con\forall$ above.

Derivation of (RG$\exists$)

$+Mon(wg)$
$-\forall X, y.[[\exists x{:}N].wg((\lambda v.X(x,v)), y) \to wg((\lambda v.[\exists x{:}N].X(x,v)), y)]$
$-[[\exists x{:}N].wg((\lambda v.X(x,v)), y) \to wg((\lambda v.[\exists x{:}N].X(x,v)), y)]$
$+[\exists x{:}N].wg((\lambda v.X(x,v)), y)$
$-wg((\lambda v.[\exists x{:}N].X(x,v)), y)$
$+N(x)$
$+wg((\lambda v.X(x,v)), y)$
$+\forall y.[\lambda v.X(x,v))(y) \to (\lambda v.[\exists x{:}N].X(x,v))(y)]$ $\qquad \dashv$derivable
$+wg((\lambda v.[\exists x{:}N].X(x,v)), y)$ $\qquad \dashv Mon(wg)$

EXERCISES §4.5.

1. Provide a derivation for (ECCA).
2. Provide a derivation for ($RG\forall$).

4.6. Positive Occurrences of a Variable

Because of the extensive use that will be made of the properties of $Lt(RG)$ and $Gt(RG)$ for monotonic and ∃- or ∀-continuous recursion generators RG, decidable sufficient conditions for these properties are defined here. The conditions are that the occurrence of Z in the formula $RG(Z, \overline{x})$ be *positive*, respectively *e-positive*; necessarily $Z{:}[\overline{\tau}]$ and $\overline{x}{:}\overline{\tau}$ have no free occurrences in RG by the definition of recursion generator given in §4.1. This result is established in §4.6.3 after appropriate definitions are provided. It generalizes theorem 3.2 of [111]. Note that the properties are established for recursion generators *without parameters* - recursion generators with parameters must be supplied with appropriate parameters that are constant terms before they can be tested.

4.6.1. Subformula path.

DEFINITION 56. *Subformula Path.*
A subformula path is a sequence $sg_1F_1, \ldots, sg_nF_n$, $n \geq 1$, *where each* sg_i *is* $+$ *or* $-$, *and each* F_i *is a formula satisfying the following conditions*:

1. *If* sg_iF_i *is* $\pm[F \downarrow G]$, *then* $sg_{i+1}F_{i+1}$ *is either one of* $-F$ *or* $-G$, *or, respectively, either one of* $+F$ *or* $+G$.
2. *If* sg_iF_i *is* $\pm\exists P$, *then* $sg_{i+1}F_{i+1}$ *is respectively* $\pm P(x)$, *where* x *is a variable without a free occurrence in any* F_j, $j \leq i$, *and is said to be respectively a* $\pm\exists$ *variable.*
3. *If* sg_iF_i *is* $\pm(\lambda u.P)(t, \overline{s})$, *then* $sg_{i+1}F_{i+1}$ *is respectively* $\pm([t/u]P)(\overline{s})$.

A *maximal* subformula path for a signed formula $\pm H$ is one in which the first signed formula sg_1F_1 is $\pm H$ and the last sg_nF_n is a signed elementary formula; that is sg_nF_n is $\pm CV(\overline{s})$ for some constant or variable CV and terms $\overline{s}$. The occurrence of $CV(\overline{s})$ in the last signed formula sg_nF_n of a maximal subformula path is said to be a *positive* instance of $CV(\overline{s})$ in $\pm H$ if sg_n is $+$, and said to be an *existentially-positive*, or just e-positive instance if in addition no free variable of $\overline{s}$ is a $-\exists$ variable.

The usual connectives ¬, ∧, ∨, →, and ↔ have been defined in terms of ↓, and the universal quantifier ∀ in terms of ∃ and ¬. The definition of subformula path can be extended in the obvious way to accommodate these connectives and quantifier.

4.6.2. Positive and e-positive occurrences of variables. The constant or variable CV is the *head* of an elementary formula $CV(\overline{s})$. A variable V that is the head of an elementary formula $CV(\overline{s})$ is said to be a *strict* head if V has no free occurrence in any of the arguments $\overline{s}$. Thus, for example, Z is the strict head of the elementary formula $Z(x, y)$.

Let Z be a variable with a free occurrence in H. It is said to be a *positive*, respectively *e-positive*, variable of a signed formula sgH if each elementary subformula of sgH in which Z has a free occurrence is a positive, respectively an e-positive subformula with Z as strict head.

Let Z be a positive variable of sgH. The *depth* of Z in sgH is defined to be the maximum of the integers n for which $sg_n F_n$ is the last signed formula of a maximal subformula path of sgH for which Z has a free occurrence in F_n.

EXAMPLES.

1. The occurrence of the variable wg in the signed formula $+Lt(wg)(x)$ or $+Gt(wg)(x)$ is positive but not e-positive. Here is a confirming subformula path for $+Lt(wg)(x)$:

   $+\forall Z.[\forall x.[wg(Z,x) \rightarrow Z(x)] \rightarrow Z(u)]$
   $-\forall x.[wg(Z,x) \rightarrow Z(x)]$
   $-[wg(Z,x) \rightarrow Z(x)]$
   $+wg(Z,x)$

   Each of the signed quantifiers $+\forall Z$ and $-\forall x$ encountered on the path has an eigenvariable with a free occurrence in $+wg(Z,x)$. The first of these means that the occurrence of wg in $+Lt(wg)(x)$ cannot be e-positive.

2. Define the recursion generator RIt with the parameters wg:$[[\tau],\tau]$ and ws:$[\tau]$:

   $$RIt \stackrel{\text{df}}{=} \lambda wg, ws, Z, x, y.[[x = 0 \wedge ws(y)] \\ \vee [\exists u{:}N].[x = S(u) \wedge wg(Z(u), y)]].$$

   Here $Z{:}[1,[\tau]]$, $x{:}1$, and $y{:}\tau$ so that $RIt{:}[[[\tau],\tau],[\tau],[1,\tau],1,\tau]$, and $RIt(wg, ws){:}[[1,\tau],1,\tau]$, the type of a recursion generator. Let Z have a positive, respectively e-positive, occurrence in $RG(Z(u), y)$, where RG is a recursion generator and let $D{:}[\tau]$ be without free variables. That then Z has a positive, respectively $\exists$-positive, occurrence in the signed formula $+RIt(RG, D, Z, x, y)$ is confirmed by the following subformula path:

   $+[[x = 0 \wedge D(y)] \vee [\exists u{:}N].[x = S(u) \wedge RG(Z(u), y)]]$
   $+[\exists u{:}N].[x = S(u) \wedge RG(Z(u), y)]$
   $+[x = S(u) \wedge RG(Z(u), y)]$
   $+RG(Z(u), y)$

3. Define the recursion generator $RF\forall$ with parameters $w1, w2{:}[\tau]$:

   $$RF\forall \stackrel{\text{df}}{=} \lambda w1, w2, Z, Y.[\neg\exists u.Y(u) \vee \exists U.[Z(w1, w2, U) \wedge \\ [\exists u{:}w1].[w2(u) \wedge \neg U(u) \wedge Y{=_e}U \cup \{u\}]]].$$

   Here $w1, w2, Y{:}[\tau]$, $Z{:}[[\tau],[\tau],[\tau]]$ and $RF\forall{:}[[\tau],[\tau],[[\tau],[\tau],[\tau]],[\tau]]$, the type of a recursion generator with two parameters of type $[\tau]$. The occurrence of the variable Z in $RF\forall(w1, w2, Z, Y)$ is e-positive as the following subformula path confirms:

   $+RF\forall(w1, w2, Z, Y)$
   $+[\neg\exists u.Y(u) \vee \exists U.[Z(w1, w2, Y) \wedge [\exists u{:}w1].[w2(u) \wedge \neg U(u) \wedge Y{=_e}U \cup \{u\}]]]$

$+\exists U.[Z(w1, w2, Y) \wedge [\exists u{:}w1].[w2(u) \wedge \neg U(u) \wedge Y{=_e}U \cup \{u\}]]]$
$+Z(w1, w2, Y)$

4.6.3. Monotonicity and continuity. Monotone recursion generators are defined in §4.3 and $\exists$- and $\forall$-continuous in §4.5. Here related properties of formulas H for which Z is a positive variable of $+H$ will be proved. The results are proved for variables Z of type $[\tau]$ but can be generalizable for variables of type $[\overline{\tau}]$.

The properties are:

M $\quad \vdash \forall X, Y.[\forall x.[X(x) \rightarrow Y(x)] \rightarrow [[X/Z]H \rightarrow [Y/Z]H]].$
C$\exists$ $\quad \vdash [\forall X{:}UCh].[[\lambda v.[\exists x{:}N].X(x, v)/Z]H \rightarrow [\exists x{:}N].[\lambda v.X(x, v)/Z]H].$
C$\forall$ $\quad \vdash [\forall X{:}DCh].[[\forall x{:}N].[\lambda v.X(x, v)/Z]H \rightarrow [\lambda v.[\forall x{:}N].X(x, v)/Z]H].$

where none of X, Y or x has a free occurrence in H, and UCh and DCh are defined in exercise 7 of §4.4.

The properties of formulas H for which Z is a positive variable of $+H$ can be established more easily at the same time as related properties of formulas H for which Z is a positive variable of $-H$. These latter properties are:

M$-$ $\quad \vdash \forall X, Y.[\forall x.[X(x) \rightarrow Y(x)] \rightarrow [[Y/Z]H \rightarrow [X/Z]H]].$
C$\exists-$ $\quad \vdash [\forall X{:}UCh].[[\forall x{:}N].[\lambda v.X(x, v)/Z]H \rightarrow [\lambda v.[\exists x{:}N].X(x, v)/Z]H].$
C$\forall-$ $\quad \vdash [\forall X{:}DCh].[[\lambda v.[\forall x{:}N].X(x, v)/Z]H \rightarrow [\exists x{:}N].[\lambda v.X(x, v)/Z]H].$

where in each case X, Y, and x have no free occurrences in H. These properties can be obtained from M, C$\exists$, and C$\forall$ respectively by replacing H by $\neg H$, for Z has a positive occurrence in $-H$ if and only if it has a positive occurrence in $\neg H$.

The following theorem establishes the desired properties:

THEOREM 41. *Let Z be a positive variable of H, respectively of $\neg H$. Then the sequent* (M), *respectively* (M$-$) *is derivable. Should Z be an e-positive variable of H, respectively of $\neg H$, then the sequents* (C$\exists$) *and* (C$\forall$), *respectively* (C$\exists-$) *and* (C$\forall-$) *are also derivable.*

PROOF. For this proof it is easiest to assume that $\neg$, $\wedge$, and $\exists$ are the primitives for the logic. The proof of the theorem is by induction on the depth of Z in H measured using these primitives. Let the depth be 0. Then H is $Z(\overline{s})$ and necessarily Z is a positive and e-positive variable of $+H$. The sequents (M), (C$\exists$) and (C$\forall$) are immediately derivable. Should Z be a positive and e-positive variable of $\neg H$, then M$-$, C$\exists-$, and C$\forall-$ are easily derived.

Assume the conclusion of the theorem whenever the depth of Z in H does not exceed d, $d \geq 0$. Consider the possibilities for H when the depth is $d + 1$.

1. It is not difficult to establish that (M$-$), (C$\exists-$) and (C$\forall-$) are each equivalent to (M), (C$\exists$) and (C$\forall$) respectively when H in one is replaced by $\neg H$ in the other. Thus the case that H is $\neg F$ follows immediately by the induction assumption.

2. Let H be $[F \wedge G]$ and let Z be a positive variable of $[F \wedge G]$, respectively of $\neg[F \wedge G]$. Let it have a free occurrence in each of F and G. Derivations for the cases when Z does not have a free occurrence in both F and G can be obtained by adapting those provided below.
   (a) Consider first the case that Z is a positive variable of $[F \wedge G]$. Then it is a positive variable of both F and G. By the induction assumption (M) is derivable when H is either F or G. A derivation of (M) for $[F \wedge G]$ in this case follows:

$-\forall X, Y.[\forall x.[X(x) \to Y(x)] \to [[X/Z][F \wedge G] \to [Y/Z][F \wedge G]]]$
$+\forall x.[X(x) \to Y(x)]$
$-[[X/Z][F \wedge G] \to [Y/Z][F \wedge G]]$
$+[X/Z][F \wedge G]$
$+[X/Z]F$
$+[X/Z]G$
$-[Y/Z][F \wedge G]$

$-[Y/Z]F$	$-[Y/Z]G$
$+[Y/Z]F$	$+[Y/Z]G$

Let now Z be an e-positive variable of $[F \wedge G]$ so that it is an e-positive variable of both F and G. By the induction assumption each of (C∃) and (C∀) is derivable when H is either F or G. A derivation of (C∃) for $[F \wedge G]$ in this case follows. A derivation of (C∀) is left as an exercise.

$-[\forall X{:}UCh].[[\lambda v.[\exists x{:}N].X(x,v)/Z][F \wedge G] \to$
$\qquad [\exists x{:}N].[\lambda v.X(x,v)/Z][F \wedge G]]$
$+UCh(X)$
$+[\lambda v.[\exists x{:}N].X(x,v)/Z][F \wedge G]$
$-[\exists x{:}N].[\lambda v.X(x,v)/Z][F \wedge G]$ (a)
$+[\lambda v.[\exists x{:}N].X(x,v)/Z]F$
$+[\lambda v.[\exists x{:}N].X(x,v)/Z]G$
$+[\forall X{:}UCh].[[\lambda v.[\exists x{:}N].X(x,v)/Z]F \to$
$\qquad [\exists x{:}N].[\lambda v.X(x,v)/Z]F]$ ⊣(C∃)
$+[\forall X{:}UCh].[[\lambda v.[\exists x{:}N].X(x,v)/Z]G \to$
$\qquad [\exists x{:}N].[\lambda v.X(x,v)/Z]G]$ ⊣(C∃)
$+[[\lambda v.[\exists x{:}N].X(x,v)/Z]F \to [\exists x{:}N].[\lambda v.X(x,v)/Z]F]$
$+[[\lambda v.[\exists x{:}N].X(x,v)/Z]G \to [\exists x{:}N].[\lambda v.X(x,v)/Z]G]$
$+[\exists x{:}N].[\lambda v.X(x,v)/Z]F]$
$+[\exists x{:}N].[\lambda v.X(x,v)/Z]G]$
$+N(x1) \quad +N(x2)$
$+[\lambda v.X(x1,v)/Z]F$
$+[\lambda v.X(x2,v)/Z]G$

$+[\forall x1, x2{:}N].[x1 \leq x2 \rightarrow \forall y.[X(x1, y) \rightarrow X(x2, y)]$ $\dashv$(UCh)
$+[x1 \leq x2 \vee x2 < x1]$ $\dashv$(O.5)

L R
$+x1 \leq x2$
$+\forall y.[X(x1, y) \rightarrow X(x2, y)]$
$+[\lambda v.X(x1, v)/Z]G$
$-[\lambda v.X(x1, v)/Z][F \wedge G]$ $\dashv$(a)

R
$+x2 < x1$

---

similar to L.

(b) Consider now the case that Z is a positive variable of $\neg[F \wedge G]$ so that it is a positive variable of both $\neg F$ and $\neg G$. By the induction assumption (M−) is derivable when H is either F or G.
A derivation of (M−) for $\neg[F \wedge G]$ can be obtained in this case by appropriate changes to the derivation of (M) for $[F \wedge G]$.
Let Z be an e-positive variable of $\neg[F \wedge G]$, and therefore of both $\neg F$ and $\neg G$. A derivation of (C∃−) for $[F \wedge G]$ in this case follows. A derivation of (C∀−) for $[F \wedge G]$ is left as an exercise.

$-[\forall X{:}UCh].[\,[[\forall x{:}N].[\lambda v.X(x, v)/Z][F \wedge G] \rightarrow$
$[\lambda v.[\exists x{:}N].X(x, v)/Z][F \wedge G]\,]$
$+UCh(X)$
$+[[\forall x{:}N].[\lambda v.X(x, v)/Z][F \wedge G]$
$-[\lambda v.[\exists x{:}N].X(x, v)/Z][F \wedge G]$

L R
$-[\lambda v.[\exists x{:}N].X(x, v)/Z]F$
$+[\forall X{:}UCh].[[\forall x{:}N].\lambda v.X(x, v)/Z]F \rightarrow$
$[\lambda v.[\exists x{:}N].X(x, v)/Z]F$ $\dashv$(C∃−)
$-[[\forall x{:}N].\lambda v.X(x, v)/Z]F$
$+N(x)$
$-[\lambda v.X(x, v)/Z]F$
$+[\lambda v.X(x, v)/Z][F \wedge G]$
$+[\lambda v.X(x, v)/Z]F$

R
$-[\lambda v.\exists x.X(x, v)/Z]G$

---

similar to F

3. Let H be $\exists z.F$ and let Z be a positive variable of H, respectively of $\neg H$. It may be assumed that z has a free occurrence in F, since otherwise the desired conclusion is immediate.

(a) Consider first the case that Z is a positive variable of $\exists z.F$. Then it is a positive and e-positive variable of F. By the induction assumption each of (M), (C$\exists$), and (C$\forall$) is derivable when H is F. Derivations of (M) and (C$\exists$) for $\exists z.F$ are left as exercises. A derivation of (C$\forall$) follows.

$-[\forall X{:}DCh].[[\forall x{:}N].[\lambda v.X(x,v)/Z]\exists z.F \rightarrow$
$\quad [\lambda v.[\forall x{:}N].X(x,v)/Z]\exists z.F]$
$+DCh(X)$ (a)
$+[\forall x{:}N].[\lambda v.X(x,v)/Z]\exists z.F$ (b)
$-[\lambda v.[\forall x{:}N].X(x,v)/Z]\exists z.F$
$+[\forall x{:}N].\exists z.[\lambda v.X(x,v)/Z]F$
$-\exists z.[\lambda v.[\forall x{:}N].X(x,v)/Z]F$ (c)
$+\forall z.[\forall X{:}DCh].[[\forall x{:}N].[\lambda v.X(x,v)/Z]F \rightarrow$
$\quad [\lambda v.[\forall x{:}N].X(x,v)/Z]F]]$ (d)$\dashv$(C$\forall$)
$+[\forall X{:}DCh].[[\forall x{:}N].[\lambda v.X(x,v)/Z]F \rightarrow$
$\quad [\lambda v.[\forall x{:}N].X(x,v)/Z]F]$
$+[[\forall x{:}N].[\lambda v.X(x,v)/Z]F \rightarrow$
$\quad [\lambda v.[\forall x{:}N].X(x,v)/Z]F]$ (e)$\dashv$(a)
$-[\lambda v.[\forall x{:}N].X(x,v)/Z]F$ $\dashv$(c)
$-[\forall x{:}N].[\lambda v.X(x,v)/Z]F$ $\dashv$(e)
$+N(x)$ (f)
$+[\lambda v.X(x,v)/Z]\exists z.F$ $\dashv$(b), (f)
$+\exists z.[\lambda v.X(x,v)/Z]F$
$+[\lambda v.X(x,v)/Z]F'$ where F' is $[z'/z]F$, z' new (g)
$+[\forall X{:}DCh].[[\forall x{:}N].[\lambda v.X(x,v)/Z]F' \rightarrow$
$\quad [\lambda v.[\forall x{:}N].X(x,v)/Z]F']]$ $\dashv$(d)
$+[[\forall x{:}N].[\lambda v.X(x,v)/Z]F' \rightarrow$
$\quad [\lambda v.[\forall x{:}N].X(x,v)/Z]F']]$ $\dashv$(a)
$-[\lambda v.[\forall x{:}N].X(x,v)/Z]F']$ $\dashv$(c)
$-[\lambda v.X(x,v)/Z]F'$ $\dashv$(f), Z +ve in F

$\dashv$(g)

(b) Consider the case that Z is a positive variable of $\neg\exists z.F$. Then it is a positive variable of $\neg F$. By the induction assumption (M$-$) is derivable when H is F. A derivation of (M$-$) for $\exists z.F$ in this case is left as an exercise.

Let Z be an e-positive variable of $\neg\exists z.F$, and therefore of $\neg F$. Derivations of (C$\exists-$) and (C$\forall-$) are left as exercises.

4. Let H be $(\lambda u.P)(t,\overline{s})$ and let Z be a positive variable of $+H$, respectively $\neg H$. It follows that Z is a positive variable of $([t/u]P)(\overline{s})$, respectively of $\neg([t/u]P)(\overline{s})$, and (M), respectively (M$-$), are derivable when H is $([t/u]P)(\overline{s})$. Derivations of (M), respectively (M$-$), when H is $(\lambda u.P)(t,\overline{s})$ are left as exercises.

Let now Z be an e-positive variable of $+H$, respectively $\neg H$. Z is an e-positive variable of $+([t/u]P)(\overline{s})$, respectively $\neg([t/u]P)(\overline{s})$, so that by the induction assumption (C∃) and (C∀), respectively (C∃−) and (C∀−) are derivable. The required derivations when H is $(\lambda u.P)(t, \overline{s})$ are left as exercises. ⊣

DEFINITION 57. *Positive and e-Positive Recursion Generators.* *Let RG be a recursion generator of type $[[\overline{\tau}], \overline{\tau}]$ and Z and $\overline{x}$ variables without free occurrences in RG of types $[\overline{\tau}]$ and $\overline{\tau}$ respectively. Then RG is a* positive, *respectively* e-positive, *recursion generator if the free occurrence of Z in $RG(Z, \overline{x})$ is positive, respectively e-positive.*

COROLLARY 42. *A positive recursion generator is monotonic, and an e-positive one is ∃- and ∀-continuous.*

EXERCISES §4.6.

1. Provide derivations for the omitted cases in the proof of the Theorem 41.
2. Given H without free variables, let $H^* \stackrel{\mathrm{df}}{=} (\lambda Z.\neg H)(\lambda u.\neg Z(u))$. Prove that H has the property (M) if and only if H^* does.
3. The properties $C\exists$ and $C\forall$ are independent properties of formulas H. To prove this it is sufficient to find counter-examples for each of the sequents

$$C\exists \vdash C\forall \text{ and } C\forall \vdash C\exists.$$

That is, it is sufficient to find in each case a formula H for which the antecedent is derivable but the succedent is not. For the first of the sequents $\exists u.Z(u)$ is such a formula and for the second $\forall u.Z(u)$. To complete the proofs in each case it is sufficient to find in each case a formula P:[1, 1] for which the succedent is not derivable. Find such formulas.

4.7. Horn Sequents and Recursion Generators

In [108], higher order Horn clauses are defined for the simple theory of types of [28]. Here a higher order version of Horn clauses for ITT called *Horn sequents* is defined in §4.7.1 and generalized for simultaneous Horn sequents in §4.7.2.

4.7.1. Horn sequents. Let RG:$[\overline{\rho}, [\overline{\sigma}], \overline{\sigma}]$ be a recursion generator with parameters of type $\overline{\rho}$; thus no variable has a free occurrence in RG; here $\overline{\rho}$ may be empty but $\overline{\sigma}$ may not. A *Horn sequent* defined by RG is any sequent of the form

$$HS)\qquad RG(\overline{w})(C(\overline{w}), \overline{x}) \vdash C(\overline{w})(\overline{x}),$$

where C:$[\overline{p}, \overline{\sigma}]$ is a constant with no occurrence in RG, and $\overline{w}$:$\overline{p}$ is a sequence of distinct variables distinct from each of the distinct variables of $\overline{x}$:$\overline{\sigma}$. C is called the *defined constant* of the Horn sequent. The *converse* of the Horn sequent (HS) is the sequent

$$C(\overline{w})(\overline{x}) \vdash RG(\overline{w})(C(\overline{w}), \overline{x}).$$

If it has not been previously defined, the constant C can be defined by

$$C \stackrel{\text{df}}{=} \lambda\overline{w}.Lt(RG(\overline{w})) \quad \text{or} \quad C \stackrel{\text{df}}{=} \lambda\overline{w}.Gt(RG(\overline{w})).$$

The following theorem is an immediate consequence of (Lt.2), (Lt.3), (Gt.2, and (Gt.3) in §4.3.1.

THEOREM 43. *Let $\overline{r}$ be a sequence of terms $\overline{r}$:$\overline{p}$ for which $RG(\overline{r})$ is a monotonic recursion generator; that is, $RG(\overline{r})$ is without free variables and the sequent $\vdash Mon(RG(\overline{r}))$ is derivable. Let C be defined to be either $\lambda\overline{w}.Lt(RG(\overline{w}))$ or $\lambda\overline{w}.Gt(RG(\overline{w}))$. Then each of the following sequents, obtained from (HS) and its converse, is derivable*:

$$RG(\overline{r})(C(\overline{r}), \overline{x}) \vdash C(\overline{r})(\overline{x}),$$
$$C(\overline{r})(\overline{x}) \vdash RG(\overline{r})(C(\overline{r}), \overline{x}).$$

Note that a recursion generator can be defined from a sequence of recursion generators $RG_1, \ldots, RG_k$ of the same type by disjunction or conjunction:

$$\lambda\overline{w}, Z, \overline{x}.[RG_1(\overline{w})(Z, \overline{x}) \vee \cdots \vee RG_k(\overline{w})(Z, \overline{x})]$$
$$\lambda\overline{w}, Z, \overline{x}.[RG_1(\overline{w})(Z, \overline{x}) \wedge \cdots \wedge RG_k(\overline{w})(Z, \overline{x})].$$

A Horn sequent defined by the disjunction of recursion generators can be equivalently expressed as a sequence of Horn sequents with the same defined constant:

$$RG_1(\overline{w})(C(\overline{w}), \overline{x}) \vdash C(\overline{w})(\overline{x})$$
$$\ldots$$
$$RG_k(\overline{w})(C(\overline{w}), \overline{x}) \vdash C(\overline{w})(\overline{x}).$$

Sometimes a single Horn sequent can be better understood when it is presented as a sequence of Horn sequents in this manner.

4.7.2. Simultaneous Horn sequents. The presentation will be for two simultaneous Horn sequents, but can be generalized to any number.

Two sequents

HS1) $\quad RG1(\overline{w})(C1(\overline{w}), \overline{x_1}, C2(\overline{w})) \vdash C1(\overline{w})(\overline{x_1})$
HS2) $\quad RG2(\overline{w})(C1(\overline{w}), C2(\overline{w}), \overline{x_2}) \vdash C2(\overline{w})(\overline{x_2})$

are said to be *simultaneous* Horn sequents with parameters $\overline{w}$ and defined constants $C1$ and $C2$ provided the terms $RG1$ and $RG2$ are *simultaneous* recursion generators; that is they satisfy the following conditions:

1. No variable has a free occurrence in $RG1$ or $RG2$.
2. $\overline{x_1}$ and $\overline{x_2}$ are sequences of distinct variables distinct from the parameters $\overline{w}$.
3. $RG1{:}[\overline{p}, [\overline{\sigma_1}], \overline{\sigma_1}, [\overline{\sigma_2}]]$ and $RG2{:}[\overline{p}, [\overline{\sigma_1}], [\overline{\sigma_2}], \overline{\sigma_2}]$, where $\overline{p}$ may be empty but not $\overline{\sigma_1}$ or $\overline{\sigma_2}$.
4. The occurrence of $Z2$ in $RG1(\overline{w})(Z1, \overline{x_1}, Z2)$ and the occurrence of $Z1$ in $RG2(\overline{w})(Z1, Z2, \overline{x_2})$ may not be vacuous.

The condition (4) ensures that the recursion is *simultaneous* rather than *serial*. Should the occurrence of $Z2$ in $RG1(\overline{w})(Z1, \overline{x_1}, Z2)$ be vacuous then there is a recursion generator $RG1'$ in which no variable has a free occurrence for which

$$\vdash RG1(\overline{w})(Z1, \overline{x_1}, Z2) \leftrightarrow RG1'(\overline{w})(Z1, \overline{x_1})$$

is derivable. Thus $C1$ can be either $\lambda\overline{w}.Lt(RG1'(\overline{w}))$ or $\lambda\overline{w}.Gt(RG1'(\overline{w}))$ and then $C2$ can be defined using the recursion generator

$$\lambda\overline{w})(Z2, \overline{x_2}.RG2(\overline{w})(C1, Z2, \overline{x_2}).$$

From the simultaneous recursion generators $RG1$ and $RG2$ a single recursion generator $RGS{:}[\overline{p}, [\overline{\sigma_1}, \overline{\sigma_2}], \overline{\sigma_1}, \overline{\sigma_2}]$ is defined as follows:

$$\begin{aligned} RGS \stackrel{\text{df}}{=} \lambda\overline{w})(Z, \overline{x_1}, \overline{x_2}.[& \\ & RG1(\overline{w})((\lambda\overline{x_1}.\exists\overline{x_2}.Z(\overline{x_1}, \overline{x_2})), \overline{x_1}, (\lambda\overline{x_2}.\exists\overline{x_1}.Z(\overline{x_1}, \overline{x_2}))) \wedge \\ & RG2(\overline{w})((\lambda\overline{x_1}.\exists\overline{x_2}.Z(\overline{x_1}, \overline{x_2})), (\lambda\overline{x_2}.\exists\overline{x_1}.Z(\overline{x_1}, \overline{x_2})), \overline{x_2}) \quad]. \end{aligned}$$

Here it is assumed that the members of $\overline{x_1}$ and $\overline{x_2}$ are distinct; a change of variables in (HS1) and (HS2) can ensure this. RGS defines a Horn sequent

$$\text{HSS)} \qquad RGS(\overline{w})(C(\overline{w}), \overline{x_1}, \overline{x_2}) \vdash C(\overline{w})(\overline{x_1}, \overline{x_2}).$$

By Theorem 43 the sequents

$$RGS(\overline{r})(C(\overline{r}), \overline{x_1}, \overline{x_2}) \vdash C(\overline{r})(\overline{x_1}, \overline{x_2})$$
$$C(\overline{r})(\overline{x_1}, \overline{x_2}) \vdash RGS(\overline{r})(C(\overline{r}), \overline{x_1}, \overline{x_2})$$

are derivable when $\overline{r}$ is a sequence of terms $\overline{r}{:}\overline{p}$ for which $RGS(\overline{r})$ is a monotonic recursion generator and C is defined to be either $\lambda\overline{w}.Lt(RGS(\overline{w}))$ or $\lambda\overline{w}.Gt(RGS(\overline{w}))$. Let the constants $C1$ and $C2$ be defined:

$$C1 \stackrel{\text{df}}{=} \lambda\overline{w}, \overline{x_1}.\exists\overline{x_2}.C(\overline{w})(\overline{x_1}, \overline{x_2}) \text{ and } C2 \stackrel{\text{df}}{=} \lambda\overline{w}, \overline{x_2}.\exists\overline{x_1}.C(\overline{w})(\overline{x_1}, \overline{x_2}).$$

Then the following theorem can be proved:

THEOREM 44. *Let* $\vdash Mon(RGS(\overline{r}))$ *be derivable. Then the sequents* (HS1) *and* (HS2) *and their converses are derivable when* $\overline{r}$ *replaces* $\overline{w}$ *provided the sequent*

$$\text{a)} \qquad \vdash \exists\overline{x_1}, \overline{x_2}.C(\overline{w})(\overline{x_1}, \overline{x_2})$$

is also derivable.

PROOF. Since (HSS) and its converse is derivable when C is defined as suggested, the following sequent is also derivable because of the definitions of $C1$, $C2$, $RG1$, and $RG2$:

HSS.1) $\vdash \forall \overline{x_1}, \overline{x_2}.[\,[RG1(\overline{w}, C1(\overline{w}), \overline{x_1}, C2(\overline{w}))$
$\wedge RG2(\overline{w}, C1(\overline{w}), C2(\overline{w}), \overline{x_2})] \leftrightarrow C(\overline{w})(\overline{x_1}, \overline{x_2})]$.

Therefore if the sequent (a) is derivable, then so are the sequents

b) $\vdash \exists \overline{x_1}.RG1(\overline{w}, C1(\overline{w}), \overline{x_1}, C2(\overline{w}))$ and
$\vdash \exists \overline{x_2}.RG2(\overline{w}, C1(\overline{w}), C2(\overline{w}), \overline{x_2})$

The sequents (HS1) and (HS2) and their converses follow immediately. ⊣

EXAMPLE. In the introduction to [103], Milnor and Tufte write

> The purpose of this note is to present one instance among several we have encountered where the use of non-well-founded sets, maximum fixed points of monotone operators and a proof method, which we call *co-induction*, are essential tools in studying the semantics of programming languages.
>
> A set is *non-well-founded* if there is an infinite sequence $A_1, A_2, \ldots$ such that A_{n+1} is a member of A_n for all $n, n \geq 1$. Otherwise it is said to be *well-founded*. Although it is often assumed in set theory that all sets are well-founded, Aczel's anti-foundation axiom [5] leads to an alternative set theory which is very useful in computer science.

The paper defines a language Exp of expressions that includes recursive functions. A relational semantics for Exp is defined in the form of a set of inference rules involving three sets Val, $Clos$, and Env that are defined by the following set equations in terms of three given sets $Const$, Var, and Exp:

$v \in Val = Const \uplus Clos$ *Values*

$E \in Env = Var \xrightarrow{\text{fin}} Val$ *Environments*

cl or $\langle x, exp, E\rangle \in Clos = Var \times Exp \times Env$ *Closures.*

A set such as each of these given and defined sets is a *domain* when it is supplied with an identity on the set as will be done in §8.4; for example, intensional identity '=' serves as an identity for each of the given sets. These set equations are rewritten as a simultaneous recursion as follows:

a)
$$RG1(Clos)(x) \vdash Val(x)$$
$$RG2(Val)(y) \vdash Env(y)$$
$$RG3(Env)(z) \vdash Clos(z)$$

where x, y, z:1; $Val, Env, Clos$:[1]; and $RG1, RG2, RG3$:[[1], 1] are *simultaneous* recursion generators for which the free occurrences of $Z3$ in $RG1(Z3)(x)$, $Z1$ in $RG2(Z1)(y)$, and $Z2$ in $RG3(Z2)(z)$ are positive and not vacuous.

The predicates Val, Env, and $Clos$ are defined in terms of $RG1$, $RG2$, and $RG3$ by this simultaneous recursion.

The recursion generator RGS is defined from $RG1$, $RG2$, and $RG3$:

$$\begin{aligned} RGS \stackrel{\text{df}}{=} \lambda Z, x, y, z.[&RG1(\lambda z.\exists x, y.Z(x, y, z))(x) \wedge \\ &RG2(\lambda x.\exists y, z.Z(x, y, z))(y) \wedge \\ &RG3(\lambda y.\exists x, z.Z(x, y, z))(z)]. \end{aligned}$$

It defines a Horn sequent

HSS) $$RGS(C, x, y, z) \vdash C(x, y, z)$$

for which both it and its converse are derivable when

$$C \stackrel{\text{df}}{=} Lt(RGS) \text{ or } C \stackrel{\text{df}}{=} Gt(RGS).$$

Define the predicates

$$\begin{aligned} Val &\stackrel{\text{df}}{=} \lambda x.\exists y, z.C(x, y, z) \\ Env &\stackrel{\text{df}}{=} \lambda y.\exists x, z.C(x, y, z) \\ Clos &\stackrel{\text{df}}{=} \lambda z.\exists x, y.C(x, y, z). \end{aligned}$$

Then by Theorem 44 generalized to three from two, the sequents (a) and their converses are derivable provided the sequent

$$\vdash \exists x, y, z.C(x, y, z)$$

is derivable.

This example is returned to again in §8.4.1.

EXERCISES §4.7.

1. Find a Horn sequent defined by the recursion generator RN of §4.4.4 and one defined by $R\leq$ of exercise (4) of §4.4.
2. Complete derivations of the sequent (HS.1) and its converse from the sequent (HSS.1) of the proof of Theorem 44.
3. The recursion generator RIt with parameters wg:[[τ], τ] and ws:[τ] was defined in Example 2 of §4.6.2:

$$\begin{aligned} RIt \stackrel{\text{df}}{=} \lambda wg, ws, Z, x, y.[&[x = 0 \wedge ws(y)] \\ &\vee[\exists u{:}N].[x = S(u) \wedge wg(Z(u), y)]]. \end{aligned}$$

A Horn sequent with defined constant It defined from $RLst$ is after contraction of its antecedent

HIt) $$\begin{aligned} &[\,[x = 0 \wedge ws(y)]\vee \\ &[\exists u{:}N].[x = S(u) \wedge wg(It(wg, ws)(u), y)]] \vdash It(wg, ws)(x, y) \end{aligned}$$

'*It*' abbreviates '*Iterate* and is defined by

$$It \stackrel{\text{df}}{=} \lambda wg.ws.Lt(RIt(wg, ws)).$$

$It(wg, ws)(x)$ is intended to be the predicate that results from iterating a recursion generator wg x times beginning with a predicate ws. Thus

$$\begin{array}{l} It(wg, ws)(0) =_e ws \\ It(wg, ws)(S(0)) =_e wg(It(wg, ws)(0)) \\ It(wg, ws)(S(S(0))) =_e wg(It(wg, ws)(S(0))), \\ \quad\vdots \end{array}$$

A *cumulative* iteration predicate CIt is defined by

$$CIt \stackrel{\text{df}}{=} \lambda wg, ws, x, y.[\exists u{:}N].[u \leq x \wedge It(wg, ws)(u, y)].$$

Provide derivations for

It.1) $\vdash \forall y.[It(wg, ws)(0, y) \leftrightarrow ws(y)]$.
It.2) $\vdash \forall x, y.[It(wg, ws)(S(x), y) \leftrightarrow [N(x) \wedge wg(It(wg, ws)(x), y)]]$.
It.3) $\vdash \forall x, y.[It(wg, ws)(x, y) \rightarrow N(x)]$.
It.4) $Mon(wg) \vdash DCh(It(wg, \mathsf{V}))$, where $\mathsf{V} \stackrel{\text{df}}{=} \lambda u.u = u$.
CIt.1) $\vdash \forall y.[[\exists x{:}N].CIt(wg, ws)(x, y) \rightarrow [\exists x{:}N].It(wg, ws)(x, y)]$.
CIt.2) $Mon(wg) \vdash UCh(CIt(wg, ws))$.
CIt.3) $\vdash \forall x, y.[[N(x) \wedge wg(CIt(wg, ws)(x), y)]$
$\rightarrow CIt(wg, ws)(S(x), y)]$.

4. Let the recursion generator *RStream* with parameter w be defined by

$$RStream \stackrel{\text{df}}{=} \lambda w, Z, z.\exists x, y.[Z(y) \wedge w(x) \wedge z{=}\langle x, y\rangle].$$

The Horn sequent defined by *RStr* after contraction of the antecedent is

$$Stream(w)(y) \wedge w(x) \wedge z = \langle x, y\rangle \vdash Stream(w)(z).$$

Necessarily $x, y{:}1$ so that $\langle x, y\rangle$ has secondary type 1 and $w{:}[1]$. *Stream* is the defined constant with parameter w. Clearly $Lt(RSteam))$ is the empty predicate. The members of $Gt(RStream))$ form what in computer science is called a *stream* of members of w. This example has been adapted from §3.1 of [16].

Review §4.4.5 and then describe the members of $Gt(RStream))$.

4.8. Definition by Iteration

Throughout this section, $wg{:}[[\tau], \tau]$ but the presentation can be easily generalized. In Exercise 2 of §4.6.2 the recursion generator RIt was defined with parameters a recursion generator wg and a predicate ws:

$$RIt \overset{\text{df}}{=} \lambda wg, ws, Z, x, y.[\\ [x = 0 \wedge ws(y)] \vee [\exists u{:}N].[x = S(u) \wedge wg(Z(u), y)]].$$

Here $wg{:}[[\tau], \tau]$ and $ws{:}[\tau]$, so that $RIt(wg, ws){:}[[1, \tau], 1, \tau]$. In Exercise 3 of §4.7 the predicate It is defined in terms of RIt:

$$It \overset{\text{df}}{=} \lambda wg.ws.Lt(RIt(wg, ws)).$$

As defined, $It(wg, ws)(x)$ is the predicate that results from iterating wg x times beginning with ws. Here it is proved that the predicates $Lt(RG)$ and $Gt(RG)$ defined in terms of the least and greatest predicate operators Lt and Gt, can also be defined in terms of It and RG provided RG is monotonic and respectively $\exists$-continuous and $\forall$-continuous. $Lt(RG)$ and $Gt(RG)$ can then alternatively be defined to be

$$(\lambda v.[\exists x{:}N].It(RG, \oslash)(x, v))$$
$$(\lambda v.[\forall x{:}N].It(RG, \mathsf{V})(x, v))$$

where $\oslash$ and V are respectively the empty and the universal predicate for the type $[\tau]$ of ws:

$$\oslash \overset{\text{df}}{=} \lambda u.\neg u = u \text{ and } \mathsf{V} \overset{\text{df}}{=} \lambda u.u = u.$$

Although iteration is suggestive of computation, these alternative definitions are not computations as is explained in Chapter 8. They are however sometimes used to derive properties of the predicates $Lt(RG)$ and $Gt(RG)$ since they permit the use of the rule (NInd) of ordinary mathematical induction derived in §4.4.4.

4.8.1. Defining $Lt(RG)$ by iteration. That the predicate $Lt(wg)$ can be defined by iterating wg beginning with $\oslash$, provided it is monotone and $\exists$-continuous, is expressed in the following derivable sequents:

ItLt.1) $Mon(wg) \vdash \forall y.[[\exists x{:}N].It(wg, \oslash)(x, y) \rightarrow Lt(wg)(y)]$.

ItLt.2) $[Mon(wg) \wedge Con\exists(wg)] \vdash$
$\forall y.[Lt(wg)(y) \rightarrow [\exists x{:}N].It(wg, \oslash)(x, y)]$.

The derivations of these sequents make use of the derivable sequents listed in Exercise 3 of §4.7.

Derivation of (ItLt.1). By Exercise (2) of §1.7, to derive (ItLt.1) it is sufficient to derive

$$Mon(wg) \vdash [\forall x{:}N].\forall y.[It(wg, \oslash)(x, y) \rightarrow Lt(wg)(y)].$$

This change is made to make use of the rule (NInd) of §4.4.4. The abbreviation

$$P \overset{\text{df}}{=} (\lambda u.\forall y.[It(wg, \oslash)(u, y) \rightarrow Lt(wg)(y)])$$

is used in the following derivation of the sequent:

$+Mon(wg)$
$-[\forall x{:}N].P(x)$

$\dashv$NInd

L R

$-P(0)$
$-\forall y.[It(wg,\oslash)(0,y) \rightarrow Lt(wg)(y)]$ $\dashv$df P
$+It(wg,\oslash)(0,y)$
$+\oslash(y)$ $\dashv$It.1

R

$-\forall x.[P(x) \rightarrow P(S(x))]$
$+P(x)$
$-P(S(x))$
$+\forall y.[It(wg,\oslash)(x,y) \rightarrow Lt(wg)(y)]$ (a)$\dashv$df P
$-\forall y.[It(wg,\oslash)(S(x),y) \rightarrow Lt(wg)(y)]$ $\dashv$df P
$+It(wg,\oslash)(S(x),y)$
$-Lt(wg)(y)$
$+wg(It(wg,\oslash)(x),y)$ $\dashv$It.2
$+wg(Lt(wg),y)$ $\dashv$(a) $Mon(wg)$
$+Lt(wg,y)$ $\dashv$Lt.2

Derivation of (ItLt.2). To derive (ItLt.2) it is sufficient by (CIt.1) of exercise (3) of §4.7 to derive

$$[Mon(wg) \wedge Con\exists(wg)] \vdash \forall y.[Lt(wg)(y) \rightarrow [\exists x{:}N].CIt(wg,\oslash)(x,y)].$$

A derivation follows.

$+[Mon(wg) \wedge Con\exists(wg)]$
$-\forall y.[Lt(wg)(y) \rightarrow [\exists x{:}N].CIt(wg,\oslash)(x,y)]$
$-\forall y.[wg([\exists x{:}N].CIt(wg,\oslash)(x),y) \rightarrow [\exists x{:}N].CIt(wg,\oslash)(x,y)]$ $\dashv$LtInd
$+wg([\exists x{:}N].CIt(wg,\oslash)(x),y)$
$-[\exists x{:}N].CIt(wg,\oslash)(x,y)$
$+UCh(CIt(wg,\oslash))$ $\dashv$CIt.2
$+\forall y.[wg([\exists x{:}N].CIt(wg,\oslash)(x),y)$
$\rightarrow [\exists x{:}N].wg(CIt(wg,\oslash)(x),y)]$ $\dashv$Con∃(wg)
$+[\exists x{:}N].wg(CIt(wg,\oslash)(x),y)$
$+N(x)$
$+wg(CIt(wg,\oslash)(x),y)$
$+CIt(wg,\oslash)(S(x),y)$ $\dashv$CIt.3
$-CIt(wg,\oslash)(S(x),y)$

4.8.2. Defining $Gt(RG)$ by iteration. That It can be used to define the predicate $Gt(wg)$ is expressed in the following two sequents:

ItGt.1. $Mon(wg) \vdash \forall y.[Gt(wg)(y) \to [\forall x{:}N].It(wg, \mathsf{V})(x, y)]$.

ItGt.2. $[Mon(wg) \wedge Con\forall(wg)] \vdash$
$$\forall y.[[\forall x{:}N].It(wg, \mathsf{V})(x, y) \to Gt(wg)(y)].$$

Derivation of (ItGt.1). By Exercise (2) of §1.7, to derive (ItGt.1) it is sufficient to derive

$$Mon(wg) \vdash [\forall x{:}N].\forall y.[Gt(wg)(y) \to It(wg, \mathsf{V})(x, y)].$$

The abbreviation

$$P \stackrel{\text{df}}{=} (\lambda u.\forall y.[Gt(wg)(y) \to It(wg, \mathsf{V})(u, y)])$$

is used in the following derivation:

$+Mon(wg)$
$-[\forall x{:}N].P(x)$
——————— $\dashv$NInd

L R
$-P(0)$
$-\forall y.[Gt(wg)(y) \to It(wg, \mathsf{V})(0, y)]$
$-It(wg, \mathsf{V})(0, y)$
$-V(y)$ $\dashv$It.1
———

R
$-\forall x.[P(x) \to P(S(x))]$
$+P(x)$
$-P(S(x))$
$+\forall y.[Gt(wg)(y) \to It(wg, \mathsf{V})(x, y)]$ (a)$\dashv$df P
$-\forall y.[Gt(wg)(y) \to It(wg, \mathsf{V})(S(x), y)]$ $\dashv$df P
$+Gt(wg)(y)$
$-It(wg, \mathsf{V})(S(x), y)$
$+It(wg, \mathsf{V})(x, y)$
$+N(x)$ (It.3)
$-wg(It(wg, \mathsf{V})(x), y)$ $\dashv$It.2
$-wg(Gt(wg), y)$ $\dashv$(a) & Mon(wg)
$-Gt(wg, y)$ $\dashv$Gt.2
———

Derivation of (ItGt.2)

$+[Mon(wg) \wedge Con\forall(wg)]$
$-\forall y.[[\forall x{:}N].It(wg, \mathsf{V})(x, y) \to Gt(wg)(y)]$
$-\forall y.[[\forall x{:}N].It(wg, \mathsf{V})(x, y) \to wg([\forall x{:}N].It(wg, \mathsf{V})(x), y)]$ $\dashv$GtInd
$+[\forall x{:}N].It(wg, \mathsf{V})(x, y)$

$-wg([\forall x{:}N].It(wg,\vee)(x),y)]$
$+DCh(It(wg,\vee))$ $\dashv$It.5
$+\forall y.[[\forall x{:}N].wg(It(wg,\vee)(x),y)$
$\quad\quad\rightarrow wg([\forall x{:}N].It(wg,\vee)(x),y)]$ $\dashv$Con$\forall(wg)$
$-[\forall x{:}N].wg(It(wg,\vee)(x),y)$
$+N(x)$
$-wg(It(wg,\vee)(x),y)$
$-It(wg,\vee)(S(x),y)$ $\dashv$It.2
$+It(wg,\vee)(S(x),y)$

4.9. Potentially Infinite Predicates

In what some regard as the first paper in computer science, namely Turing's description of Turing machines [145], the concept of a *potentially infinite* tape was introduced. This is a tape on which only finitely many symbols could be written at any given time but which could be expanded in finite amounts to accommodate any finite number of symbols. Thus there is no bound on the number of symbols that could be written, but the number is always finite. This is a typical limitation appearing often in the description of computing devices or processes or languages. For example, Scott in [129] remarks on the domain L of *locations*

> Concerning L, all we would know is that there are infinitely many locations, because new ones can be produced indefinitely.

The number of such locations is sometimes called *potentially* infinite, because at no time are infinitely many locations actually used but there is always the potential for using finitely many more. Potentially infinite domains must be treated differently than a truly infinite domain such as the natural numbers. A function with source a potentially infinite domain can, at any given time, be defined for only finitely many arguments while a function with source the domain of natural numbers is generally defined for every natural number. The study of potentially infinite domains within a formal logic requires special definitions. Here a number of such definitions are given and illustrated with applications from computer science in preparation for Chapter 8 where an attempt is made to justify ITT as a logical foundation for computer science.

4.9.1. Characters and strings. The terms of a formal language consist of finite strings of finitely many characters satisfying given grammatical rules; see for example [4]. The characters must be syntactically distinct and therefore their representations in a logic must be *derivably* distinct. For example, the ten digits '0', '1', ... , '8', and '9' are the characters from which the customary notation for the numerals are constructed. A representation of them in ITT can be given by the following definitions:

$$
\begin{aligned}
0 &\stackrel{\text{df}}{=} \lambda u.\neg u = u \\
1 &\stackrel{\text{df}}{=} S(0) \\
2 &\stackrel{\text{df}}{=} S(S(0)) \\
&\cdots \\
8 &\stackrel{\text{df}}{=} S(\cdots S(S(0))\cdots) \quad \text{8 S's} \\
9 &\stackrel{\text{df}}{=} S(S(\cdots S(S(0))\cdots)).
\end{aligned}
$$

Here S is defined in §4.4. That these representations of the digits are derivably distinct follows from (S.1) and (S.2) of that section. The success of these terms as representations of characters depends essentially on the secondary typing allowed in ITT.

Numerals with more than one digit, which are strings of digits not beginning with the digit '0', can be represented using the ordered pair $\langle\,\rangle$ predicate and the predicate *Nil* defined in §4.4.2. For example, the numeral '123' is represented by the term

$$\langle 1, \langle 2, \langle 3, \mathit{Nil}\rangle\rangle\rangle.$$

This term is an example of what in computer science is called a *list data structure*. Such lists are used to represent finite sequences of characters as described next in §4.9.2. Because of the properties (0P.1) and (OP.2) of §4.4.2, distinct strings of characters are represented by *derivably* distinct terms of ITT. The success of these terms as representations of strings of characters also depends essentially on secondary typing.

4.9.2. Lists. Let the recursion generator $RLst$ with parameter w be defined by

$$RLst \stackrel{\text{df}}{=} \lambda w, Z, z.[z = \mathit{Nil} \vee \exists u, v.[w(u) \wedge Z(v) \wedge z = \langle u, v\rangle]]$$

Here it is assumed that u, v:1, so that $\langle u, v\rangle$ has secondary type 1, and that w:[1]. Thus $RLst$:[[1], [1], 1]. Two predicates defined from $RLst$ are

$$Lst \stackrel{\text{df}}{=} (\lambda w.Lt(RLst(w))) \quad \text{and} \quad GLst \stackrel{\text{df}}{=} (\lambda w.Gt(R[Lst](w)).$$

Since a Horn sequent defined by $RLst$ and its converse are each derivable when the constant C of (HS) of §4.7.1 is Lst or $GLst$, each of the following sequents is derivable:

Lst) $\vdash \forall x.[Lst(w)(x)$
$\qquad \leftrightarrow [x = \mathit{Nil} \vee \exists u, v.[w(u) \wedge Lst(w)(v) \wedge x = \langle u, v\rangle]]].$

GLst) $\vdash \forall x.[GLst(w)(x)$
$\qquad \leftrightarrow [x = \mathit{Nil} \vee \exists u, v.[w(u) \wedge GLst(w)(v) \wedge x = \langle u, v\rangle]]].$

Given a predicate P:[1], the terms to which the predicate $Lst(P)$ applies represent list data structures with elements members of the predicate P. A list data structure is a way of representing finite sequences of members of any parameter predicate P. *Nil* represents the empty sequence; see any elementary textbook on data structures, such as [150].

Consider, for example, the following definitions of the predicates *Digit*, *Digit*1, *NumNil* and *Numeral*:

$$Digit \stackrel{\text{df}}{=} \lambda u.[u = 0 \vee u = 1 \vee u = 2 \vee \cdots \vee u = 9]$$
$$Digit1 \stackrel{\text{df}}{=} \lambda u.[u = 1 \vee u = 2 \vee \cdots \vee u = 9]$$
$$NumNil \stackrel{\text{df}}{=} \lambda x.Lst(Digit)(x)$$
$$Numeral \stackrel{\text{df}}{=} \lambda x.\exists u, v.[Digit1(u) \wedge NumNil(v) \wedge x = \langle u, v\rangle].$$

The members of *Digit*1 are the digits that may begin a string of digits that are numerals, while members of *Digit* may be any other digit in the string. The members of *NumNil* may follow the first digit of a string with *Nil* in this case representing the empty string of digits. Some members of *NumNil* are

$$Nil, \langle 0, Nil\rangle, \cdots, \langle 9, Nil\rangle, \langle 0, \langle 1, Nil\rangle\rangle, \langle 0, \langle 1, \langle 2, Nil\rangle\rangle\rangle.$$

Some members of *Numeral* are

$$\langle 1, Nil\rangle, \cdots, \langle 9, Nil\rangle, \langle 1, \langle 0, \langle 1, Nil\rangle\rangle\rangle, \langle 9, \langle 1, \langle 1, Nil\rangle\rangle\rangle.$$

As discussed in §4.4.5 for the non-well-founded GnN, the members of $GnLst$, where

$$GnLst \stackrel{\text{df}}{=} (\lambda w, z.[GLst(w)(z) \wedge \neg Lst(w)(z)])$$

are non-well-founded. For this predicate the following sequent is derivable:

GnLst) $\quad \vdash \forall w, z.[GnLst(w)(z) \rightarrow [\exists x{:}w].[\exists y{:}GnLst(w)].z = \langle x, y\rangle].$

Thus, for example, no member of $GnLst(N)$ terminates as $\langle 1, \langle 0, \langle 1, Nil\rangle\rangle\rangle$ does in a pair $\langle 1, Nil\rangle$ with second member *Nil*; members of $GnLst(N)$ can always be continued by a pair with second element a member of $GnLst(N)$. Although "list" is usually reserved for the well-founded members of $Lst(w)$, when the parameter w has been assigned a predicate $P{:}[1]$, the members of $GnLst(P)$ can also be called lists when there is no danger of confusion.

4.9.3. Universal quantification. The occurrences of symbols on a Turing tape [145] is the extension of a potentially infinite predicate; at any given time there are at most finitely many occurrences, but there is no bound on the number. Here the importance of distinguishing between universal quantification over *potentially* infinite, as opposed to infinite, predicates is demonstrated.

Let the recursion generator *RFP* with parameter w be defined by

$$RFP \stackrel{\text{df}}{=} \lambda w, Z, Y.[\neg\exists u.Y(u) \vee \\ \exists U.[Z(U) \wedge [\exists u{:}w].[\neg U(u) \wedge Y =_e U \cup \{u\}]\,]\,],$$

where $w, Y, U{:}[\tau]$. When $FP \stackrel{\text{df}}{=} \lambda w.Lt(RFP(w))$, the formula $FP(w, Y)$ is intended to hold for predicates $Y{:}[\tau]$ with finitely many members from the extension of w:$[\tau]$. Thus *FP* abbreviates *Finite Predicate*, or for set minded readers, *Finite Power* set. The following sequent is derivable by (Lt.2):

$$\text{FP)} \quad \vdash \forall w, Y.[FP(w,Y) \leftrightarrow [\neg\exists u.Y(u) \vee \exists U.[FP(w,U) \wedge [\exists u{:}w].[\neg U(u) \wedge Y =_e U \cup \{u\}]\,]\,].$$

Let $w1, w2{:}[\tau]$ and let P be a finite subpredicate of $w1$; that is, assume that the sequent

$$\text{a)} \quad \vdash FP(w, P)$$

is derivable. In some cases members $p_1, \ldots, p_k$ of $w1$ may be known for which

$$\text{b)} \quad \vdash P =_e \lambda x.[x = p_1 \vee \cdots \vee x = p_k]$$

is also derivable. Then

$$\vdash [\,[\forall x{:}P].w2(x) \leftrightarrow [w2(p_1) \wedge \cdots \wedge w2(p_k)]\,]$$

can be derived. Thus universal quantification over a predicate with a given finite number of members can be expressed by conjunction. But not all finite predicates are so neatly defined.

Consider the following example that has been adapted from the example of §3.1.3, greatest fixed points, of [117]. Let St:[1] and R:[1, 1] be given predicates. St can be understood to be the states of a process that can move from a state x to a state y whenever $R(x, y)$ holds. Let the recursion generator RE be defined by

$$RE \stackrel{\text{df}}{=} \lambda Z, x.[St(x) \wedge \forall y.[R(x,y) \to Z(y)].$$

Note that because of the quantifier $\forall y$, it is only possible to conclude from Corollary 42 of §4.6.3 that RE is a positive recursion generator but not an e-positive one.

A Horn sequent with defined constant E defined from RE is after contraction of the antecedent

$$St(x), \forall y.[R(x,y) \to E(y)] \vdash E(x).$$

Clearly $Lt(RE)$ is empty but $Gt(RE)$ is of interest.

Since the formula $R(x, y)$ is defined by a process, it is to be expected that there are at most finitely many states y that the process can move to from a given state x, although the number is potentially infinite. Formally this is expressed by

$$\text{c)} \quad \vdash [\forall x{:}St].FP(St, \lambda y.R(x,y)).$$

Thus RE is not the correct recursion generator needed to describe the process since the quantifier $\forall y$ in RE is over a finite but potentially infinite domain $\lambda y.R(x, y)$ that can vary with x.

Let the formula $R\forall(w1, w2, Y)$ express that Y is a finite subpredicate of $w1$ and that $[\forall u{:}Y].w2(u)$ holds, where $w1, w2, Y{:}[\tau]$. Then RE should be replaced by REF:

$$REF \stackrel{\text{df}}{=} \lambda Z, x.[St(x) \wedge F\forall(St, Z, \lambda y.R(x, y))]$$

provided $F\forall$ can be correctly defined; that is provided that the sequent

F∀. $\forall Y, w1, w2.[F\forall(w1, w2, Y) \leftrightarrow [FP(w1, Y) \wedge [\forall u{:}Y].w2(u)]]$

can be derived.

A recursion generator $RF\forall$ for which

$$F\forall \stackrel{\text{df}}{=} \lambda w1, w2, Y.Lt(RF\forall(w1, w2), Y)$$

is defined in Example 3 of §4.6.2. It is

$$RF\forall \stackrel{\text{df}}{=} \lambda w1, w2, Z, Y.[\neg\exists u.Y(u) \vee \exists U.[Z(w1, w2, U) \wedge [\exists u{:}w1].[w2(u) \wedge \neg U(u) \wedge Y =_e U \cup \{u\}]]].$$

With $F\forall$ defined by $RF\forall$, an important difference between REF and RE must be noted: Although Z is positive but not e-positive in RE it is both positive and e-positive in REF. Thus by Corollary 42 of §4.6.3 RE is ∀-continuous so that by (ItGt.1) and (ItGt.2) of §4.8.2 $Gt(REF)$ can be defined by iteration:

d) $\forall y.[Gt(REF)(y) \leftrightarrow [\forall x{:}N].It(REF, \vee)(x, y)]$.

Readers are asked to provide a derivation for (F∀.1). A direct derivation using (LtInd) for $F\forall$ in one direction and also (LtInd) for FP in the other direction should succeed. But another possibility presents itself. It is demonstrated in Example 3 of §4.6.2 that the occurrence of Z in $RF\forall(w1, w2, Z, Y)$ is e-positive so that $RF\forall$ is monotonic and ∃-continuous by Corollary 42 of §4.6.3. It is easily seen for the same reason that RFP is also monotonic and ∃-continuous. Therefore by (ItLt.1) and (ItLt.2) of §4.8.1, both FP and $F\forall$ can be defined by iteration:

ItFP. $\forall w1, w2, Y.[FP(w1, Y) \leftrightarrow [\exists x{:}N].It(RFP(w1), \oslash)(x, Y)]$

ItF∀. $\forall w1, w2, Y.[F\forall(w1, w2, Y) \leftrightarrow [\exists x{:}N].It(RF\forall(w1, w2), \oslash)(x, Y)]$.

A derivation of (F∀.1) using these sequents and (NInd) should also succeed.

EXERCISES §4.9.

1. Redefine $Rlst$ of §4.9.2 so that no member of N appears more than once in a list from either $Lt(RLst(N))$ or $Gt(RLst(N))$.
2. Provide a derivation for (GnLst) of §4.9.2.
3. Lst and $GLst$ are defined in §4.9.2. Each of the following terms has the type of a recursion generator when $Z{:}[1]$:

$$\lambda Z, z.Lst(Z)(z) \quad \text{and} \quad \lambda Z, z.GLst(Z)(z).$$

Describe the membership of the predicates

$$Lt(\lambda Z, z.Lst(Z)(z)),\ Gt(\lambda Z, z.Lst(Z)(z)),$$
$$Lt(\lambda Z, z.GLst(Z)(z)), \text{ and } Gt(\lambda Z, z.GLst(Z)(z)).$$

4. A list that is a member of $Lst(P)$ or $GLst(P)$ can represent a finite set of members of P:[1]. Thus for example $\langle 1, \langle 0, \langle 1, Nil\rangle\rangle\rangle$ represents the subset of N with members 0 and 1 as does $\langle 1, \langle 0, \langle 1, z\rangle\rangle\rangle$ if it is a member of $GLst(N)$. Nil represents the empty subset of P and each other list of $Lst(P)$ or $GLst(P)$ represents a subset with members determined by the list. A given member x of P may appear more than once in a given list, but that can be prevented in the definition of the lists. Define a recursion generator $RFPLst$ with parameter w:[1] for which the members of $Lt(RFPLst(w))$ are represented as lists without repetitions.
5. When $w1$:[σ] and $w2$:[τ], then $\langle u, v\rangle$:[[σ, τ]] when u and v are members of $w1$ and $w2$, and thus $FF(w1, w2)$:[[[[σ, τ]]]]. However, should $w1, w2$:[1], then by secondary typing $\langle u, v\rangle$:1 and thus $FF(w1, w2)$:[[1]]. In this case each member X of $FF(w1, w2)$ can be represented as a member of $Lst(\otimes(w1, w2))$ satisfying the restriction for functions, where

$$\otimes \stackrel{\text{df}}{=} \lambda z.[\exists u{:}w1].[\exists u{:}w1].z = \langle u, v\rangle.$$

The advantage of this representation of finite functions when $w1, w2$:[1] is that each finite function with domain $w1$ and range $w2$ is represented by a term of secondary type 1. If $w1 \xrightarrow{\text{fin}} w2$ is the predicate that applies to such representations then $w1 \xrightarrow{\text{fin}} w2$:[1]. Provide a definition for $\xrightarrow{\text{fin}}$.
6. Provide a derivation for (FP) in §4.9.3.
7. Provide a derivation for (F∀).

CHAPTER 5

CHOICE AND FUNCTION TERMS

5.1. Introduction

The types of the logics TT and ITT, apart from their notation, differ from the types of Church's Simple Theory of Types [28] only in the absence of types for functions with values of type other than the type [] of the truth values. This chapter is concerned primarily with filling that gap.

A notation for functions is not a necessity for a predicate logic since a function of n arguments, $n \geq 0$, can be represented as a predicate of $n+1$ arguments with the $n+1$'st argument being the value of the function. But a functional notation can greatly simplify some assertions. Consider, for example, an assertion that an arity 2 function $+$ is associative. Using an infix functional notation this is expressed in the sequent

$$\vdash \forall x, y, z.(x + (y + z))=((x + y) + z).$$

If, on the other hand, the function $+$ is expressed as an arity 3 predicate, then the statement of associativity becomes much more complicated. It is necessary to introduce variables $v1$ and $v2$ to record the values of $(y + z)$ and $(x + v1)$ respectively, and variables $w1$ and $w2$ to record the values of $(x + y)$ and $(w1 + z)$ respectively, and then conclude that $v2$ is $w2$:

$$\begin{aligned}&\vdash \forall x, y, z, v1, v2, w1, w2.\\ &\quad [+(y, z, v1) \wedge +(x, v1, v2) \wedge +(x, y, w1) \wedge +(w1, z, w2) \rightarrow v2{=}w2].\end{aligned}$$

But a functional notation that is admitted as primitive must of necessity be interpreted as a notation for total functions, not partial functions, over the domain of the function; thus, for example, the sequent

$$\vdash \forall x, y.\exists z.(x + y)=z$$

is necessarily a derivable sequent of the logic.

An implicit notation for functions with values that are extensionally defined, and therefore not of type 1, is available in ITT through the syntactic sugaring of application. For example, the terms Lt and Gt have been used in this way. Thus $Lt(RN)$, where RN is the recursion generator defined in §4.4, is a predicate of type [1] that is the value of the function Lt for the argument RN of type [[1], 1]. Another example is the Cartesian product, defined

$$\otimes \stackrel{\mathrm{df}}{=} \lambda X, Y, z_1, \ldots, z_n.\exists \overline{x}, \overline{y}.[z_1{=}\langle x_1, y_1\rangle \wedge \cdots \wedge z_n{=}\langle x_n, y_n\rangle \wedge X(\overline{x}) \wedge Y(\overline{y})]$$

where $X{:}[\overline{\sigma}]$ and $Y{:}[\overline{\tau}]$, with $\overline{\sigma}$ and $\overline{\tau}$ sequences $\sigma_1, \ldots, \sigma_n$ and $\tau_1, \ldots, \tau_n$, $n \geq 1$. Thus $\otimes(X, Y)$ is the value of the Cartesian product of X and Y.

A third example appears in §4.1 where a map c in the sense of [111] or an operator F in the sense of [103] is defined to be

$$(\lambda X, x.RG(X, x))$$

where RG is a recursion generator. This term behaves like a notation for a monotonic function F when RG is monotonic. This use of application sugaring makes it unnecessary to introduce a special notation into ITT for functions with values that are extensionally defined.

An implicit notation for functions with arguments and values that are of type 1 is available in ITT through the secondary typing of terms. The term S defined in §4.4.1 is used as a notation for a function with type 1 arguments and values in the derivations of (S.1) and (S.2): $S(t)$ is a first order term when t is. Similarly the ordered pair may be regarded as a notation for a function of two arguments of type 1 and value type 1. But in each of these cases the functions implicitly defined are total. A notation for partial functions in ITT with arguments of any type and intensionally defined values must be explicitly introduced. The notation is defined in terms of a notation for what are known as *choice* terms.

5.1.1. A functional notation from choice terms. Introducing a notation for partial functions into a logic requires dealing with the syntactic and semantic problems associated with *non-denoting* terms; that is properly typed terms that cannot be given a value by a valuation. Consider, for example, the defined predecessor predicate *Pred*:

$$Pred \stackrel{\mathrm{df}}{=} \lambda u, v.S(v){=}u.$$

The following sequent is derivable from (S.1) of §4.4.1:

$$\vdash \forall x, y.[Pred(S(x), y) \rightarrow x = y].$$

Thus $Pred(S(x), y)$ can be understood to define a function with value x for argument $S(x)$. But that function has no value for argument 0; the sequent

$$\vdash \exists v.Pred(0, v)$$

is not derivable; the function is necessarily *partial* since the value of the function for argument 0 is necessarily *non-denoting*.

In [42] a number of ways of dealing with partial functions and non-denoting terms are considered; the paper's extensive bibliography provides further references. A method is described there for the formulation of the simple theory of types of [9], a modification of the logic of [28]. The method involves modifying the axioms of the logic to accomplish the goal of allowing non-denoting terms to be introduced and reasoned about.

The traditional method of dealing with partial functions and the possible resulting non-denoting terms uses Russell's definite description operator $(\iota v)F$ [149, 147]; it is read "the v such that F"; see for example [28] and [120]. The value of the predecessor function for an argument x is then $(\iota v)Pred(x, v)$ when there is a unique v for which $Pred(x, v)$.

With λ abstraction available, it is not necessary to have the ι-operator as a variable binding operator like λ. The operator can be understood to be applicable to any predicate term M of type $[\tau]$ so that ιM is a *function term*. For example, a term $(\iota v)Pred(x, v)$ in Russell's notation is replaced by the function term $\iota(\lambda v.Pred(x, v))$ since $(\lambda v.Pred(x, v))$:[t[v]].

For a function term ιM to be denoting, the predicate term M must have an extension with a single member; that is

$$\exists v.[M(v) \wedge \forall u.[M(u) \rightarrow u = v]]$$

must be satisfied. This formula is traditionally abbreviated to

$$\exists! v.M(v).$$

The ι-operator is called the *definite* description operator because if this formula is satisfied, the single member of the extension of M is *described* by M. An *indefinite* description operator when applied to M on the other hand selects an arbitrary member of its extension. Following Hilbert and Bernays [78], this operator is generally denoted by ε and referred to as the *choice* operator. Thus for a *choice term* εM to be denoting it is sufficient for $\exists v.M(v)$ to be satisfied. The ε-operator can, in some respects, be thought of as the inverse of the λ-operator since $\varepsilon(\lambda v.F)$:t[v].

Clearly if a function term ιM is denoting, then different occurrences of it in a formula have the same denotation. Thus the sequent

$$\vdash [\exists v.[M(v) \wedge \forall u.[M(u) \rightarrow u = v]] \rightarrow \iota M = \iota M]$$

should be derivable. But that is not the case for the choice term εM; that is the sequent

$$\vdash [\exists v.M(v) \rightarrow \varepsilon M = \varepsilon M]$$

will not in general be derivable since the two occurrences of εM may denote different members of the extension of M. On the other hand the sequent

$$\vdash [\exists v.M(v) \rightarrow (\lambda v.v = v)\varepsilon M]$$

should be derivable, since there is but one occurrence of εM. The application term $(\lambda v.v = v)\varepsilon M$ cannot therefore contract to $\varepsilon M = \varepsilon M$; the definition of β-contraction for a logic with choice terms must exclude a β-contraction

$$(\lambda v.P)\varepsilon M > [\varepsilon M/v]P.$$

For a choice term must always be accompanied by a *scope* within which the term is intended to be denoting. In the last of the displayed sequents $(\lambda v.v = v)$ is the scope of εM.

The ι-operator can be defined in terms of the ε-operator with an additional restriction that ensures uniqueness of choice:

$$\iota M \stackrel{\text{df}}{=} \varepsilon(\lambda v.[M(v) \wedge \forall u.[M(u) \rightarrow u = v]])$$

where u and v are distinct variables without a free occurrence in M. Although here a function term ιM is defined as a choice term without a scope, such terms must have a scope as all choice terms must have.

In this chapter the ε-operator is introduced into an extension ITTε of ITT. The syntax for ITTε is defined in §5.2. The proof theory is described in §5.3. A semantics for ITTε is described in §5.4 in terms of a mapping of some terms of ITTε into terms of ITT. This mapping gives meaning to all the terms of ITTε for which it is defined, since a semantics has been defined for ITT. Using this mapping a completeness and Cut-elimination theorem similar to Theorem 33 of §3.5 is proved along with a theorem from which it can be concluded that ITTε is a conservative extension of ITT; that is a sequent of formulas of ITT is derivable in ITTε if and only if it is derivable in ITT. The consistency of ITTε is an immediate consequence.

In §5.5 the ι-operator is used to introduce a functional notation for total or partial functions with arguments and values of any type. Thus ITTε can be seen as supplying the function terms of the type theory of [28] that are missing from TT.

The presentation of choice terms in ITTε has similarities with the method of dependent types discussed in [124] and employed in PVS, the system described in [110]. Dependent types allow for the definition of a domain for the arguments of a partial function over which the function becomes total. In §5.6 the domains of partial functions are suggested as dependent types.

5.2. Introducing Choice Terms

The types, variables, and constants of ITTε are those of ITT. In addition ITTε makes use of the untyped symbol ε that is used in choice terms.

DEFINITION 58. *Terms of ITTε.*

1. *A constant or variable cv is a term of type* t$[cv]$. *A term cv of type other than* 1 *is a* predicate *term.*
2. *Let M be a predicate term of type* $[\tau]$. *Then εM is a* choice *term and term of type τ, but not a predicate term.*
3. *Let P be a predicate term of type* $[\tau, \overline{\tau}]$ *and Q any term of type τ. Then* (PQ) *is a* predicate *term of type* $[\overline{\tau}]$. *If Q is a choice term, then P is the* scope *of its occurrence in* (PQ).

4. *Let P be a predicate term of type $[\overline{\tau}]$ and v a variable of type τ. Then $(\lambda v.P)$ is a* predicate *term of type $[\tau, \overline{\tau}]$.*
5. *Let P be a predicate term of type $[\overline{\tau}]$ in which at most variables of type 1 have a free occurrence. Then P has* primary *type $[\overline{\tau}]$ and* secondary *type 1.*

A *formula* of ITTε is a predicate term of type [ ]; a choice term of that type is not a formula. Further, the substitution operator $[Q/v]P$ is only defined for a term Q that is not a choice term.

Given a predicate M of type $[\tau]$, the notation εM denotes an arbitrary member of the extension of the predicate M, if one exists. But note that εM is a term of ITTε only if $M{:}[\tau]$; no type is assigned to εM if $M{:}[\tau, \overline{\sigma}]$, when $\overline{\sigma}$ is not empty. However the effect of having choice terms formed from such M can be achieved with the terms $(\lambda \overline{u}.(P\varepsilon M(\overline{u})))$ when $\overline{u}{:}\overline{\sigma}$.

An occurrence of a choice term εM in a term P is *free* if each free occurrence of a variable in M is a free occurrence in P. An occurrence of a choice term is said to be free in itself.

5.2.1. The Relation $>_\varepsilon$ on terms of ITTε. The relation $>$ was first defined for terms of TT and extended in §3.1.3 for terms of ITT as follows: $P > P'$ if P' is the result of replacing an occurrence of a contractable subterm of P by its contraction, or results from P by a change of bound variable. To define $>_\varepsilon$ for terms of ITTε therefore requires the prior definition of the contractable terms of ITTε.

DEFINITION 59. *Contractable terms and their contractions.*
The contractable *terms of ITTε and their* contractions *take the following forms*:

(β) *$(\lambda v.P)Q$ contracts to $[Q/v]P$, when Q is not a choice term.*
(η) *$(\lambda v.Pv)$ contracts to P, when v has no free occurrence in P.*
(ε) *$(P\varepsilon M)$ contracts to $(\lambda \overline{u}.[\exists v{:}M].P(v, \overline{u}))$, when $P{:}[\tau, \overline{\sigma}]$, $M{:}[\tau]$, and $\overline{u}{:}\overline{\sigma}$ and $v{:}\tau$ are distinct variables without free occurrences in P or M.*

Note that P in an η-contraction cannot be a choice term since P must be a predicate term for (Pv) to be a term when $P(v, \overline{u})$ is written in its unsugared form $((\ldots(((Pv))u_1)\ldots)u_n)$, where $\overline{u}$ is $u_1, \ldots, u_n$.

A term $(\lambda x.P)\varepsilon M$ is not β-contractable but is ε-contractable:

$$((\lambda x.P)\varepsilon M) >_\varepsilon (\lambda \overline{u}.[\exists v{:}M].((\lambda x.P)v)(\overline{u})]) > (\lambda \overline{u}.[\exists v{:}M].[v/x]P(\overline{u})])$$

A premature contraction, such as $(\lambda x.P)\varepsilon M$ to $[\varepsilon M/x]P$, requiring an improper substitution can assign the wrong scope to a choice term and result in undesirable changes of meaning as demonstrated in the following examples.

EXAMPLES. *The Importance of Scope.*

1. Let P be $\forall z.C(x, z)$. Then

$$((\lambda x.P)\varepsilon M) >_\varepsilon [\exists v{:}M].(\lambda x.P)v > [\exists v{:}M].[v/x]P > [\exists v{:}M].\forall z.C(v, z).$$

In this case the scope of εM is $(\lambda x.\forall z.C(x,z))$. On the other hand, since $C(\varepsilon M, z)$ in its unsugared form is $((C\varepsilon M)z)$, it follows that

$$\forall z.C(\varepsilon M, z) >_\varepsilon \forall z.((\lambda u.[\exists v{:}M].(Cv)u)z) > \forall z.[\exists v{:}M].C(v,z)$$

reversing the order of the quantifiers. In this second case the scope of εM is C.

2. This example illustrates the effect of premature contraction on negated formulas. Consider the formula $((\lambda x.\neg C(x))\varepsilon M)$ in which εM has scope $(\lambda x.\neg C(x))$. Then

$$((\lambda x.\neg C(x))\varepsilon M) >_\varepsilon [\exists v{:}M].(\lambda x.\neg C(x))v > [\exists v{:}M].\neg C(v)$$

while $[\varepsilon M/x]\neg C(x)$ is $\neg C(\varepsilon M)$ so that εM has scope C and

$$\neg C(\varepsilon M) >_\varepsilon \neg[\exists v{:}M]C(v).$$

3. This last example has been adapted from [125]. Let C be understood to be the predicate of being bald, and M the predicate of being a present king of France. Then $C(\varepsilon M)$ expresses that a present king of France is bald. Since there is no present king of France, $C(\varepsilon M)$ is false which concurs with the contraction $[\exists v{:}M].C(v)$. On the other hand the sentence "a present king of France is not bald" has two meanings, one corresponding to $\neg C(\varepsilon M)$, which is true since $C(\varepsilon M)$ is false, and one corresponding to $(\lambda x.\neg C(x))(\varepsilon M)$ which translates to $[\exists v{:}M].\neg C(v)$ and is false.

These examples illustrate the importance of determining the correct scope for a choice term in the context of a given formula. It requires consideration of the sugaring of terms.

Consider now the predicate *Den* of any type $[\tau]$:

$$Den \overset{\text{df}}{=} (\lambda u.u{=}u).$$

If Q is not a choice term, then

$$Den(Q) > Q = Q.$$

But

$$Den(\varepsilon M) >_\varepsilon [\exists v{:}M].Den(v) > [\exists v{:}M].v = v.$$

Thus $Den(Q)$ properly expresses that Q is a denoting term. Beeson in [17] and Farmer in [42] use the notation $Q\downarrow$ to express the same thing.

EXERCISES §5.2.

1. The definite description operator in [28] is assigned a type. Is it possible for the constant ε to be treated like the constants $\downarrow$ and $\forall$ and be assigned a type?
2. Give a justification of the contraction of $P(\overline{r}, \varepsilon M, \overline{s})$ to $[\exists v{:}M].P(\overline{r}, v, \overline{s})$, where $P{:}[\overline{p}, \tau, \overline{\sigma}]$, $M{:}[\tau]$, $\overline{r}{:}\overline{p}$, $\overline{s}{:}\overline{\sigma}$, $v{:}\tau$, and v has no free occurrence in P, M, $\overline{r}$ and $\overline{s}$.

3. State and prove a substitution lemma for ITTε similar to Lemma 23 of ITT in §3.1.2.

5.3. Proof Theory for ITTε

Here all the rules of ITTε are listed. Note that each rule of ITT is a rule of ITTε since a formula of ITT is a formula of ITTε. But since a choice term of type [] is not a formula, they can appear only as subterms of formulas of ITTε. For example, in the first of the $+\downarrow$-rules, G may be a choice term but F may not, while in the second F may be a choice term but G may not. Neither F nor G may be a choice term in the $-\downarrow$-rule. Further, since P in a term (PQ) cannot be a choice term, the P in the $\exists$-rules is necessarily a predicate term.

Note in the conclusion of the $-\exists$ rule, the scope discussion of the examples of §5.2.1 is relevant if Q is εM.

$+\downarrow$ $\quad \dfrac{+[F\downarrow G]}{-F} \quad \dfrac{+[F\downarrow G]}{-G}$ $\qquad$ $-\downarrow$ $\quad \dfrac{-[F\downarrow G]}{+F \quad +G}$

$+\exists$ $\quad \dfrac{+\exists P}{+P(y)}$ y new to branch $\qquad$ $-\exists$ $\quad \dfrac{-\exists P}{-P(Q)}$ Q any term

$+\lambda$ $\quad \dfrac{+F}{+G}$ $F >_\varepsilon G$ $\qquad$ $-\lambda$ $\quad \dfrac{-F}{-G}$ $F >_\varepsilon G$

$+=$ $\quad \dfrac{+P = Q}{+\forall Z.[Z(P) \rightarrow Z(Q)]}$ $\qquad$ $-=$ $\quad \dfrac{-P = Q}{-\forall Z.[Z(P) \rightarrow Z(Q)]}$

$+Int$ $\quad \dfrac{+P=^1 Q}{+P=^\tau Q}$ $\qquad$ $-Int$ $\quad \dfrac{-P=^1 Q}{-P=^\tau Q}$

P and Q have primary type τ and secondary type 1.

Cut $\quad \dfrac{}{+F \quad -F}$

Note that a choice term appearing in the premiss of an application of one of the $+\downarrow$ rules may be absent from the conclusion. Thus these rules can remove a non-denoting term; no other rule can do so. The first of the following examples illustrates how this property can be exploited in an application of the $+\wedge$ rule, derived from a the $+\downarrow$ rule.

EXAMPLE DERIVATIONS.

1. A conditional expression is frequently used in programming languages. It will be expressed here as $\varepsilon Cond(x, y_1, y_2)$ where $Cond$ is defined:

$$Cond \stackrel{\text{df}}{=} \lambda u, v_1, v2, v.[[u{=}1 \wedge v{=}v_1] \vee [\neg u{=}1 \wedge v{=}v2]].$$

The basic property of $\varepsilon Cond$ is expressed in the following derivable sequent:

Cond) $\vdash \forall x, y_1, y_2.[[x{=}1 \rightarrow \varepsilon Cond(x, y_1, y_2){=}y_1] \wedge [\neg x{=}1 \rightarrow \varepsilon Cond(x, y_1, y_2){=}y_2]].$

A choice term $\varepsilon Cond(Q, Q1, Q2)$ may be denoting even though one of $Q1$ or $Q2$ is not. For example, the following sequent is derivable

$$\vdash \forall y_1.\varepsilon Cond(1, y_1, \varepsilon Pred(0)){=}y_1.$$

Recall that an alternative to the infix notation used here for $=$ is the prefix notation $= (\varepsilon Cond(1, y_1, \varepsilon Pred(0)), y_1)$ in which it is clear that the scope of $\varepsilon Cond(1, y_1, \varepsilon Pred(0))$ is $=$. An abbreviated derivation of the sequent making use of this change of notation follows:

$-\forall y_1. = (\varepsilon Cond(1, y_1, \varepsilon Pred(0)), y_1)$
$- = (\varepsilon Cond(1, y_1, \varepsilon Pred(0)), y_1)$
$-[\exists v{:}Cond(1, y_1, \varepsilon Pred(0))]. = (v, y_1)$
$-\exists v.[Cond(1, y_1, \varepsilon Pred(0), v) \wedge = (v, y_1)]$
$-[Cond(1, y_1, \varepsilon Pred(0), y_1) \wedge = (y_1, y_1)]$
$-Cond(1, y_1, \varepsilon Pred(0), y_1)$
$-(\lambda u, v_1, v_2, v.[[u{=}1 \wedge v{=}v_1] \vee [\neg u = 1 \wedge v{=}v_2]])(1, y_1, \varepsilon Pred(0), y_1)$
$-[[1 = 1 \wedge y_1 = y_1] \vee [\neg 1 = 1 \wedge y_1 = \varepsilon Pred(0)]]$
$-[1 = 1 \wedge y_1 = y_1]$

2. The scope discussion of §5.2.1 is relevant to the form of a derivable rule $+\forall_\varepsilon$ for ITTε. Here is the rule and its justification:

$$+\forall_\varepsilon \qquad \frac{+\forall P}{+P(\varepsilon M) \quad - \exists v.M(v)}$$

A derivation of the rule follows together with justifications for some lines. Letters (a), (b), and (c) designate three nodes in the derivation that are referenced in justifications $\dashv$(a) and $\dashv$(b),(c); other justifications such as the first $\dashv$df$\forall$ indicates how the node has been obtained.

$+\forall P$
$+(\lambda U.\neg\exists(\lambda v.\neg U(v)))P$ $\quad\dashv$df$\forall$
$+\neg\exists(\lambda v.\neg P(v))$
$-\exists(\lambda v.\neg P(v))$
$-(\lambda v.\neg P(v))\varepsilon M$ $\quad\dashv -\exists$

$$
\begin{array}{lll}
-\exists v.[M(v) \wedge (\lambda v.\neg P(v))v] & & \text{(a)} \\
\hline
& & \dashv\text{Cut} \\
+\exists v.M(v) & -\exists v.M(v) & \\
+M(v) & & \text{(b)} \\
-[M(v) \wedge (\lambda v.\neg P(v))v] & & \dashv\text{(a)} \\
-(\lambda v.\neg P(v))v & & \\
-\neg P(v) & & \\
+P(v) & & \text{(c)} \\
\hline
& & \dashv\text{Cut} \\
-P(\varepsilon M) & +P(\varepsilon M) & \\
-\exists v.[M(v) \wedge P(v)] & & \\
-[M(v) \wedge P(v)] & & \\
\hline
& & \dashv\text{(b),(c)}
\end{array}
$$

EXERCISES §5.3.

1. Provide a derivation for the sequent:

$$\vdash \forall x.\varepsilon Pred(S(x))=x.$$

2. Let $Q{:}[\sigma, \tau]$. Provide a derivation for the following sequent:

$$\vdash \forall x.[Den(\varepsilon Q(x)) \rightarrow Q(x, \varepsilon Q(x))].$$

But for the replacement of ι by ε, this sequent is similar to the axiom scheme 9^α of [28].

5.4. Consistency and Completeness

Rather than developing a semantics for ITTε, a mapping ‡ of some terms of ITTε to terms of ITT is defined. The semantics of ITTε is then understood in terms of the semantics of ITT.

DEFINITION 60. *The mapping ‡.*
A term $P^\ddagger$ of ITT is defined as follows for some terms P of ITTε.

1. *$P^\ddagger$ is P, when no choice term has an occurrence in P.*
2. *$(PQ)^\ddagger$ is $(P^\ddagger Q^\ddagger)$, when P is a predicate term and Q is not a choice term.*
3. *$(P\varepsilon M)^\ddagger$ is $(\lambda\overline{u}.[\exists v{:}M^\ddagger].P^\ddagger(v,\overline{u})])$, when $P{:}[\tau,\overline{\sigma}]$ is a predicate term, and $\overline{u}{:}\overline{\sigma}$ and $v{:}\tau$ are distinct variables without free occurrences in P or M.*
4. *$(\lambda x.P)^\ddagger$ is $(\lambda x.P^\ddagger)$, when P is a predicate term.*

Note that $P^\ddagger$ is not defined when P is a choice term or a term of type 1 in which a choice term has an occurrence.

LEMMA 45. *‡ maps to ITT.*
Let P be any term of ITTε that is not a choice term. Then $P^\ddagger$ is a term of ITT of the same type as P and with the same free variables.

PROOF. The proof is by induction on the Definition 58 in §5.2 of term of ITTε. For a constant or variable cv, $cv^\ddagger$ is cv. $(PQ)^\ddagger$ is $(P^\ddagger Q^\ddagger)$ if Q is not a choice term. $(P\varepsilon M)^\ddagger$ is $(\lambda\overline{u}.[\exists v{:}M^\ddagger].P^\ddagger(v,\overline{u})])$. $(\lambda x.P)^\ddagger$ is $(\lambda x.P^\ddagger)$.

A term P that is not a choice term but in which a choice term has an occurrence can only have the type 1 as a secondary type and a predicate type as primary. Thus if P has secondary type 1 in ITTε, then $P^\ddagger$ has that type in ITT. ⊣

The following lemma states an important substitution property for the ‡ mapping

LEMMA 46. *Let P and Q be any predicate terms of ITTε. Then*

$$([Q/x]P)^\ddagger \quad is \quad [Q^\ddagger/x]P^\ddagger.$$

PROOF. The proof is by induction on $ct[P]$; that is the number of uses made of application and abstraction in the construction of P. The conclusion is immediate if that number is 0. Assume the conclusion if the number is less than ct, and let $ct[P]$ be ct. There are three possibilities for P; it may be (P_1P_2), $(P_1\varepsilon M)$, or $(\lambda v.P_1)$. Only the second case presents any difficulties. But $([Q/x](P_1\varepsilon M))^\ddagger$ is $([Q/x]P_1\varepsilon[Q/x]M)^\ddagger$ while the latter is

$$[\exists v{:}([Q/x]M)^\ddagger].([Q/x]P_1)^\ddagger(v).$$

By the induction assumption this is

$$[\exists v{:}([Q^\ddagger/x]M^\ddagger].([Q^\ddagger/x]P_1^\ddagger(v)$$

which is $([Q^\ddagger/x](P_1\varepsilon M)^\ddagger$ as required. ⊣

That ITTε is a conservative extension of ITT, and therefore consistent, is a consequence of the next theorem.

THEOREM 47. *A derivable rule of ITT is obtained from each rule of ITTε by replacing each premiss and conclusion $\pm F$ of the rule by $\pm F^\ddagger$.*

PROOF. The only difficult cases are the $-\exists$ rule with Q a choice term and the $\pm\lambda$ rules.

- A derivation for the $-\exists$ case follows:

  $-(\exists P)^\ddagger$
  $-\exists P^\ddagger$
  $-P^\ddagger(\varepsilon M^\ddagger)$
  $-\exists v.[M^\ddagger(v) \wedge P^\ddagger(v)]$
  $-\exists v.[M(v) \wedge P(v)]^\ddagger$
  $-(\exists v.[M(v) \wedge P(v)])^\ddagger$
  $-(P(\varepsilon M))^\ddagger$

- The rules to be derived for the $\pm\lambda$ case are

$$\frac{\pm F^\ddagger}{\pm G^\ddagger}$$

where $F >_\varepsilon G$. It is sufficient to prove that if $F >_\varepsilon G$ holds in ITTε then $F^\ddagger >_\varepsilon G^\ddagger$ holds in ITT, where in ITTε $>_\varepsilon$ is defined using Definition 59 of contractable term in §5.2.1.

Let $F >_\varepsilon G$ hold in ITTε. Then G is obtained from F by replacing a contractable subterm CT of F by its contraction CT'. The three possibilities for CT and CT', (β), (η), and (ε), are described in Definition 59. For the case (η) when CT is $(\lambda v.Pv)$ and CT' is P the conclusion is immediate since $(\lambda v.Pv)^\ddagger$ is $(\lambda v.P^\ddagger v)$. For the case (β) when CT is $(\lambda v.P)Q$ and CT' is $[Q/v]P$ the conclusion follows from Lemma 46 since $((\lambda v.P)Q)^\ddagger$ is $(\lambda v.P^\ddagger)Q^\ddagger$ and $([Q/v]P)^\ddagger$ is $[Q^\ddagger/v]P^\ddagger$. For the case (ε), when CT is $(P\varepsilon M)$ and CT' is $(\lambda\overline{u}.[\exists v{:}M].P(v,\overline{u})])$, the conclusion follows from Definition 60 of the mapping ‡.

The remaining cases are left to the reader. ⊣

Since the translation by ‡ of each rule of ITTε is a rule of ITT, the following corollary is immediate.

COROLLARY 48. *If $\Gamma \vdash \Theta$ is derivable in ITTε, then $\Gamma^\ddagger \vdash \Theta^\ddagger$ is derivable in ITT.*

In particular, therefore, a sequent $\Gamma \vdash \Theta$ of ITT that is derivable in ITTε is derivable in ITT since $\Gamma^\ddagger \vdash \Theta^\ddagger$ is $\Gamma \vdash \Theta$. Thus ITTε is a conservative extension of ITT and consistent.

5.4.1. Completeness of ITTε. Although no semantics in the style provided for ITT has been provided for ITTε, the mapping ‡ provides a semantics for ITTε directly in terms of the semantics for ITT: A sequent $\Gamma \vdash \Theta$ is valid in ITTε if its translation $\Gamma^\ddagger \vdash \Theta^\ddagger$ is valid in ITT, or equivalently is derivable in ITT with or without Cut. Interpreted in this way Corollary 48 to Theorem 47 corresponds directly to Theorem 29, the consistency theorem of ITT. Corresponding to the completeness theorem of ITT, Theorem 33, is then the following theorem:

THEOREM 49. *If $\Gamma^\ddagger \vdash \Theta^\ddagger$ is derivable in ITT, then $\Gamma \vdash \Theta$ is derivable in ITTε.*

PROOF. Let $\Gamma \vdash \Theta$ be an underivable sequent of ITTε. It will be proved that $\Gamma^\ddagger \vdash \Theta^\ddagger$ is underivable in ITT.

Definition 48 in §3.5 for a descending chain for a given underivable sequent of ITT can be immediately extended for ITTε. From Theorem 47 it follows that from a descending chain for $\Gamma \vdash \Theta$, a descending chain for $\Gamma^\ddagger \vdash \Theta^\ddagger$ can be obtained. The sequent is therefore underivable. ⊣

COROLLARY 50. *If a seqent has a derivation in ITTε then it has a Cut-free derivation.*

PROOF. Here a sketch of the proof is offered that the reader is invited to complete. Let $\Gamma \vdash \Theta$ be a derivable sequent of ITTε without a Cut-free derivation. Then a descending chain for $\Gamma \vdash \Theta$ can be found in much the same manner as a descending chain can be found for a sequent of ITT without a Cut-free derivation. Further a translation by ‡ of the descending chain of ITTε converts it into an incomplete descending chain for ITT that can be completed to a descending chain for $\Gamma^\ddagger \vdash \Theta^\ddagger$ in ITT. But that it is impossible since $\Gamma \vdash \Theta$ has a Cut-free derivable in ITT. ⊣

EXERCISES §5.4.

1. Complete the proof of Theorem 47 for the rules not considered.
2. Complete the details of the proof of Corollary 50.

5.5. Function Terms

In §5.1.1 a function term ιM is defined to be

$$\varepsilon(\lambda v.[M(v) \wedge \forall u.[M(u) \rightarrow u = v]]).$$

Using the notation $[!v{:}M]$ to abbreviate $[M(v) \wedge \forall u.[M(u) \rightarrow u = v]]$,

$$\iota M \quad \text{is} \quad \varepsilon(\lambda v.[!v{:}M]).$$

This notation will be used throughout this section.

Since a function term is a choice term it too must have a scope; examples like those presented in §5.2.1 that stress the importance of scope for choice terms can be repeated for function terms.

EXERCISES §5.5.

1. The μ operator is a function with arguments subsets of N, the natural numbers, and value the least member. Provide a definition of μ and a derivation of the sequent:

$$\vdash \forall X.[[\forall x{:}X].N(x) \wedge \exists x.X(x) \rightarrow \iota\mu(X) = \varepsilon\mu(X)].$$

2. Let the predicate Ub, for Upper bound, of type [[1], 1] be defined:

$$Ub \stackrel{\text{df}}{=} (\lambda Z, v.[\forall x{:}Z].x \leq v)$$

where $\leq$ is the predicates defined in Exercise (4) of §4.4. Provide a derivation for the sequent

$$\vdash \forall X.[[\forall x{:}X].N(x) \rightarrow [\forall x{:}X].x \leq \varepsilon Ub(X)].$$

Define a predicate LUb, for Least Upper bound, for which the following sequent is derivable:

$$\vdash \forall X.[[\forall x{:}X].N(x) \wedge \exists x.X(x) \rightarrow \iota LUb(X) \leq \varepsilon Ub(X)].$$

Verify that the sequent is derivable for your definition of LUb.

3. An identity function for type τ terms can be defined in ITTε; it is $\iota Id(u)$. where $Id \stackrel{\text{df}}{=} (\lambda u, v.u{=}v)$ with $u, v{:}\tau$. Thus $Id{:}[\tau, \tau]$ and $\iota Id(u){:}\tau$. The scope of $\iota Id(u)$ in the formula is $(= u)$. Provide a derivation for the sequent:

$$\vdash \forall u.u{=}\iota Id(u).$$

4. For exercise 1 of §4.4 derivations are to be provided for the following sequents:

   Hd.1) $\vdash \forall x, y.Hd(\langle x, y\rangle, x)$
   Tl.1) $\vdash \forall x, y.Tl(\langle x, y\rangle, y)$
   Hd.2) $\vdash \forall z, x, y.[[Hd(z, x) \wedge Hd(z, y)] \rightarrow x{=}y]$
   Tl.2) $\vdash \forall z, x, y.[[Tl(z, x) \wedge Tl(z, y)] \rightarrow x{=}y]$
   Hd.3) $\vdash [\forall z{:}Pr].\exists!x.Hd(z, x)$
   Tl.3) $\vdash [\forall z{:}Pr].\exists!y.Tl(z, y)$

   Using the notation introduced in this section, the latter two can be written

   Hd.3) $\vdash [\forall z{:}Pr].\exists x.[!x{:}Hd(z, x)]$
   Tl.3) $\vdash [\forall z{:}Pr].\exists y.[!y{:}Tl(z, y)]$

   Provide derivations for:

   $$\vdash \forall x, y.\iota Hd(\langle x, y\rangle){=}x$$
   $$\vdash \forall x, y.\iota Tl(\langle x, y\rangle){=}y$$
   $$\vdash [\forall z{:}Pr].z = \langle \iota Hd(z), \iota Tl(z)\rangle$$

5. The inverse of the successor S is a partial function, the "destructor" for the "constructor" S, using the terminology of [113], or the predecessor function `Pred` using the terminology of [103]. Define the function as a function term and derive the sequent:

   $$\vdash \forall x.\exists!y.\iota Pred(S(x)).$$

5.6. Partial Functions and Dependent Types

It is argued in Chapter 1 that types are implicitly present in natural languages through the distinction made between the subject and predicate of an elementary assertion. Adherence to the structure imposed by the types ensures coherence for the terms and formulas of the logic. A similar benefit can accrue by defining domains for partial functions and treating the domains as defined types: A function is defined for arguments in the domain. Such domains are referred to as *dependent types* in computer science; see, for example, §A.11.2 of [124] or [118] for a discussion of dependent types in category theory. For although they play a role similar to the given types of the logic, the determination of whether an expression adheres to dependent typing restrictions is fundamentally different; in general it requires providing a derivation for a sequent.

The predicate $FUNC$ is defined

$$FUNC \stackrel{\text{df}}{=} (\lambda Z.\forall v, v'.[Z(v) \wedge Z(v') \rightarrow v{=}v']).$$

A predicate P for which

$$\vdash FUNC(P)$$

is derivable is said to be *functional*, since then

a) $\vdash [\exists v.P(v) \to \exists! v.P(v)]$

is derivable.

Let $Q{:}[\overline{\tau}, \sigma]$ and let $\overline{u}{:}\overline{\tau}$. Assume that

$$\vdash \forall \overline{u}.FUNC(Q(\overline{u}))$$

is derivable. Then $Q(\overline{u}){:}[\sigma]$ and $\varepsilon Q(\overline{u})$ is a function term that defines a partial function with domain

$$(\lambda \overline{u}.Den(\varepsilon Q(\overline{u}))).$$

To determine if a sequence $\overline{t}$ of terms has the appropriate dependent types for an argument of the function, it is necessary to derive the sequent

$$\vdash (\lambda \overline{u}.Den(\varepsilon Q(\overline{u})))(\overline{t})$$

and sufficient by (a) to derive

$$\vdash (\lambda \overline{u}.\exists v.Q(\overline{x}, v))(\overline{t}).$$

Further simplifications are possible in some cases. For example the functions defined from the sum and product predicates of the members of N can be proved to have as their domain

$$(\lambda u, v.[N(u) \wedge N(v)]).$$

In this case the determination of whether arguments for the functions are of the appropriate dependent types is somewhat simplified.

CHAPTER 6

INTUITIONIST LOGIC

6.1. Intuitionist/Constructive Mathematics

The nominalism that motivates ITT is similar in some respects to the views expressed by Martin-Löf in the introduction to [96] and developed into the logic of [97] also described by Beeson in [17]. It is therefore appropriate to consider an intuitionist formulation of ITT.

A sequent calculus version ELG [EL*Gentzen*] of the classical elementary logic EL is described in §1.7.6. As noted in §1.7.7, Gentzen described in [52] sequent calculus formulations of both classical and intuitionist first order logic, the latter as formalized by Heyting in [77]. The intuitionist sequent calculus results from the classical by restricting each sequent in a derivation to having at most one formula in its succedent. These formulations of first order logic can also be found in [88] and in [144]. Looked at from this point of view intuitionist logic can be seen to be a restricted form of the classical. But conversely, results are proved in [88] that provide formulations of classical first order logic as a sublogic of intuitionist logic.

Analogous results are proved in §6.3.1 for a Genzen sequent calculus formulation ITTG [ITT*Gentzen*] of ITT described in §6.2 and a Genzen sequent calculus formulation of an intuitionist version HITTG [*Heyting*ITTG] of ITTG described in §6.3. That HITTG may be a reasonable foundation for intuitionist and/or constructive mathematics then follows from a result established in §6.3, namely that if $\Gamma \vdash [F \vee G]$ is derivable in HITTG, then so is either $\Gamma \vdash F$ or $\Gamma \vdash G$, and if $\Gamma \vdash \exists R$ is derivable, then so is $\Gamma \vdash R(t)$ for some term t.

A semantic tree version HITT of HITTG is described in §6.4 and proved to be equivalent to HITTG. It has been suggested in part by the semantic tableaux for first order intuitionist logic defined by Beth in [20]. To provide for counter-examples for sequents of HITT it is natural to consider *forests* of semantic trees, not just semantic trees, a forest being a collection of semantic trees with common initial nodes defined by a sequent. Examples are provided in §6.5 that illustrate special features of HITT derivations. In [64], Kripke's model structures, with respect to which first-order intuitionistic logic has been proved complete [90], are extended for HITT and completeness and Cut elimination theorems proved for the logic and for HITTε, the extension of HITT with choice terms.

The question as to how much of the recursion theory developed in Chapter 4 for ITT can also be developed in HITT has a happy, perhaps surprising answer: The reader is asked to confirm in Exercise 1 of §6.4 that each of the displayed sequents of Chapter 4 is a derivable sequent of HITT. In particular Peano's axioms are derivable in HITT. As it is demonstrated in [64] that the main body of intuitionist mathematics can be developed in HITT, it is fair to say that logicism provides a foundation for both classical and intuitionist mathematics. In this regard it would be of interest to know the relationship between HITT and Martin-Löf's logic described in [96]; a preliminary study of that relationship appears in [64].

6.2. A sequent calculus formulation ITTG of ITT

In §1.7.6 a Gentzen sequent calculus formulation ELG of the elementary logic EL is described. Here that formulation is adapted and extended to provide a sequent calculus formulation ITTG of ITT. To smooth the transition to the intuitionist version HITTG of ITTG the logical connectives $\wedge$ and $\neg$ are taken to be primitive in ITTG instead of joint denial $\downarrow$. The definition of the *nonidentity* predicate $\neq$ in ITTG is:

$$\neq \overset{\text{df}}{=} \lambda x, y.\exists Z.[Z(x) \wedge \neg Z(y)].$$

This definition could be replaced by two additional rules of deduction $\pm \neq$. A type superscript may be added to $\neq$ to indicate the type of the arguments of the defined predicate.

As with ITT, Γ and Θ are finite *sets* of formulas and not the finite sequences of Gentzen's sequent calculus, so that the two *Thin* rules stated below are the only structural rules needed. The rules of deduction of ITTG are expressed in the same notation used in §1.7.6 in the formulation of ELG: The expressions 'Γ, F' and 'Θ, F' are understood to be $\Gamma \cup \{F\}$ and $\Theta \cup \{F\}$.

The axioms are all sequents

$$F \vdash F.$$

The rules are:

$\vdash \wedge$ $$\frac{\Gamma \vdash \Theta, F \quad \Gamma \vdash \Theta, G}{\Gamma \vdash \Theta, [F \wedge G]}$$

$\wedge \vdash$ $$\frac{\Gamma, F \vdash \Theta}{\Gamma, [F \wedge G] \vdash \Theta} \qquad \frac{\Gamma, G \vdash \Theta}{\Gamma, [F \wedge G] \vdash \Theta}$$

$\vdash \neg$ $$\frac{\Gamma, F \vdash \Theta}{\Gamma \vdash \Theta, \neg F}$$

$\neg \vdash$ $$\frac{\Gamma \vdash \Theta, F}{\Gamma, \neg F \vdash \Theta}$$

$\vdash\exists$ $\dfrac{\Gamma \vdash \Theta, P(t)}{\Gamma \vdash \Theta, \exists P}$ *eigen* term t

$\exists\vdash$ $\dfrac{\Gamma, P(y) \vdash \Theta}{\Gamma, \exists P \vdash \Theta}$ *eigen* variable y not free below premiss

$\vdash\lambda$ $\dfrac{\Gamma \vdash \Theta, G}{\Gamma \vdash \Theta, F}$ $F > G$

$\lambda\vdash$ $\dfrac{\Gamma, G \vdash \Theta}{\Gamma, F \vdash \Theta}$ $F > G$

$\vdash Int_c$ $\dfrac{\Gamma \vdash \Theta, P \neq^{\tau} Q}{\Gamma \vdash \Theta, P \neq^{1} Q}$

$Int_c\vdash$ $\dfrac{\Gamma, P \neq^{\tau} Q \vdash \Theta}{\Gamma, P \neq^{1} Q \vdash \Theta}$

P and Q have primary type τ and secondary type 1

$\vdash Thin$ $\dfrac{\Gamma \vdash \Theta}{\Gamma \vdash \Theta, F}$

$Thin\vdash$ $\dfrac{\Gamma \vdash \Theta}{\Gamma, F \vdash \Theta}$

Cut $\dfrac{\Gamma \vdash \Theta, F \quad \Gamma, F \vdash \Theta}{\Gamma \vdash \Theta}$

It is left to the reader to prove that a sequent is derivable in ITTG if and only if it is derivable in ITT. The proof of Theorem 8 of §1.7.6 provides a good beginning.

A sequent calculus formulation ITTGε of the extension ITTε of ITT described in Chapter 5 can be provided by adding the rules described in §5.3 to those of ITTG.

Although the connectives $\vee$ and $\rightarrow$ and the quantifier $\forall$ are not primitive in ITTG, classical versions of them can be defined:

$$\vee_c \stackrel{\text{df}}{=} \lambda X, Y.\neg[\neg X \wedge \neg Y] \qquad X, Y{:}[\,]$$

$$\rightarrow_c \stackrel{\text{df}}{=} \lambda X, Y.\neg[X \wedge \neg Y] \qquad X, Y{:}[\,]$$

$$\forall_c \stackrel{\text{df}}{=} \lambda X.\neg\exists(\lambda u.\neg X(u)) \qquad X{:}[\tau]$$

As with the primitive binary connective $\wedge$, the infix notation will be used with the two defined connectives.

EXERCISES §6.2.

1. Prove that a sequent is derivable in ITTG if and only if it is derivable in ITT.

6.3. An intuitionist formulation HITTG of ITTG

For HITTG the full roster of common logical connectives and quantifiers are primitive: $\neg$, $\wedge$, $\vee$, $\rightarrow$, $\forall$, and $\exists$. Thus HITTG shares with ITTG the connectives $\neg$ and $\wedge$ and the quantifier $\exists$. Therefore a formula of ITTG is a formula of HITTG and the definitions of $=_c$, $\vee_c$, $\rightarrow_c$ and the quantifier $\forall_c$ defined in §6.2 in terms of $\neg$, $\wedge$ and $\exists$ can be used in HITTG.

The identity predicate $=$ of HITTG is defined as it was in §2.5, but now the connective $\rightarrow$ is the intuitionist implication and the quantifier $\forall$ the intuitionist universal quantifier:

$$= \stackrel{\mathrm{df}}{=} \lambda x, y.\forall Z.[Z(x) \rightarrow Z(y)].$$

The axioms of ITTG, extended to include all formulas of HITTG, are the axioms of HITTG. The most important difference between HITTG and ITTG is in the definition of their sequents: A sequent $\Gamma \vdash \Theta$ of formulas of HITTG is required to satisfy the additional requirement:

H) The succedent Θ is either empty or a single formula.

As a consequence the rules of deduction of HITTG, which include all the rules of ITTG other than the Int_c rules, are affected. Two rules of ITTG in particular are of direct concern for they may have a premiss satisfying (H) but have a conclusion that does not; they are $\vdash \neg$ and $\vdash$ *Thin* in which Θ must be empty. If these conditions are imposed on applications of these two rules then it is an easy matter to establish that every derivable sequent of HITTG satisfies the condition (H).

The following are the rules for the additional connectives and quantifier of HITTG.

$$\vdash\vee \quad \frac{\Gamma \vdash F}{\Gamma \vdash [F \vee G]} \quad \frac{\Gamma \vdash G}{\Gamma \vdash [F \vee G]} \qquad \vee\vdash \quad \frac{\Gamma, F \vdash \Theta \quad \Gamma, G \vdash \Theta}{\Gamma, [F \vee G] \vdash \Theta}$$

$$\vdash\rightarrow \quad \frac{\Gamma, F \vdash G}{\Gamma \vdash [F \rightarrow G]} \qquad \rightarrow\vdash \quad \frac{\Gamma \vdash F \quad \Gamma, G \vdash \Theta}{\Gamma, [F \rightarrow G] \vdash \Theta}$$

$$\vdash\forall \quad \frac{\Gamma \vdash P(y)}{\Gamma \vdash \forall P} \qquad \forall\vdash \quad \frac{\Gamma, P(t) \vdash \Theta}{\Gamma, \forall P \vdash \Theta}$$

eigen variable y not free below premiss; eigen term t

The Int_c rules of ITTG are replaced by the following rules

$$\vdash Int \quad \frac{\Gamma \vdash P =^\tau Q}{\Gamma \vdash P =^1 Q} \qquad Int \vdash \quad \frac{\Gamma, P =^\tau Q \vdash \Theta}{\Gamma, P =^1 Q \vdash \Theta}$$

P and Q have primary type τ and secondary type 1

The following theorem is characteristic for an intuitionist/constructive logic.

THEOREM 51. *Witnesses for disjunctions and existential formulas.*
Let $\Gamma \vdash [F \vee G]$, *respectively* $\Gamma \vdash \exists R$, *be derivable in HITTG. Then in the first case so is either* $\Gamma \vdash F$ *or* $\Gamma \vdash G$ *derivable, and in the second case so is* $\Gamma \vdash R(t)$ *for some term* t.

PROOF. It may be assumed that the formula F in an axiom is elementary. The proof is by induction on the length of the derivation of the sequent. No formula of an axiom is either a disjunction or an existential formula. Assume the theorem holds for derivations of length n and consider a sequent $\Gamma \vdash [F \vee G]$ with a derivation of length $n + 1$. The sequent is necessarily the conclusion of an antecedent rule, of $\vdash$ *Thin*, of $\vdash \vee$ rule, or of *Cut*. In each case the theorem holds.

Consider now a sequent $\Gamma \vdash \exists R$ with a derivation of length $n + 1$. The sequent is necessarily the conclusion of an antecedent rule, of $\vdash$ *Thin*, or of $\vdash \exists$. In each of these cases the theorem follows immediately. $\dashv$

6.3.1. Derivability in ITTG and HITTG. The next definition provides a means for representing a non-empty succedent Θ of a sequent $\Gamma \vdash \Theta$ of ITTG by a formula Θ^H of ITTG and therefore of HITTG.

DEFINITION 61. *A succedent of a sequent of* ITTG *as a formula of* ITTG.

1. *If* Θ *is* $\{G\}$ *then* Θ^H *is* G.
2. *If* Θ *is* $\{G_1, \ldots, G_n, G_{n+1}\}$ *then* Θ^H *is* $[\{G_1, \ldots, G_n\}^H \vee_c G_{n+1}\}]$, $1 \leq n$.

Using this definition, if Θ is non-empty then a succeedent Θ, G is represented by $[\Theta^H \vee_c G]$.

LEMMA 52. *The following are derivable rules of* HITTG *when* F *and* G *are formulas and* Θ *a non-empty set of formulas of* ITTG, *and* $P{:}[\tau]$ *and* $t{:}\tau$ *are terms of* ITTG. *In addition the rules remain derivable in HITTG when the succedent of each premiss and conclusion is replaced by its double negation.*

$\vdash \neg^H$
$$\frac{\Gamma, F \vdash \Theta^H}{\Gamma \vdash [\Theta^H \vee_c \neg F]}$$

$\neg^H \vdash$
$$\frac{\Gamma \vdash [\Theta^H \vee_c F]}{\Gamma, \neg F \vdash \neg\neg\Theta^H}$$

$\vdash \wedge^H$
$$\frac{\Gamma \vdash [\Theta^H \vee_c F] \quad \Gamma \vdash [\Theta^H \vee_c G]}{\Gamma \vdash [\Theta^H \vee_c [F \wedge G]]}$$

$\vdash \forall^H$
$$\frac{\Gamma \vdash [\Theta^H \vee_c P(y)]}{\Gamma \vdash [\Theta^H \vee_c \forall P]}$$
y not free below premiss

$\vdash Thin^H$ $\dfrac{\Gamma \vdash \Theta^H}{\Gamma \vdash [\Theta^H \vee_c F]}$

$\vdash \lambda^H$ $\dfrac{\Gamma \vdash [\Theta^H \vee_c G]}{\Gamma \vdash [\Theta^H \vee_c F]}$
where $F > G$

$\vdash Int^H$ $\dfrac{\Gamma \vdash [\Theta^H \vee_c P \neq^\tau Q]}{\Gamma \vdash [\Theta^H \vee_c P \neq^1 Q]}$ $Int^H \vdash$ $\dfrac{\Gamma, P \neq^\tau Q \vdash \theta^H}{\Gamma, P \neq^1 Q \vdash \Theta^H}$

PROOF. A derivation of the $\vdash \neg^H$ rule follows.

$\Gamma, F \vdash \Theta^H$
$\Gamma, \neg\Theta^H, F \vdash$
$\Gamma, \neg\Theta^H \vdash \neg F$
$\Gamma, \neg\Theta^H, \neg\neg F \vdash$
$\Gamma, [\neg\Theta^H \wedge \neg\neg F] \vdash$
$\Gamma \vdash \neg[\neg\Theta^H \wedge \neg\neg F]$
$\Gamma \vdash [\Theta^H \vee_c \neg F]$

A derivation of the $\neg^H\vdash$ rule can be obtained from one application of Cut to the premiss of the derived rule and the following derivable sequent.

$$[\Theta^H \vee_c F], \neg F \vdash \neg\neg\Theta^H.$$

A derivation of this sequent follows.

$\neg\Theta^H \vdash \neg\Theta^H$ $\neg F \vdash \neg F$

$\neg\Theta^H, \neg F \vdash [\neg\Theta^H \wedge \neg F]$
$\neg[\neg\Theta^H \wedge \neg F], \neg\Theta^H, \neg F \vdash$
$[\Theta^H \vee_c F], \neg\Theta^H, \neg F \vdash$
$[\Theta^H \vee_c F], \neg F \vdash \neg\neg\Theta^H$

A derivation of the $\vdash \wedge^H$ rule can be obtained from applications of Cut to the two premisses and the derivable sequent

$$[\Theta^H \vee_c F], [\Theta^H \vee_c G] \vdash [\Theta^H \vee_c [F \wedge G]].$$

A derivation of this sequent follows:

$F \vdash F$ $G \vdash G$

$F, G \vdash [F \wedge G]$
$F, G, \neg[F \wedge G] \vdash$
$F, \neg[F \wedge G] \vdash \neg G$ $\neg\Theta^H \vdash \neg\Theta^H$

$F, \neg\Theta^H, \neg[F \wedge G] \vdash [\neg\Theta^H \wedge \neg G]$
$F, \neg\Theta^H, \neg[F \wedge G], \neg[\neg\Theta^H \wedge \neg G] \vdash$
$F, \neg\Theta^H, \neg[F \wedge G], [\Theta^H \vee_c G] \vdash$
$\neg\Theta^H, \neg[F \wedge G], [\Theta^H \vee_c G] \vdash \neg F \qquad \neg\Theta^H \vdash \neg\Theta^H$

$\neg\Theta^H, \neg[F \wedge G], [\Theta^H \vee_c G] \vdash [\neg\Theta^H \wedge \neg F]$
$\neg\Theta^H, \neg[F \wedge G], \neg[\neg\Theta^H \wedge \neg F], [\Theta^H \vee_c G] \vdash$
$\neg\Theta^H, \neg[F \wedge G], [\Theta^H \vee_c F], [\Theta^H \vee_c G] \vdash$
$[\neg\Theta^H \wedge \neg[F \wedge G]], [\Theta^H \vee_c F], [\Theta^H \vee_c G] \vdash$
$[\Theta^H \vee_c F], [\Theta^H \vee_c G] \vdash \neg[\neg\Theta^H \wedge \neg[F \wedge G]]$
$[\Theta^H \vee_c F], [\Theta^H \vee_c G] \vdash [\Theta^H \vee_c [F \wedge G]]$

A derivation of the $\vdash \exists^H$ rule can be obtained from an application of Cut to the derivable sequent

$$[\Theta^H \vee_c P(t)] \vdash [\Theta^H \vee_c \exists P].$$

A derivation of the sequent follows:

$P(t) \vdash P(t)$
$P(t) \vdash \exists P$
$P(t), \neg\exists P \vdash$
$\neg\exists P \vdash \neg P(t) \qquad \neg\Theta^H \vdash \neg\Theta^H$

$\neg\Theta^H, \neg\exists P \vdash [\neg\Theta^H \wedge \neg P(t)]$
$\neg[\neg\Theta^H \wedge \neg P(t)], \neg\Theta^H, \neg\exists P \vdash$
$[\Theta^H \vee_c P(t)], \neg\Theta^H, \neg\exists P \vdash$
$[\Theta^H \vee_c P(t)], [\neg\Theta^H \wedge \neg\exists P] \vdash$
$[\Theta^H \vee_c P(t)] \vdash \neg[\neg\Theta^H \wedge \neg\exists P]$
$[\Theta^H \vee_c P(t)] \vdash [\Theta^H \vee_c \exists P]$

A derivation of the $\vdash Thin^H$ rule follows:

$\Gamma \vdash \Theta^H$
$\Gamma, \neg F \vdash \Theta^H$
$\Gamma, \neg\Theta^H, \neg F \vdash$
$\Gamma, [\neg\Theta^H \wedge \neg F] \vdash$
$\Gamma \vdash \neg[\neg\Theta^H \wedge \neg F]$
$\Gamma \vdash [\Theta^H \vee_c F]$

Each application of the $\vdash \lambda^H$ rule is an application of the $\vdash \lambda$ rule of HITTG since if $F > G$ then $[\Theta^H \vee_c F] > [\Theta^H \vee_c G]$.

Derivations of the Int^H rules follow respectively from one application of Cut to the sequents

a) $[\theta^H \vee_c P{\neq}^\tau Q] \vdash [\theta^H \vee_c P{\neq}^1 Q]$

b) $P{\neq}^1 Q \vdash P{\neq}^\tau Q.$

Derivations for these can be obtained from the sequents

c) $P{\neq}Q \vdash \neg P{=}Q$

d) $\neg P{=}Q \vdash P{\neq}Q.$

A derivation of (c) follows:

$$\frac{Z(P) \vdash Z(P) \qquad Z(Q) \vdash Z(Q)}{Z(P), [Z(P) \rightarrow Z(Q)] \vdash Z(Q)}$$

$Z(P), \forall Z.[Z(P) \rightarrow Z(Q)] \vdash Z(Q)$
$Z(P), P{=}Q \vdash Z(Q)$
$Z(P), \neg Z(Q), P{=}Q \vdash$
$[Z(P) \wedge \neg Z(Q)], P{=}Q \vdash$
$\exists Z.[Z(P) \wedge \neg Z(Q)], P{=}Q \vdash$
$P{\neq}Q, P{=}Q \vdash$
$P{\neq}Q \vdash \neg P{=}Q$

A derivation of (d) follows, where $t \stackrel{\mathrm{df}}{=} \lambda x.P{=}x$:

$P{=}Q \vdash P{=}Q$
$\neg P{=}Q, P{=}Q \vdash$
$\neg P{=}Q, t(Q) \vdash$

$$\frac{\neg P{=}Q \vdash \neg t(Q) \qquad \vdash t(P)}{\neg P{=}Q \vdash [t(Q) \wedge \neg t(Q)]}$$

$\neg P{=}Q \vdash \exists Z.[Z(Q) \wedge \neg Z(Q)]$
$\neg P{=}Q \vdash P{\neq}Q$

A derivation of (a) follows:

$P{=}^{\tau}Q \vdash P{=}^{\tau}Q$
$P{=}^{1}Q \vdash P{=}^{\tau}Q$
$P{=}^{1}Q, \neg P{=}^{\tau}Q \vdash$
$\neg P{=}^{\tau}Q \vdash \neg P{=}^{1}Q$
$P{\neq}^{\tau}Q \vdash P{\neq}^{1}Q$ From (c) and (d)
$P{\neq}^{\tau}Q, \neg P{\neq}^{1}Q \vdash$

$$\frac{\neg P{\neq}^{1}Q \vdash \neg P{\neq}^{\tau}Q \qquad \neg\Theta^{H} \vdash \neg\Theta^{H}}{\neg\Theta^{H}, \neg P{\neq}^{1}Q \vdash [\neg\Theta^{H} \wedge \neg P{\neq}^{\tau}Q]}$$

$[\neg\Theta^{H} \wedge \neg P{\neq}^{1}Q] \vdash [\neg\Theta^{H} \wedge \neg P{\neq}^{\tau}Q]$
$\neg[\neg\Theta^{H} \wedge \neg P{\neq}^{\tau}Q], [\neg\Theta^{H} \wedge \neg P{\neq}^{1}Q] \vdash$
$\neg[\neg\Theta^{H} \wedge \neg P{\neq}^{\tau}Q] \vdash \neg[\neg\Theta^{H} \wedge \neg P{\neq}^{1}Q]$
$[\Theta^{H} \vee_{c} P{\neq}^{\tau}Q] \vdash [\Theta^{H} \vee_{c} P{\neq}^{1}Q]$

A similar but simpler derivation of (b) is left to the reader.

Consider now the double negation modification to the succedents of the premisses and conclusions of these derived rules.

When the succedent Θ^H of the premiss of the succedent $\neg^H$ or $Thin^H$ rule is replaced by $\neg\neg\Theta^H$ the rule remains derivable since

$$[\neg\neg\Theta^H \vee_c \neg F] \vdash \neg\neg[\Theta^H \vee_c \neg F]$$

is derivable in HITTG. The other rules remain derivable when changed because when Θ is not empty and F is any formula both the sequents

$$[\Theta^H \vee_c F] \vdash \neg\neg[\Theta^H \vee_c F] \text{ and } \neg\neg[\Theta^H \vee_c F] \vdash [\Theta^H \vee_c F]$$

are derivable. ⊣

THEOREM 53. *Derivability Relationships between* ITTG *and* HITTG.

1. *If* $\Gamma \vdash \Theta$ *is derivable in* ITTG *then* $\Gamma \vdash \neg\neg\Theta^H$ *is derivable in* HITTG.
2. *If* $\Gamma \vdash \Theta$ *is derivable in* HITTG *then* $\Gamma^c \vdash \Theta^c$ *is derivable in* ITTG, *where* $\Gamma^c \vdash \Theta^c$ *is obtained from* $\Gamma \vdash \Theta$ *by replacing each occurrence of* $\vee$, $\rightarrow$, *and* $\forall$ *in a formula of the sequent by* $\vee_c$, $\rightarrow_c$, *and* $\forall_c$.

PROOF. 1. Note that if no application of the $\neg\vdash$ rule is used in a derivation of a sequent $\Gamma \vdash \Theta$ in ITTG, then the sequent $\Gamma \vdash \Theta^H$ is derivable in HITTG. However, because of the conclusion of the $\neg^H\vdash$ rule, if an application of the $\neg\vdash$ rule is used then the sequent $\Gamma \vdash \neg\neg\Theta^H$ is derivable in HITTG. This is possible because of the double negation modification allowed to the succedents of the premisses and conclusions of the derived rules.

2. Each axiom and each rule of deduction of HITTG is transformed to an axiom, rule, or derived rule of ITTG when a sequent $\Gamma \vdash \Theta$ of HITTG is replaced by $\Gamma^c \vdash \Theta^c$. ⊣

6.4. Semantic Tree Formulation HITT of HITTG

A sequent is derivable in the semantic tree formulation of ITT if and only if it is derivable in ITTG. This permits a derivation in ITTG to be presented in a condensed more easily comprehended form. The derivations supporting the conclusions of Chapter 4 on recursions are all of ITT. To simplify a determination of whether these conclusions can also be supported in HITTG, a semantic tree formulation HITT of HITTG is described next in which the rules for $=$ are replaced with a direct definition.

$+\neg \quad \dfrac{+\neg F}{-F} \qquad\qquad -\neg \quad \dfrac{-\neg F}{+F}$

$+\wedge \quad \dfrac{+[F \wedge G]}{+F} \quad \dfrac{+[F \wedge G]}{+G} \qquad\qquad -\wedge \quad \dfrac{-[F \wedge G]}{-F \quad -G}$

$+\vee \quad \dfrac{+[F \vee G]}{+F \quad +G} \qquad\qquad -\vee \quad \dfrac{-[F \vee G]}{-F} \quad \dfrac{-[F \vee G]}{-G}$

$+\rightarrow \quad \dfrac{+[F \rightarrow G]}{-F \quad +G}$ $\qquad -\rightarrow \quad \dfrac{-[F \rightarrow G]}{+F} \quad \dfrac{-[F \rightarrow G]}{-G}$

$+\forall \quad \dfrac{+\forall P}{+P(t)}$ $\qquad -\forall \quad \dfrac{-\forall P}{-P(y)}$
y new to branch

$+\exists \quad \dfrac{+\exists P}{+P(y)}$
y new to branch
$\qquad -\exists \quad \dfrac{-\exists P}{-P(t)}$

$+\lambda \quad \dfrac{+F}{+G}$
$F > G$
$\qquad -\lambda \quad \dfrac{-F}{-G}$
$F > G$

$+Int \quad \dfrac{+P =^1 Q}{+P =^\tau Q}$ $\qquad -Int \quad \dfrac{-P =^1 Q}{-P =^\tau Q}$
P and Q have primary type τ and secondary type 1.

$Cut \quad \dfrac{}{+F \quad -F}$

The semantic rules for ITT in §3.4 determine the semantic trees that can be constructed for a sequent $\Gamma \vdash \Theta$ of ITT; the trees are generalizations of the trees described in §1.7.2 for EL. As a consequence of the restriction (H) on the membership of the succedent Θ of a sequent of HITT, a semantic tree of HITT based on a given sequent is a restricted form of a semantic tree based on the same sequent in ITT:

DEFINITION 62. *An HITT Semantic Tree Based on a Sequent.*
An HITT semantic tree based on a sequent $\Gamma \vdash \Theta$, where Γ is $\{F_1, \ldots, F_m\}$ and Θ is empty, or respectively $\{G\}$, is defined as follows:

1. *The tree consisting of a single branch with nodes $+F_1, \ldots, +F_m$, respectively $+F_1, \ldots, +F_m, -G$ is a semantic tree based on the sequent. The node $-G$, if it exists, is* available *on the branch.*
2. *Let T be a semantic tree based on a sequent $\Gamma \vdash \Theta$. Then any* immediate extension *T' of T is a semantic tree based on the sequent if it is obtained from T in one of the following ways from an open branch of T; that is a branch that is not* closed *by a pair of nodes $+F$ and $-F$, where $-F$ is* available.
   (a) *T' is obtained from T by extending the branch by a node that is the*

single conclusion of a rule with a possible *premiss on the branch; that is a node of the branch that is either a + node or an* available *– node. If the conclusion is a – node, then any previously available – node on the branch is* unavailable *on the extended branch and the conclusion is* available.

(b) *$\mathcal{T}'$ is obtained from $\mathcal{T}$ by extending the branch to two branches using the two conclusions of a rule with a possible premiss. If a conclusion is a – node, then any previously available – node on the branch defined by that conclusion is* unavailable *on the extended branch and the conclusion is* available.

(c) *$\mathcal{T}'$ is obtained from $\mathcal{T}$ by extending the branch to two branches using the two conclusions $+F$ and $-F$ of an application of Cut. Any previously available – node on the branch defined by $-F$ is* unavailable *on the extended branch and $-F$ is* available.

This definition ensures that at most one available – node appears on a branch of a semantic tree based on a sequent of HITT, reflecting the restriction (H) on the succedents of HITTG.

A tree $\mathcal{T}''$ is an *extension* of $\mathcal{T}$ if $\mathcal{T}''$ is $\mathcal{T}$ or is the immediate extension of a tree $\mathcal{T}'$ that is an extension of $\mathcal{T}$. A tree $\mathcal{T}$ is a *predecessor* of $\mathcal{T}''$ if the latter is an extension of the former. A tree $\mathcal{T}''$ is said to be *Cut-free* if no predecessor $\mathcal{T}'$ is an immediate extension as defined in clause (2c). A tree based on a sequent is a *derivation* of the sequent if each branch of the tree is closed. An example derivation that is Cut-free follows.

EXAMPLE DERIVATION. $\vdash \neg\neg[F \vee \neg F]$

1	$-\neg\neg[F \vee \neg F]$	initial tree. (1) is available
2	$+\neg[F \vee \neg F]$	$-\neg$ from (1).
3	$-[F \vee \neg F]$	$+\neg$ from (2). (1) is unavailable and (3) is available
4	$-\neg F$	$-\vee$ from (3). (3) is unavailable and (4) is available
5	$+F$	$-\neg$ from (4).
6	$-[F \vee \neg F]$	$+\neg$ from (2). (4) is unavailable and (6) is available
7	$-F$	$-\vee$ from (6). (6) is unavailable and (7) is available
8	——	Branch closes because of (5) and availability of (7)

Note that the duplication of the node (3) at (6) cannot be removed as it could be for a derivation for ITT: (4) is the conclusion of $-\vee$ with premiss (3); (7) is the conclusion of $-\vee$ with premiss (6). (7) cannot be the conclusion of $-\vee$ with premiss (3) since the conclusion (4) has made (3) unavailable. This illustrates a fundamental difference in the uses that can be made of the $-\vee$ rule in ITT and HITT: The two possible conclusions of an application of the rule with the same premiss cannot appear on the same branch since the first such application makes the premiss unavailable. Similarly it is not possible to have two or more conclusions of the $-\exists$ rule from the same premiss on the same branch since the first conclusion makes the premiss unavailable.

OTHER EXAMPLES. In §4.2 derivations of ITT are given for the sequents (Lt.1) and (Gt.1). As can be easily confirmed, these derivations are also derivations of HITT. Similarly the derivations provided in §4.3 for the sequents (Lt.2) and (Gt.2) are also derivations of HITT. Indeed all of the displayed derivations of Chapter 4 are derivations of HITT as the reader is asked to confirm.

6.4.1. Properties of HITT derivations. Lemma 7 in §1.7.3 states properties of EL tree derivations and Lemma 30 in §3.4.4 properties of ITT. Here the lemma is updated for HITT. As with the earlier logics the properties can narrow the choices that must be considered while attempting to construct a derivation.

LEMMA 54. *Properties of HITT Derivations.*
If a sequent $\Gamma \vdash \Theta$ has a derivation, then it has a derivation with the following properties:

1. *Each branch of the derivation has exactly one closing pair.*
2. *Each variable with a free occurrence in the eigen term t of an application of $+\forall$ or of $-\exists$ has a free occurrence in a node above the conclusion of the rule. With the possible exception of a single constant for each type, each constant occurring in the eigen term t of an application of $+\forall$ or of $-\exists$ has an occurrence in a node above the conclusion of the rule.*
3. *Let the sequent $\Gamma \vdash \Theta$ have a derivation. Then there is a derivation for a sequent $\Gamma' \vdash \Theta'$, where $\Gamma' \subseteq \Gamma$ and $\Theta' \subseteq \Theta$, for which each node of the derivation has a descendent that is a closing node.*
4. *If $\pm G$ is the conclusion of an application of one of the λ-rules from the premiss $\pm F$, then F may be assumed to be either a formula $(\lambda v.P)(t)$ or a formula $cv(\overline{s})$ where cv is either a constant or variable and $\overline{s}$ is a non-empty sequence of terms.*
5. *A $-$ node can be used at most once on a branch as the premiss of a rule with a $-$ conclusion on the branch; however, a $+$ node with a $-$ conclusion on the branch can be used any number of times as a premiss.*
6. *On a given branch there need be at most one application of a rule with a given $+$ conclusion.*
7. *If both $+F$ and $-G$ are conclusions of the same node $-[F \rightarrow G]$, then the node $-G$ may be assumed to occur immediately after the node $+F$.*

PROOF. Proofs of the first three properties are left to the reader. (4) is a consequence of the fact that any application of a λ-rule can be postponed until after the application of any other logical rule. (5) is an immediate consequence of the availability restrictions.

• Consider property (6). Let sF be the premiss for an application of a rule with conclusion $+G$. The result is easily established for every rule with conclusion(s) a $+$ node. The $+\rightarrow$ rule presents complications because one of its conclusions is $-$. Let $+[F \rightarrow G]$ be the premiss for two applications of

$+\rightarrow$, with the second conclusions extending the branch defined by the nodes $+[F \rightarrow G]$ and $+G$ as illustrated here:

$$
\begin{array}{llll}
\ldots & & & \\
+[F \rightarrow G] & & & \\
\ldots & & & \\
\hline
-F & +G & & \\
\ldots & \ldots & & \\
\cline{2-3}
 & -F & +G & \\
 & \ldots & \ldots (a) &
\end{array}
\quad \Rightarrow \quad
\begin{array}{ll}
\ldots & \\
+[F \rightarrow G] & \\
\ldots & \\
\hline
-F & +G \\
\ldots & \ldots \\
 & \ldots (a)
\end{array}
$$

Since any conclusion with premiss the second $+G$ could equally well have the first $+G$ as premiss, the second application of the $+\rightarrow$ rule can be dropped.

• Consider property (7). Let both $+F$ and $-G$ be conclusions of the same node $-[F \rightarrow G]$. That the node $-G$ must be below the node $+F$ is immediate; otherwise $-[F \rightarrow G]$ would be unavailable for the conclusion of $+F$. Assume that $-G$ does not immediately follow $+F$. This means that one or more $+$ nodes appear between $+F$ and $-G$ on a branch connecting them. There cannot be any $-$ nodes between them since such a node would make the node $-[F \rightarrow G]$ unavailable. Further each of the $+$ nodes between $+F$ and $-G$ is necessarily either a $+$ conclusion of Cut, or a $+$ conclusion of an application of a rule with a $+$ node as premiss. It is sufficient to prove therefore that any such application can be postponed until *after* the second application of $-\rightarrow$.

It is not difficult to prove that an application of Cut with $+$ conclusion immediately preceding $-G$ can be postponed. Consider therefore the rules with premiss and a conclusion a $+$ node. One such rule is, for example, the $+\rightarrow$ rule with premiss $+[H1 \rightarrow H2]$ and conclusions $-H1$ and $+H2$. The reversal of the applications of the two rules is illustrated here:

$$
\begin{array}{ll}
\ldots & \\
-[F \rightarrow G] & \\
\ldots & \\
+F & \\
\ldots & \\
\hline
-H1 & +H2 \\
 & -G
\end{array}
\quad \Rightarrow \quad
\begin{array}{ll}
\ldots & \\
-[F \rightarrow G] & \\
\ldots & \\
+F & \\
\ldots & \\
-G & \\
\hline
-H1 & +H2
\end{array}
$$

The fact that $-G$ is unavailable in the branch determined by $-H1$ is of no importance since it is only available in the branch determined by $+H2$ in the original derivation. The application of any other possible rule can be postponed in a similar fashion. ⊣

6.4.2. Equivalence of HITT and HITTG. The following theorem confirms that the restriction (H) on the sequents of HITTG has been properly treated in the definition of derivation of HITT.

THEOREM 55. *Equivalence of HITT and HITTG.*
A sequent is derivable in HITT if and only if it is derivable in HITTG.

PROOF. To prove that a sequent derivable in HITTG is also derivable in HITT is the easier of the two parts of the theorem and is proved first.

• Let $\Gamma \vdash \Theta$ be a derivable sequent of HITTG. The proof that it is also a derivable sequent of HITT is by induction on the number of applications of rules of deduction in the derivation. If the number is 0 then the sequent is $F \vdash F$ for some formula F and is immediately derivable in HITT. Assume that a sequent derivable in HITTG using n or less applications of rules is also derivable in HITT. Consider a sequent $\Gamma \vdash \Theta$ with a derivation using $n+1$ applications and consider one of the possibilities for the last rule applied.

$\vdash \wedge$:

$$\frac{\Gamma \vdash F \quad \Gamma \vdash G}{\Gamma \vdash [F \wedge G]}$$

$-\wedge$:

$$\frac{\begin{array}{c} +\Gamma \\ -[F \wedge G] \end{array}}{-F \quad -G}$$

In the tree of signed formulas headed $-\wedge$ on the right $+\Gamma$ is to be understood to be a $+$ node for every formula in Γ. By the induction assumption there exist derivations in HITT for the two premisses $\Gamma \vdash F$ and $\Gamma \vdash G$ of the $\vdash \wedge$ rule. Therefore these derivations, without their initial nodes $+\Gamma$, $-F$ and $-G$, can be added below $-F$ and $-G$ providing a derivation for $\Gamma \vdash [F \wedge G]$ in HITT as required.

Each of the other possibilities for the last rule applied in the derivation of HITTG can be dispensed with in a similar fashion. Consider for example the case that the Cut rule is the last rule applied in a derivation of $\Gamma \vdash \Theta$.

$$\frac{\Gamma \vdash F \quad \Gamma, F \vdash \Theta}{\Gamma \vdash \Theta}$$

$$\frac{\begin{array}{c} +\Gamma \\ -\Theta \end{array}}{+F \quad -F}$$

Here $-\Theta$ is either empty or a single $-$ node. By the induction assumption there are derivations of the sequents $\Gamma \vdash F$ and $\Gamma, F \vdash \Theta$ in HITT. Since the conclusion $-F$ makes that node $-\Theta$ unavailable in the right branch, if it exists, the tree on the right can be extended to a derivation in HITT. The cases corresponding to the other rules of HITTG are left to the reader.

• Let now $\Gamma \vdash \Theta$ be a derivable sequent of HITT. The initial tree of the derivation is a single branch

$$\begin{array}{c} +\Gamma \\ -\Theta \end{array}$$

with $+$ nodes from formulas of Γ and possibly a $-$ node from Θ. Should the initial tree be closed then $\Gamma \vdash \Theta$ can be derived in HITTG by applications of *Thin* $\vdash$.

Assume now that a derivation of $\Gamma \vdash \Theta$ in HITT exists with $n+1$ applications of the rules. Consider the first application of a rule; that is an application with conclusion(s) immediately below the initial nodes. It may be assumed that a conclusion of this rule is an ancestor for a member of a closing pair on some branch of the derivation, for otherwise the application of the rule could be dropped. Consider some of the possibilities.

Cut:

$$\frac{\begin{array}{l}+\Gamma \\ -\Theta\end{array}}{+F \quad -F}$$

Cut:

$$\frac{\Gamma \vdash F \quad \Gamma, F \vdash \Theta}{\Gamma \vdash \Theta}$$

The given HITT derivation of $\Gamma \vdash \Theta$ can be understood to be derivations of the sequents $\Gamma, F \vdash \Theta$ and $\Gamma \vdash F$. The latter because the conclusion $-F$ makes $-\Theta$ unavailable; hence there can be no node below $-F$ that is a conclusion from a premiss in $-\Theta$. Each of these derivations has n or fewer applications of the semantic rules. By the induction assumption, therefore, each has a derivation in HITTG so that $\Gamma \vdash \Theta$ does also by one application of *Cut*.

$-\wedge$:

$$\frac{\begin{array}{l}+\Gamma \\ -[F \wedge G]\end{array}}{-F \quad -G}$$

$\vdash \wedge$:

$$\frac{\Gamma \vdash F \quad \Gamma \vdash G}{\Gamma \vdash [F \wedge G]}$$

Since the node $-[F \wedge G]$ is unavailable in every branch of the tree, there must exist derivations in HITT for the sequents $\Gamma \vdash F$ and $\Gamma \vdash G$. By the induction assumption there exists derivations in HITTG for these sequents and therefore by one application of $\vdash \wedge$ for $\Gamma \vdash [F \wedge G]$.

$+\wedge$:

$$\begin{array}{l}+\Gamma' \\ +[F \wedge G] \\ -\Theta \\ +F\end{array}$$

$\wedge \vdash$:

$$\frac{\Gamma', [F \wedge G], F \vdash \Theta}{\Gamma', [F \wedge G] \vdash \Theta}$$

Here $+\Gamma'$ is $+\Gamma$ without $+[F \wedge G]$. The derivation on the left is also a derivation for $\Gamma', [F \wedge G], F \vdash \Theta$ with n applications of rules. Therefore by the induction assumption there is a derivation for this sequent in HITTG. But then by one application of $\wedge \vdash$ there is also a derivation for the sequent $\Gamma, [F \wedge G] \vdash \Theta$ as required.

It might be thought that the cases for the two $\neg$ rules are difficult but they are similar to the two cases for $\wedge$.

$-\neg$:	$\vdash \neg$:
$+\Gamma$	$\Gamma, F \vdash$
$-\neg F$	———
$+F$	$\Gamma \vdash \neg F$

By item (6) of Lemma 54 in §6.4.1 the node $-\neg F$ need never be used again as a premiss so that it follows that a derivation with n applications can be provided for $\Gamma, F \vdash$. It follows that $\Gamma, F \vdash$ is derivable in HITTG by the induction assumption and therefore also $\Gamma \vdash \neg F$ by one application of $\vdash \neg$.

$+\neg$:	$\vdash \neg$:
$+\Gamma'$	$\Gamma', \neg F \vdash F$
$+\neg F$	———
$-F$	$\Gamma', \neg F \vdash$

The sequent $\Gamma', \neg F \vdash F$ has a derivation with n applications. Therefore by the induction assumption it is derivable in HITTG. $\Gamma', \neg F \vdash$ is then derivable by one application of $\vdash \neg$.

There are two cases for $-\rightarrow$ corresponding to the two possible conclusions $+F$ and $-G$. The second of the two cases is similar to the case for $-\neg$:

$-\rightarrow$:	$\vdash\rightarrow$:
$+\Gamma$	$\Gamma \vdash G$
$-[F \rightarrow G]$	———
$-G$	$\Gamma, F \vdash G$
	———
	$\Gamma \vdash [F \rightarrow G]$

Since the conclusion $-G$ makes the premiss $-[F \rightarrow G]$ unavailable, the derivation is a derivation of the sequent $\Gamma \vdash G$ with n applications. Thus by the induction assumption $\Gamma \vdash G$ is derivable in HITTG, as is $\Gamma \vdash [F \rightarrow G]$ by one application of *Thin*$\vdash$ and one application of $\vdash\rightarrow$.

The first case for $-\rightarrow$ is the most complicated of all the cases. The complication arises from the fact that the node $-[F \rightarrow G]$ may be the premiss for a second application of $-\rightarrow$ with conclusion $-G$. But by item (7) of Lemma 54 in §6.4.1 it may be assumed that the conclusion $-G$ of the second application of $-\rightarrow$ follows immediately after the conclusion $+F$ of the first.

$-\rightarrow$:	$\vdash\rightarrow$:
$+\Gamma$	$\Gamma, F \vdash G$
$-[F \rightarrow G]$	———
$+F$	$\Gamma \vdash [F \rightarrow G]$
$-G$	

Thus the derivation on the left is also a derivation of $\Gamma, F \vdash G$ with $n - 1$ applications since $-G$ makes $-[F \rightarrow G]$ unavailable. By the induction assumption, therefore, there is a derivation in HITTG of the sequent. One application of $\vdash\rightarrow$ then provides the desired result.

The remaining cases are left for the reader. $\dashv$

EXERCISES §6.4.

1. Confirm that each of the displayed sequents of Chapter 4 is a derivable sequent of HITT.
2. Complete the proof of Theorem 55.

6.5. Semantics for HITT

In [90], Kripke provided a semantics for first order intuitionistic logic [77] and proved it complete with respect to a class of models called *Kripke model-structures*; this first order logic is obtained from HITT by dropping the $\pm\lambda$ and $\pm Int$ rules. In [64], the Kripke model structures are extended for HITT and HITTε and completeness and cut-elimination proved for the logics. To provide some insights into how HITT differs from the logic ITT, some examples of derivable and underivable sequents of HITT are given.

6.5.1. Forests of semantic trees. A major difference between ITT and HITT becomes evident when derivations are systematically sought for underivable sequents. In ITT a single semantic tree defines a counter-model, while in HITT, because of the restriction on the availability of negative formulas, *forests* of semantic trees are used. Forests of semantic trees are indicated in the following examples by a horizontal dotted line, in effect describing two semantic trees instead of one.

The first example illustrates that an unavailable node may not be a member of a closing pair of nodes for a branch:

EXAMPLE 1. $\neg\neg F \vdash F$

$+\neg\neg F$
$-F \cdots\cdots -\neg F$
$\qquad\qquad +F$

The node $-F$ is made unavailable by the addition of the node $-\neg F$ so that it cannot be paired with $+F$ to close the branch. To emphasize this fact, the node $-\neg F$ is shown as an *alternative* to $-F$; that is, either the node $-F$ or the node $-\neg F$ is to be understood as the node immediately following $+\neg\neg F$. Thus two semantic trees are described in the diagram, namely:

$+\neg\neg F$	$+\neg\neg F$
$-F$	$-\neg F$
	$+F$

The next example illustrates that the $-\vee$ rule compels a choice of conclusion that also results in a second semantic tree.

EXAMPLE 2. $\vdash [F \vee \neg F]$

$-[F \vee \neg F]$
$-F$ .... $-\neg F$
$+F$

The dotted line between the two nodes $-F$ and $-\neg F$ again indicates that each is to tbe understood as the node immediately below the root node $-[F \vee \neg F]$. Here the dotted line is necessary because each of the conclusions $-F$ and $-\neg F$ of $-[F \vee \neg F]$ makes that node unavailable on each of the branches. Two semantic trees result:

$-[F \vee \neg F]$	$-[F \vee \neg F]$
$-F$	$-\neg F$
	$+F$

An example involving quantification is offered next; it is taken from [90].

EXAMPLE 3. $\forall x.[P(x) \vee F] \vdash [\forall x.P(x) \vee F]$

$+\forall x.[P(x) \vee F]$
$-[\forall x.P(x) \vee F]$
$+[P(x) \vee F]$ x is an eigenterm for the $+\forall$ rule.

$+P(x)$	$+F$
$-\forall x.P(x)$	$-F$
$-P(y)$	———
$+[P(y) \vee F]$	

$+P(y)$	$+F$
———	

The Kripke model structure in Figure 3 of Kripke's [90], which is a counter-model for the sequent $\forall x.[P(x) \vee F] \vdash [\forall x.P(x) \vee F]$, can be obtained from this semantic tree.

The next example is included because the Kripke model structure in Figure 1 of Kripke's [90] is a counter-model for the sequent. Here, following Kripke, P, Q, and R are constants of type [ ].

EXAMPLE 4. $P \vdash [\neg R \vee \neg[Q \wedge \neg[R \vee \neg R]]]$

$+P$
$-[\neg R \vee \neg[Q \wedge \neg[R \vee \neg R]]]$
$-\neg R$ .... $-\neg[Q \wedge \neg[R \vee \neg R]]$
$+R$ $\quad$ $+[Q \wedge \neg[R \vee \neg R]]$
$+Q$
$+\neg[R \vee \neg R]$
$-[R \vee \neg R]$
$-R$ .... $-\neg R$
$+R$

The nodes G, H_1, H_2, H_3, and H_4 of the Kripke model structure in Figure 1 of [90], correspond to the sets of nodes of this example as follows:

G to $\{+P\}$ H_1 to $\{+P, +R\}$ H_2 to $\{+P, +Q\}$
H_3 to $\{+P, +Q, -R\}$ H_4 to $\{+P, +Q, +R\}$

EXAMPLE 5. $\neg\forall x.\neg P(x) \vdash \exists x.P(x)$

$+\neg\forall x.\neg P(x)$
$-\exists x.P(x)$ ··············· $-\forall x.\neg P(x)$
$-P(x)$ $-\neg P(x)$
$+P(x)$

The last example, unlike the first five, is derivable; its derivation is the example derivation given in §6.4 and is the second tree in a forest of two trees.

EXAMPLE 6. $\vdash \neg\neg[F \vee \neg F]$

$-\neg\neg[F \vee \neg F]$
$+\neg[F \vee \neg F]$
$-[F \vee \neg F]$
$-F$ ··············· $-\neg F$
$+F$
$-[F \vee \neg F]$
$-F$
———

Two branches are defined, the second of which is a derivation:

$-\neg\neg[F \vee \neg F]$	$-\neg\neg[F \vee \neg F]$
$+\neg[F \vee \neg F]$	$+\neg[F \vee \neg F]$
$-[F \vee \neg F]$	$-[F \vee \neg F]$
$-F$	$-\neg F$
	$+F$
	$-[F \vee \neg F]$
	$-F$
	———

The first of the two trees is not closed but can be by extending it. Thus two different derivations can be described from the forest. The second tree results from the 'best' choices of alternatives. The first when extended to a derivation has unnecessary repetitions.

The construction of the counter-models used in the proof of the completeness of HITT in [64] does not rely on forests of trees. As has been remarked before, the completeness proof offers little insight into the motivation for intuitionistic logic.

6.6. Recursions in HITT

In §6.4 the derivations of the sequents (Lt.1), (Gt.1), (Lt.2), and (Gt.2) in §4.2 and §4.3 are offered as examples of derivations of ITT that are also derivations of HITT. The conclusion of an examination of other derivable sequents of ITT cited in Chapter 4 is that recursion theory as it is developed in Chapter 4 for ITT serves equally well as a development for HITT. Thus since appropriate Kripke model-structures are definable for HITT, it is reasonable to conclude that the fundamental logic for recursion theory is intuitionistic and that the classical logic version is a convenient simplification.

CHAPTER 7

LOGIC AND MATHEMATICS

7.1. Introduction

It is unlikely that anything new can be added to the discussion of the relationship between logic and mathematics one hundred and twenty years after the publication of Frege's [49] and seventy years after that of Gödel's [69]. But since the logic ITT differs in several important respects from previously considered logics it is perhaps worthwhile restating some old positions in the light of the new logic.

As expressed by Steen in the introduction to [137],

> Mathematics is the art of making vague intuitive ideas precise and then studying the result.

The tool available for practicing the art is logic; thus mathematics is applied logic, a view consistent with Russell's logicism as expressed in the introduction. The subject matter of mathematics is defined predicates, and the goal of mathematics is the discovery of properties, including membership, of these predicates. This echoes the view expressed by Ayer in the quote displayed in the introduction. Evidence has already been supplied in Chapter 4 for this view of mathematics. Further evidence is supplied in this chapter. But first some implications of Gödel's theorems are discussed in §7.2.

MacLane describes in [95] an argument used in category theory that he claims cannot be formalized in any standard set theory or logic. That it cannot be formalized in either of the standard set theories is immediate. That it cannot be formalized in a logic is a direct challenge to the view that mathematics is applied logic. MacLane's claim is refuted in §7.3 for the logic ITTε using a simpler form of the argument for Abelian semi-groups suggested by Feferman in [44]. The formalization of Feferman's simplified argument makes essential use of both polytyping and secondary typing; it has been adapted from the formalization described in [65] for the inconsistent logic NaDSet.

The argument for category theory in ITTε is more complicated but essentially not different in principle from the one for Abelian semi-groups. It too depends on secondary typing, suggesting that the nominalism that motivates ITT motivates category theory as well. It must be stressed, however, that this does not mean that the goals of category theory can best be pursued through ITT; rather, it only means that the methods employed by category theory can be justified in ITT. The relationship between logic and category theory is similar in some respects to the relationship between say assembly language programming and functional programming: Category theory provides a more efficient way of reasoning about its fundamental concepts that are, nevertheless, definable in logic; but logic permits the exploration of concepts beyond the fundamental concepts of category theory such as the use and mention distinction explored in this book.

MacLane's claim correctly denies set theory a foundational role for mathematics. In §7.4 the goals of set theory are examined and it is suggested that these can be achieved in ITT. More recently Aczel has extended set theory with an anti-foundation axiom [5]. It appears that an application of non-well-founded sets can be replaced by an application of non-well-founded predicates as described in Chapter 4. A general method of translation appears possible but is not attempted here.

7.2. Self-Applied Logic

The logic ITT cannot of course avoid the implications of the two fundamental theorems of Gödel [69]. But the view that mathematics is logic applied in the study of defined predicates provides a different perspective of them. The proofs of Gödel's theorems for the logic ITT require an application of the logic to a study of the defined derivability predicate of ITT. A profound, but not disturbing, consequence of this study is that there is a formula F for which the derivability predicate applies to neither F nor $\neg F$. A second profound consequence is that a formula expressing the consistency of the derivability predicate can only be derived in ITT under additional assumptions. The conclusion as described by Feferman on page 233 of [45] is

> ··· the 'true reason' for the incompleteness phenomena is that though a formal system S may be informally recognized to be correct, we must adjoin formal expression of that recognition by means of a reflection principle in order to decide Gödel's undecidable statements.

In a paper first published in 1963 and reprinted in Chapter 10 of [143], Montague studied what he called a syntactic treatment of modality in which a sentence such as 'Man is rational' is to be understood to mean ''Man is rational' is a necessary truth'. Thus the first sentence is understood as expressing a property of the *name* 'Man is rational' of a sentence. This is a view similar to the view described as 'nominalist' in the introduction. In the paper Montague extends the conclusion of Feferman. It may be possible for ITT to be self-applied so as to extend the conclusion of Feferman in somewhat the same way.

7.3. Logic and Category Theory

MacLane's example of an argument that traditional set theories and logics cannot formalize has been simplified by Feferman in [44], to the following: An Abelian semi-group is defined by an associative and commutative binary product on a set of elements with an identity relation. The Cartesian product of Abelian semi-groups is an associative and commutative binary product on Abelian semi-groups and isomorphism between Abelian semi-groups is an identity on Abelian semi-groups. Therefore the set of Abelian semi-groups is itself an Abelian semi-group under Cartesian product and isomorphism. The self-membership inherent in the conclusion precludes it being stated in any standard set theory and appears to violate the type restrictions of a typed logic, although not necessarily those of the polytyped logic TT as the example

$$\vdash Ext(Ext)$$

of §2.5.1 demonstrates. As presented here the formalization of Feferman's example requires not only polytyping but also the secondary typing available in ITT and ITTε. The formalization is described in the remainder of this section using the logic ITTε in order that the two projection functions ιHd and ιTl for ordered pairs, introduced in Exercise 4 of §5.5, may be used.

7.3.1. Abelian semi-groups. A predicate SG with membership the Abelian semi-groups is defined

$$SG \stackrel{\text{df}}{=} (\lambda Z.\exists XA, XP, XI.[\mathsf{Axioms}(XA, XP, XI) \wedge Z = \langle XA, XP, XI\rangle])$$

where the definition of a triple is similar to that of pair as defined in §4.4.2 so that $\langle XA, XP, XI\rangle$ is $\lambda w.w(XA, XP, XI)$. $\mathsf{Axioms}(XA, XP, XI)$ is the conjunction of the following ten formulas:

1. $[\forall u{:}XA].XI(u, u)$
2. $[\forall u, v{:}XA].[XI(u, v) \rightarrow XI(v, u)]$
3. $[\forall u, v, w{:}XA].[XI(u, v) \wedge XI(v, w) \rightarrow XI(u, w)]$
4. $[\forall ua, ub, v, w{:}XA][XI(ua, ub) \wedge XP(ua, v, w) \rightarrow XP(ub, v, w)]$
5. $[\forall ua, ub, v, w{:}XA][XI(va, vb) \wedge XP(u, va, w) \rightarrow XP(u, vb, w)]$
6. $[\forall u, v, wa, wb{:}XA].[XI(wa, wb) \wedge XP(u, v, wa) \rightarrow XP(u, v, wb)]$
7. $[\forall u, v{:}XA].[\exists wa{:}XA].XP(u, v, wa)$
8. $[\forall u, v, wa, wb{:}XA].[XP(u, v, wa) \wedge XP(u, v, wb) \rightarrow XI(wa, wb)]$
9. $[\forall u, v, wa, wb{:}XA].[XP(u, v, wa) \wedge XP(v, u, wb) \rightarrow XI(wa, wb)]$
10. $[\forall u, v, w, wa, wb, za, zb{:}XA].[$
    $[XP(u, v, wa) \wedge XP(wa, w, za) \wedge XP(u, wb, zb) \wedge XP(v, w, wb)]$
    $\rightarrow XI(za, zb)]$

These formulas are suitable axioms for an Abelian semi-group, where XP is the product predicate and XI the identity predicate for the members of XA. Formulas (1)–(6) express that XI is an identity on XA with respect to XP.

Formulas (7)–(10) express that XP is an associative and commutative binary operation on XA. Thus

$$\vdash SG(\langle N, \oplus, =\rangle)$$

for example, is derivable, where N is defined in §4.4.4 and $\oplus$ is the sum predicate of N defined in §8.2. Further, if CP is defined to be the Cartesian product of members of SG, and IS the isomorphism relation between members, then the following sequent should also be derivable

SG) $$\vdash SG(\langle SG, CP, IS\rangle).$$

The definitions of CP and IS are given in §7.3.2, a type analysis of them and other terms is undertaken in §7.3.3, and a sketch of a derivation of (SG) is given in §7.3.4.

7.3.2. Definitions. It is assumed that projection functions A, P, and I for triples W have been defined in a manner similar to the manner in which projection functions ιHd and ιTl are defined in Exercise 4 of §5.5 for pairs. In particular, it is assumed that the following sequent is derivable in ITTε:

Tr) $$\vdash [\forall W{:}SG].W = \langle A[W], P[W], I[W]\rangle.$$

The use of these projections functions could be eliminated and the needed definitions and derivation could be carried out in ITT, but they would then become unnecessarily complicated.

Two members U and V of SG are *isomorphic* if $IS(U, V)$ holds, where

$$\begin{aligned}
IS \stackrel{\text{df}}{=} \lambda U, V.[&SG(U) \wedge SG(V) \wedge \exists Z.[\\
&[\forall u1{:}A[U]].[\exists u2{:}A[V]].Z(u1, u2)\wedge\\
&\quad[\forall u2{:}A[V]].[\exists u1{:}A[U]].Z(u2, u1)\wedge\\
&\quad\quad[\forall u1, v1, w1{:}A[U]].[\forall u2, v2, w2{:}A[V]].[\\
&\quad\quad\quad Z(u1, u2) \wedge Z(v1, v2) \wedge Z(w1, w2) \rightarrow\\
&\quad\quad\quad\quad [P[U](u1, v1, w1) \leftrightarrow P[V](u2, v2, w2)]\wedge\\
&\quad\quad\quad\quad\quad [I[U](u1, v1) \leftrightarrow I[V](u2, v2)]]]]].
\end{aligned}$$

The Cartesian product CP of two groups U and V is defined:

$$\begin{aligned}
&CP \stackrel{\text{df}}{=}\\
&\quad\lambda U, V, W.[SG(U) \wedge SG(V)\wedge\\
&\quad\quad W = \langle(A[U] \otimes A[V]), (P[U] \otimes P[V]), (I[U] \otimes I[V])\rangle]
\end{aligned}$$

where the following abbreviations have been used:

Ab.1. $$(A[U] \otimes A[V]) \stackrel{\text{df}}{=} \lambda z.[\exists u1{:}A[U]].[\exists u2{:}A[V]].z = \langle u1, u2\rangle.$$

Ab.2. $$\begin{aligned}
&(P[U] \otimes P[V]) \stackrel{\text{df}}{=}\\
&\quad\lambda z1, z2, z3.[\exists u1, v1, w1{:}A[U]].[\exists u2, v2, w2{:}A[V]].\\
&\quad\quad[z1 = \langle u1, u2\rangle \wedge z2 = \langle v1, v2\rangle \wedge z3 = \langle w1, w2\rangle\wedge\\
&\quad\quad\quad P[U](u1, v1, w1) \wedge P[V](u2, v2, w2)].
\end{aligned}$$

Ab.3. $(I[U] \otimes I[V]) \stackrel{\text{df}}{=}$
$\lambda z1, z2.[\exists u1, v1{:}A[U]].[\exists u2, v2{:}A[V]].$
$[z1 = \langle u1, u2\rangle \wedge z2 = \langle v1, v2\rangle \wedge$
$I[U](u1, v1) \wedge I[V](u2, v2)].$

7.3.3. Typing. Consider now the typing of the terms SG, CP, and IS and their subterms. In this discussion the notation t[cv] that designates the type of a constant or variable cv will be used to denote also the type of a term.

1. The variables appearing in the axioms. The lower case variables may have any type δ. Then t[XA], t[XP], and t[XI] are respectively [δ], [δ,δ,δ], and [δ,δ].
2. SG as defined in §7.3.1. t[SG] is [t[Z]] is [t$\langle XA, XP, XI\rangle$] is [[[t[$XA$], t[$XP$], t[$XI$]]]] is [[[[$\delta$], [$\delta,\delta,\delta$], [$\delta,\delta$]]]].
   Let the notation $\tau(\delta)$ abbreviate [[[δ], [δ,δ,δ], [δ,δ]]] so that t[SG] is [$\tau(\delta)$].
3. CP as defined in §7.3.2. In this analysis, δ is assumed to be 1. Since in this case t[SG] is [$\tau(1)$], it follows that t[U] and t[V] is each $\tau(1)$ so that t[CP] is [$\tau(1), \tau(1)$, t[W]].
   t[W] is t[$\langle(A[U] \otimes A[V]), (P[U] \otimes P[V]), (I[U] \otimes I[V])\rangle$] is [[t[$(A[U] \otimes A[V])$]], [t[$(P[U] \otimes P[V])$]], t[$(I[U] \otimes I[V])$]]].
   t[$(A[U] \otimes A[V])$] is [t[z]] is [t[$\langle u1, u2\rangle$]] is [1] by the secondary typing of $\langle u1, u2\rangle$.
   Similarly, t[$(P[U] \otimes P[V])$] is [1,1,1], and t[$(I[U] \otimes I[V])$] is [1,1].
   Thus t[W] is $\tau(1)$ and t[CP] is [$\tau(1), \tau(1), \tau(1)$].
   Note that this conclusion depends on secondary typing; if δ is of any type other than 1, then t[U] is t[V] is $\tau(\delta)$ and t[W] is not $\tau(\delta)$.
4. IS as defined in §7.3.2. t[IS] is [$\tau(1), \tau(1)$].

7.3.4. Derivation of (SG). Note that axiom (7) is the only axiom in which an existential quantifier appears. That is why it is selected for detailed treatment in the sketch of a derivation of (SG) that follows. Axiom (8) is also selected in order to examine one axiom in which both XP and XI appear.

Derivation of (SG)
$-SG(\langle SG, CP, IS\rangle)$
$-\text{Axioms}(SG, CP, IS)$

Ten branches result from applications of the $-\wedge$ rule, each branch headed by a node obtained from one of the axioms. Each branch must be closed. The branches headed by the nodes obtained from axioms (7) and (8) are selected for detailed treatment in that order.

$-[\forall u, v{:}SG].[\exists w{:}SG].CP(u, v, w)$
$+SG(U)$ (a)
$+SG(V)$
$-\exists w.[SG(w) \wedge CP(U, V, w)]$

Replace w with Cw, where

$$Cw \stackrel{\text{df}}{=} \langle (A[U] \otimes A[V]), (P[U] \otimes P[V]), (I[U] \otimes I[V]) \rangle$$

$-SG(Cw) \wedge CP(U, V, Cw)$

L R

$-SG(Cw)$

$-\text{Axioms}((A[U] \otimes A[V]), (P[U] \otimes P[V]), (I[U] \otimes I[V]))$

Ten branches result from applications of the $-\wedge$ rule, each branch headed by a node obtained from one of the axioms. Each branch must be closed. The branch headed by the node obtained from axiom (7), which is referred to as Ax.7, is selected for detailed treatment.

$-[\forall x, y:(A[U] \otimes A[V])].[\exists z:(A[U] \otimes A[V])].(P[U] \otimes P[V])(x, y, z)$
$+(A[U] \otimes A[V])(x)$ (b)
$+(A[U] \otimes A[V])(y)$
$-\exists z.[(A[U] \otimes A[V])(z) \wedge (P[U] \otimes P[V])(x, y, z)$ (c)
$+[\exists u1{:}A[U]].[\exists u2{:}A[V]].x = \langle u1, u2 \rangle$ $\dashv$df $\otimes$
$+A[U](x1)$
$+A[V](x2)$
$+x = \langle x1, x2 \rangle$
$+A[U](y1)$
$+A[V](y2)$
$+y = \langle y1, y2 \rangle$ $\dashv$similar
$+\exists XA, XP, XI.[\text{Axioms}(XA, XP, XI) \wedge U = \langle XA, XP, XI \rangle]$ $\dashv$(a)
$+U = \langle A[U], P[U], I[U] \rangle$ $\dashv$(Tr)
$+\text{Axioms}(A[U], P[U], I[U])$
$+[\forall u, v{:}A[U]].[\exists w{:}A[U]].P[U](u, v, w)$ $\dashv$Ax.7
$+[\exists w{:}A[U]].P[U](x1, y1, w)$
$+A[U](w1)$ (d)
$+P[U](x1, y1, w1)$
$+A[U](w2)$ (e)
$+P[V](x2, y2, w2)$ $\dashv$Ax.7
$+(A[U] \otimes A[V])(\langle w1, w2 \rangle)$ (f),$\dashv$(d),(e)

Replace z in $-\exists z$ with $\langle w1, w2 \rangle$

$-[(A[U] \otimes A[V])(\langle w1, w2 \rangle) \wedge (P[U] \otimes P[V])(x, y, \langle w1, w2 \rangle)$
$-(P[U] \otimes P[V])(x, y, \langle w1, w2 \rangle)$ $\dashv$(f) $-\wedge$
$-[\exists u1, v1, w1'{:}A[U]].[\exists u2, v2, w2'{:}A[V]].$
$\quad [x = \langle u1, u2 \rangle \wedge y = \langle v1, v2 \rangle \wedge \langle w1, w2 \rangle = \langle w1', w2' \rangle] \wedge$
$\quad\quad P[U](x1, y1, w1') \wedge P[V](x2, y2, w2')$ $\dashv$Ab.2

$-A[U](w1) \wedge A[V](w2) \wedge$
$[x = \langle x1, x2\rangle \wedge y = \langle y1, y2\rangle \wedge \langle w1, w2\rangle = \langle w1, w2\rangle] \wedge$
$P[U](x1, y1, w1) \wedge P[V](x2, y2, w2)$

R

$-CP(U, V, Cw)$
$-[SG(U) \wedge SG(V) \wedge Cw = (A[U] \otimes A[V]), (P[U] \otimes P[V]), (I[U] \otimes I[V])\rangle]$

The branch resulting from axiom(8) follows.

$-[\forall u, v, wa, wb{:}SG].[CP(u, v, wa) \wedge CP(u, v, wb) \rightarrow IS(wa, wb)]$
$+SG(U) \quad +SG(V) \quad +SG(Wa) \quad +SG(Wb)$
$+CP(U, V, Wa) \quad +CP(V, U, Wb)$
$-IS(Wa, Wb)$
$+Wa = \langle (A[U] \otimes A[V]), (P[U] \otimes P[V]), (I[U] \otimes I[V])\rangle]$
$+Wb = \langle (A[V] \otimes A[U]), (P[V] \otimes P[U]), (I[V] \otimes I[U])\rangle]$
$-\exists Z.[[\forall u1{:}(A[U] \otimes A[V])].[\exists u2{:}(A[V] \otimes A[U])].Z(u1, u2) \wedge$
$[\forall u2{:}(A[V] \otimes A[U])].[\exists u1{:}(A[U] \otimes A[V])].Z(u2, u1) \wedge$
$[\forall u1, v1, w1{:}(A[U] \otimes A[V])].[\forall u2, v2, w2{:}(A[V] \otimes A[U])].[$
$Z(u1, u2) \wedge Z(v1, v2) \wedge Z(w1, w2) \rightarrow$
$[(P[U] \otimes P[V])(u1, v1, w1) \leftrightarrow (P[V] \otimes P[U])(u2, v2, w2)] \wedge$
$[(I[U] \otimes I[V])(u1, v1) \leftrightarrow (I[V] \otimes I[U])(u2, v2)]]]$ $\dashv$df *IS*

Replace Z by CZ, where $CZ \stackrel{\text{df}}{=} \lambda z1, z2.\exists u1, u2.[z1 = \langle u1, u2\rangle \wedge z2 = \langle u2, u1\rangle]$.

$-[[\forall u1{:}(A[U] \otimes A[V])].[\exists u2{:}(A[V] \otimes A[U])].CZ(u1, u2) \wedge$
$[\forall u2{:}(A[V] \otimes A[U])].[\exists u1{:}(A[U] \otimes A[V])].CZ(u2, u1) \wedge$
$[\forall u1, v1, w1{:}(A[U] \otimes A[V])].[\forall u2, v2, w2{:}(A[V] \otimes A[U])].[$
$CZ(u1, u2) \wedge CZ(v1, v2) \wedge CZ(w1, w2) \rightarrow$
$[(P[U] \otimes P[V])(u1, v1, w1) \leftrightarrow (P[V] \otimes P[U])(u2, v2, w2)] \wedge$
$[(I[U] \otimes I[V])(u1, v1) \leftrightarrow (I[V] \otimes I[U])(u2, v2)]]]$

L R

$-[\forall u1{:}(A[U] \otimes A[V])].[\exists u2{:}(A[V] \otimes A[U])].CZ(u1, u2)$
$+(A[U] \otimes A[V])(z1)$
$-[\exists u2{:}(A[V] \otimes A[U])].CZ(z1, u2)$
$+[\exists u1{:}A[U]].[\exists u2{:}A[V]].z1 = \langle u1, u2\rangle$ $\dashv$df $\otimes$
$+A[U](u1) \qquad +A[V](u2) \qquad z1 = \langle u1, u2\rangle$
$+(A[V] \otimes A[U])(\langle u2, u1\rangle)$
$-(A[V] \otimes A[U])(\langle u2, u1\rangle) \wedge CZ(z1, \langle u2, u1\rangle)$

R

$-[\forall u2{:}(A[V]\otimes A[U])].[\exists u1{:}(A[U]\otimes A[V])].CZ(u2,u1)\wedge$
$\quad[\forall u1,v1,w1{:}(A[U]\otimes A[V])].[\forall u2,v2,w2{:}(A[V]\otimes A[U])].[$
$\quad\quad CZ(u1,u2)\wedge CZ(v1,v2)\wedge CZ(w1,w2)\rightarrow$
$\quad\quad\quad[(P[U]\otimes P[V])(u1,v1,w1)\leftrightarrow(P[V]\otimes P[U])(u2,v2,w2)]\wedge$
$\quad\quad\quad\quad[(I[U]\otimes I[V])(u1,v1)\leftrightarrow(I[V]\otimes I[U])(u2,v2)]]]$

RL RR

similar to L

RR

$-[\forall u1,v1,w1{:}(A[U]\otimes A[V])].[\forall u2,v2,w2{:}(A[V]\otimes A[U])].[$
$\quad CZ(u1,u2)\wedge CZ(v1,v2)\wedge CZ(w1,w2)\rightarrow$
$\quad\quad[(P[U]\otimes P[V])(u1,v1,w1)\leftrightarrow(P[V]\otimes P[U])(u2,v2,w2)]\wedge$
$\quad\quad\quad[(I[U]\otimes I[V])(u1,v1)\leftrightarrow(I[V]\otimes I[U])(u2,v2)]]]$
$+(A[U]\otimes A[V])(za1)\quad +(A[U]\otimes A[V])(za2)\quad +(A[U]\otimes A[V])(za3)$
$+(A[V]\otimes A[U])(zb1)\quad +(A[V]\otimes A[U])(zb2)\quad +(A[V]\otimes A[U])(zb3)$
$+CZ(za1,zb1)\quad +CZ(za2,zb2)\quad +CZ(za3,zb3)$ (g)
$-[[(P[U]\otimes P[V])(za1,za2,za3)\leftrightarrow(P[V]\otimes P[U])(zb1,zb2,zb3)]\wedge$
$\quad\quad[(I[U]\otimes I[V])(za1,za2)\leftrightarrow(I[V]\otimes I[U])(zb1,zb2)]]$

RRL RRR

$-[(P[U]\otimes P[V])(za1,za2,za3)\leftrightarrow(P[V]\otimes P[U])(zb1,zb2,zb3)]$

RRLL RRLR

$+(P[U]\otimes P[V])(za1,za2,za3)$
$-(P[V]\otimes P[U])(zb1,zb2,zb3)$
$+[\exists u1,v1,w1{:}P[U]].[\exists u2,v2,w2{:}P[V]].$
$\quad[za1=\langle u1,u2\rangle\wedge za2=\langle v1,v2\rangle\wedge za3=\langle w1,w2\rangle]$ $\dashv$Ab.2
$+P[U](u1)\quad +P[U](v1)\quad +P[U](w1)$
$+P[V](u2)\quad +P[V](v2)\quad +P[V](w2)$
$+za1=\langle u1,u2\rangle\quad +za2=\langle v1,v2\rangle\quad +za3=\langle w1,w2\rangle$ (h)
$-[\exists u1,v1,w1{:}P[V]].[\exists u2,v2,w2{:}P[U]].$
$\quad[zb1=\langle u1,u2\rangle\wedge zb2=\langle v1,v2\rangle\wedge zb3=\langle w1,w2\rangle]$ Ab.2
$-[zb1=\langle u2,u1\rangle\wedge zb2=\langle v2,v1\rangle\wedge zb3=\langle w2,w1\rangle]$

$\dashv$(g), (h)

RRLR

Similar to RRLL

RRR

Adapt argument for P for I.

Consider now the typing of the formulas of this derivation. The two occurrences of SG in the sequent (SG) have different types. t[SG] is $[\tau(1)]$ for the second occurrence as has been determined in item (2) in §7.3.3 and t[SG] for the first occurrence is $[[[[\tau(1)],[\tau(1),\tau(1),\tau(1)],[\tau(1),\tau(1)]]]]$. Thus for this typing the lower case variables of the axioms are all of type $\tau(1)$.

EXERCISES §7.3.

1. Provide the necessary definitions and a derivation for the sequent (Tr) of §7.3.2.
2. Extend the sketch of a derivation for (SG) by considering details for axiom (4).

7.4. Set Theory and Logic

Zermelo described in [151] the motivation for developing an axiomatization of set theory:

> ... Under these circumstances there is at this point nothing left for us to do but ... to seek out the principles required for establishing the foundation of this mathematical discipline. In solving the problem we must, on the one hand, restrict these principles sufficiently to exclude all contradictions and, on the other, take them sufficiently wide to retain all that is valuable in this theory.

Cantor's theory of transfinite numbers was part of what was to be retained in Zermelo's set theory. That theory can of course be developed within ITT by formalizing a set theory within ITT in much the same way that Abelian semigroups were formalized in §7.3.1, and developing the theory within the set theory. Because it has finitely many axioms, the Bernays-Gödel version of the theory described in [68] can be most easily formalized in this way. But a formalization of ZFC, Zermelo-Fraenkel set theory with the axiom of choice, is a more interesting challenge. A sketch of how this might be done is provided in §7.4.1.

Since an axiom of extensionality is the first axiom of Zermelo's set theory it might be thought that it is essential to the development of Cantor's theory and that it's absence from ITT precludes the direct development of Cantor's theory within that logic. But that is not the case: Cantor's diagonal argument, that is central to his theory, is formalized in ITT in §7.4.2.

Set theory has more recently been extended by Aczel in [5] with an anti-foundation axiom to permit the definition of non-well-founded sets. It is obtained from ZFC, as described in §7.4.1, by replacing the foundation axiom, which asserts that every set is wellfounded, by an anti-foundation axiom that asserts that every possible non-well-founded set exists. In Chapter 8 of [5] the theory is used to provide a semantics for the calculi of Milner described in [101, 102]. It is used also by Milner in [103] to prove the consistency of the static and dynamic semantics of a small functional programming language with recursive functions. That this application can be completed in ITT is demonstrated in Chapter 8. Barwise and Etchemendy in [15] and Barwise and Moss in [16] provide many other examples of applications of the non-well-founded set theory. It is not known whether these applications can also be completed in ITT.

Set theory, like category theory, provides a useful shorthand for expressing mathematical arguments, but ultimately its applications must be justified through formalization in a logic like ITT.

7.4.1. A set theory formalized in ITT. The non-logical axioms of ZFC are described in appendix B of [5]. To express them in ITT it is necessary to introduce two constants M:[1] and $\in$:[1, 1]. The first is the predicate that has *sets* as it members while the second is the membership relation between sets. The usual infix notation for $\in$ will be used. The convention will be maintained that all lower case variables are of type 1, that the intensional identity $=$ of ITT is of type [1,1], and that X:[1] and Y:[1, 1]. The latter two variables are needed to express two axiom schemes of the first order logic formulation of ZFC.

The non-logical axioms of ZFC expressed in ITT are:

Extensionality:

$$[\forall x, y{:}M].[\,[\forall u{:}M].[u \in x \leftrightarrow u \in y] \rightarrow x = y].$$

Pairing:

$$[\forall x, y{:}M].[\exists z{:}M].[x \in z \wedge y \in z].$$

Union:

$$[\forall x{:}M].[\exists z{:}M].[\forall u, v{:}M].[u \in v \wedge v \in x \rightarrow u \in z].$$

Powerset:

$$[\forall x{:}M].[\exists z{:}M].[\forall v{:}M].[\,[\forall u{:}M].[u \in v \rightarrow u \in x] \rightarrow v \in z].$$

Infinity:

$$\begin{aligned}&[\exists z{:}M].[[\exists x{:}M].[x \in z \ \wedge [\forall u{:}M].\neg u \in x] \wedge \\ &\quad [\forall x{:}M].[x \in z \rightarrow [\exists y{:}M].y \in z \ \wedge \ x \in y]].\end{aligned}$$

Separation:

$$\begin{aligned}&\forall X.[\,[\forall u{:}X].M(u) \rightarrow \\ &\quad [\forall x{:}M].[\exists z{:}M].[\forall u{:}M].[u \in z \leftrightarrow [u \in x \wedge X(u)]\,].\end{aligned}$$

Collection:

$$\begin{aligned}&\forall Y.[\forall u, v.[Y(u, v) \rightarrow M(u) \wedge M(v)] \rightarrow \\ &\quad [\forall x{:}M].[\,[\forall u{:}M].[u \in x \rightarrow \exists v.\,Y(u, v)] \rightarrow \\ &\qquad [\exists z{:}M].[\forall u{:}M].[u \in x \rightarrow \exists v.\,Y(u, v) \wedge v \in z]\,]].\end{aligned}$$

Choice:

$$\begin{aligned}&[\forall y{:}M].[[\forall x{:}M].[x \in y \rightarrow [\exists v{:}M].v \in x] \wedge \\ &\quad [\forall x1, x2{:}M].[x1 \in y \wedge x2 \in y \rightarrow \\ &\qquad\qquad [\,[\exists v{:}M].v \in x1 \wedge v \in x2 \rightarrow x1{=}x2]\,] \rightarrow \\ &\qquad \exists z.[\forall x{:}M].[x \in y \rightarrow \\ &\qquad\qquad [\exists v{:}M].[v \in x \wedge [\forall u{:}M].[u \in x \rightarrow [u \in z \leftrightarrow u{=}v]\,]].\end{aligned}$$

Foundation:

$$[\forall y{:}M].[[\exists x{:}M].x \in y \to [\exists x{:}M].[x \in y \wedge [\forall u{:}M].[u \in x \to \neg u \in y]]].$$

A closed formula of ITT is a ZF-formula if it is obtained from a closed formula of ZF by replacing each quantifier by the corresponding bound quantifier $[\forall v{:}M]$ or $[\exists v{:}M]$ of ITT. The predicate ZFC that applies to ZF-formulas that are theorems of the set theory ZFC as it is formalized in ITT is defined:

$$ZFC \stackrel{\text{df}}{=} \lambda Z.[\mathsf{Axioms} \to Z]$$

where Axioms is the conjunction of the nine axioms of ZFC. That

$$\text{Th)} \qquad \vdash ZFC(F)$$

is derivable in ITT for each ZF-formula F that is a translation of a theorem of the set theory ZFC is not difficult to establish. But a real challenge is to prove that (Th) cannot be derived for too many ZF-formulas F; that is that the set of ZF-formulas F for which (Th) can be derived is consistent, or at least that it is consistent if the set theory ZFC is consistent.

Fortunately it is not necessary to depend on the formalization of ZFC in ITT to develop Cantor's transfinite number theory.

7.4.2. Cantor's diagonal argument in ITT. The power-set relation PS, between a unary predicate W and a predicate X that applies only to members of W, is defined

$$PS \stackrel{\text{df}}{=} \lambda W, X.[\forall x{:}X].W(x).$$

A *Cantor* map of a predicate W is a mapping Xm of the membership of W into the membership of $PS(W)$:

$$Mp \stackrel{\text{df}}{=} \lambda W, Xm.[[\forall x{:}W].[\exists X{:}PS(W)].$$
$$[Xm(x, X) \wedge \forall Y.[Xm(x, Y) \to X =_e Y]].$$

Note that extensional identity is used for members of $PS(W)$ while intensional identity is implicitly used for members of W.

A Cantor map Xm of W defines a member XD of $PS(W)$, the *diagonal* member of $PS(W)$ determined by Xm:

$$Dg \stackrel{\text{df}}{=} \lambda W, Xm, x.[Mp(W, Xm) \wedge W(x) \wedge [\exists X{:}Xm(x)].\neg X(x)].$$

Cantor's *diagonal argument* gets its name from the use made of the "diagonal" predicate $Dg(W, Xm)$ to derive the following sequent that expresses what is known as *Cantor's Lemma*:

$$\vdash \forall W.[\forall Xm{:}Mp(W)].[\exists X{:}PS(W)].[\forall x{:}W].\neg Xm(x, X).$$

A derivation of the sequent follows:

$-\forall W.[\forall Xm{:}Mp(W)].[\exists X{:}PS(W)].[\forall x{:}W].\neg Xm(x, X)$
$+Mp(W, Xm)$ (a)
$-[\exists X{:}PS(W)].[\forall x{:}W].\neg Xm(x, X)$

Replace X by $Dg(W, Xm)$

$-[PS(W, Dg(W, Xm)) \wedge [\forall x{:}W].\neg Xm(x, Dg(W, Xm))]$

L | R

$-PS(W, Dg(W, Xm))$
$-[\forall x{:}Dg(W, Xm)].W(x)$ ⊣df PS
$+Dg(W, Xm)(x)$
$-W(x)$
$+[Mp(W, Xm) \wedge W(x) \wedge [\exists X{:}Xm(x)].\neg X(x)]$ ⊣df Dg
$+W(x)$

R
$-[\forall x{:}W].\neg Xm(x, Dg(W, Xm))]$
$+W(x)$ (b)
$-\neg Xm(x, Dg(W, Xm))$
$+Xm(x, Dg(W, Xm))$ (c)
$+\forall x.[W(x) \rightarrow \exists X.[PS(W, X) \wedge$
$\quad Xm(x, X) \wedge \forall Y.[Xm(x, X) \rightarrow X =_e Y]]]$ (a) ⊣df Mp
$+\exists X.[PS(W, X) \wedge Xm(x, X) \wedge \forall Y.[PS(W, Y) \wedge Xm(x, Y) \rightarrow X =_e Y]]$
$+[PS(W, X) \wedge Xm(x, X) \wedge \forall Y.[PS(W, Y) \wedge Xm(x, Y) \rightarrow X =_e Y]]$
$+PS(W, X)$
$+Xm(x, X)$ (d)
$+\forall Y.[Xm(x, Y) \rightarrow X =_e Y]$ (e)
$+[Xm(x, Dg(W, Xm)) \rightarrow X =_e Dg(W, Xm)]$ ⊣(e)

RL | RR

$-Xm(x, Dg(W, Xm))$

⊣(c)

RR
$+X =_e Dg(W, Xm)$
$+\forall x.[X(x) \leftrightarrow Dg(W, Xm)(x)]$
$+[X(x) \leftrightarrow Dg(W, Xm)(x)]$

RRL | RRR

$+X(x)$
$+Dg(W, Xm, x)$
$+[Mp(W, Xm) \wedge W(x) \wedge [\exists X{:}Xm(x)].\neg X(x)]$ ⊣df Dg
$+\exists X.[Xm(x, X) \wedge \neg X(x)]$

$+Xm(x, Y)$
$-Y(x)$
$+[Xm(x, Y) \rightarrow X =_e Y]$ $\dashv$(e)
$+X =_e Y$
$+Y(x)$

RRR
$-X(x)$
$-Dg(W, Xm, x)$
$-\exists X.[Xm(x, X) \wedge \neg X(x)]$ $\dashv$df Dg,(a),(b)
$-[Xm(x, X) \wedge \neg X(x)]$

$\dashv$(d)

EXERCISES §7.4.

1. A real number in the closed interval [0,1] can be represented as a sequence of *bits*, where each bit is 0 or 1. An enumeration of sequences of real numbers in the interval is then an enumeration of sequences of 0's and 1's. Provide definitions and a derivation to prove that for any such enumeration there is a real number not enumerated by it.

7.5. Why Is Mathematics Applicable?

Mathematics is described in the introduction to this chapter as applied logic. Quine remarks in the preface to [120]

> Of that which receives precise formulation in mathematical logic, an important part is already vaguely present as a basic ingredient of daily discourse.

This view is reinforced by the work of Chomsky [25], Devlin [40, 39], and others on the inherent structure of natural languages. Chomsky speculates in [25]:

> At this point one can only speculate, but it is possible that the number facility developed as a by-product of the language facility. The latter has features that are quite unusual, perhaps unique in the biological world. In technical terms it has the property of "discrete infinity." To put it simply, each sentence has a fixed number of words: one, two, three, forty-seven, ninety-three, etc. And there is no limit in principle to how many words the sentence may contain.

Work cited in [40, 39] appears to confirm the view that logical/mathematical concepts are inherent to the natural languages that mankind uses to communicate with others. Thus logic/mathematics is necessarily applicable since it has evolved with the languages of mankind to be applicable in a world with some degree of order and predictability.

CHAPTER 8

LOGIC AND COMPUTER SCIENCE

8.1. Introduction

To demonstrate that theoretical computer science, like mathematics,

> ... deals exclusively with concepts definable in terms of a very small number of fundamental logical concepts, and that all its propositions are deducible from a very small number of fundamental logical principles ...

as stated by Russell in [126], it is sufficient to demonstrate that its concepts can be defined within an intensional logic such as ITT.

There is an immediate advantage to doing this. ITT has a denotational semantics in the sense that each term P of type τ is assigned a value $\Phi(\tau, P)$ from a domain $DI(\tau)$ by each valuation Φ that is a model. Further this valuation interprets application and abstraction correctly. As a consequence if the fundamental concepts of computer science can be defined as terms of ITT, it is then unnecessary to develop a special semantics for them. The main purpose of this chapter is therefore to attempt a demonstration of this.

The topics covered in Chapter 4 are, of course, relevant to computer science; in particular the recursion generator method of defining well-founded and non-well-founded predicates provides a semantics for the *definitional* aspect of logic programming languages. But it must be stressed that no process for actually *computing* such predicates is described. This distinction between *defining* and *computing* is clearly evident in the difference between the descriptions of the polytyping algorithms for TT and ITT and the recursion generator definitions of Chapter 4. Since the concepts of a *computation* and of a *programming* language in which a process can be described for undertaking a computation on a machine are clearly fundamental to computer science, they are discussed in §8.2.

Logic programming languages, as described in [10] and [50] for example, are but a very few of the programming languages that have been developed or are under development. The sources [127] and [148] provide motivation and descriptions of early programming languages while [121] provides a theoretical basis for programming languages and references to more recent ones. As the need for more complex programs arises, styles of programming languages evolve, such as aspect oriented programming that has evolved from object-oriented programming [87]. These higher order programming languages are intended to hide complexity in simple but abstract structures that help programmers increase efficiency and avoid errors.

As suggested by Mosses in [106], the deconstruction of the meanings given to programming languages by denotational semantics can assist in the clarification and evolution of such structures:

> The characteristic feature of Denotational Semantics is that one gives semantic objects for *all* phrases—not only for complete programs. The semantic object specified for a phrase is called the *denotation* of the phrase. The idea is that the denotation of each phrase represents the contribution of that phrase to the semantics of any complete program in which it may occur.
>
> The denotations of compound phrases must depend only on the denotations of their subphrases. (Of course, the denotations of basic phrases do not depend on anything.) This is called *compositionality*.

The major difference between the denotational semantics provided for ITT and the denotational semantics for programming languages lies in the domains to which the valuations map. For ITT the domains are defined in an unformalized set theory that could be replaced by an extended ITT; an *extended* logic being necessary because of Tarski's fundamental theorem [142]. For programming languages it has been traditional since Scott's fundamental work [129] to define the domains using categorical, topological and algebraic methods; see for example [72] and [3]. The domains are defined as the fixed points of *domain equations* that are constructed from given basic domains using domain constructors.

An essential goal of this chapter must be to demonstrate that the domains needed for a semantics of programming languages can be defined in ITT using recursion generators. Three features peculiar to ITT, arising from the need to distinguish between the intension and extension of predicates, makes this possible:

D1: There are only types for individuals and predicates; there are no types for functions. In particular there are no types for functions with values of type 1.

D2: Terms of type $[\overline{\tau}]$, in which at most variables of type 1 have a free occurrence, have that type as its *primary* type and the type 1 as a *secondary* type.

D3: A rule of *intensionality* replaces an axiom of *extensionality*.

Recursive predicates are defined in ITT in the same way as any individual name or predicate:

P.1: Any constant C of the logic not previously defined can be defined to be either an individual or predicate name by a meta-logical definition of the form

$$C \stackrel{\mathrm{df}}{=} P$$

where P is any term in which C has no occurrence and no variable has a free occurrence. The type of C is the type of P. C is treated as an abbreviation for P; that is, whenever necessary an occurrence of C in a derivation can be replaced by an occurrence of P.

P.2: Given a *recursion generator* with parameters of type $\overline{p}$, that is a defined predicate RG of type $[\overline{p}, [\overline{\tau}], \overline{\tau}]$, and the defined predicates Lt and Gt of type $[\,[[\overline{\tau}], \overline{\tau}\,], \overline{\tau}]$, each of the terms

$$(\lambda\overline{u}.Lt(RG(\overline{u})) \text{ and } (\lambda\overline{u}.Gt(RG(\overline{u}))$$

can be used to define a *recursive* predicate C of type $[\overline{p}, \overline{\tau}]$, where $\overline{u}{:}\overline{p}$. Without any further assumptions on RG, a very general *induction principle* in the form of a sequent is derivable for each of the defined predicates.

P.3: If a recursion generator RG without parameters is assumed to be *monotonic*, that is if the sequent

$$\vdash Mon(RG)$$

is derivable, then the sequents

$$\vdash Lt(RG) =_e RG(Lt(RG)) \text{ and } \vdash Gt(RG) =_e RG(Gt(RG))$$

are also derivable; that is each of $Lt(RG)$ and $Gt(RG)$ is a *fixed point* X of an extensional identity $X =_e RG(X)$.

P.4: If a recursion generator $RG{:}[[\overline{\tau}], \overline{\tau}]$ without parameters is assumed to be both monotonic and $\exists$-continuous, that is if the sequent

$$\vdash [Mon(RG) \wedge Con\exists(RG)]$$

is derivable, then the sequent

$$\vdash Lt(RG) =_e \lambda\overline{v}.[\exists x{:}N].It(RG, \oslash)(x, \overline{v})$$

is derivable. If it is assumed to be both monotonic and $\forall$-continuous, that is if the sequent

$$\vdash [Mon(RG) \wedge Con\forall(RG)]$$

is derivable, then the sequent

$$\vdash Gt(RG) =_e \lambda\overline{v}.[\forall x{:}N].It(RG, \mathrm{V})(x, \overline{v})$$

is also derivable. Here *It* is an *iteration* predicate defined in §4.8 and $\oslash$ and $\mathrm{V}{:}[\overline{\tau}]$ are the empty and universal predicates respectively of their type. Further, under the assumption that RG is monotone and $\exists$-continuous, each of the following sequents are derivable:

$$\vdash [\forall x{:}N].\forall\overline{v}.[\exists u.[u \leq x \wedge It(RG,\oslash)(u,\overline{v})] \rightarrow Lt(RG)(\overline{v})].$$
$$\vdash [\forall x{:}N].\forall\overline{v}.[\exists u.[u \leq x \wedge It(RG,\oslash)(u,\overline{v})] \rightarrow \exists u.[u \leq S(x) \wedge It(RG,\oslash)(u,\overline{v})]].$$

Therefore for each natural number n, the predicate

$$\lambda\overline{v}.\exists u.[u \leq S^n(0) \wedge It(RG,\oslash)(u)]$$

is an *approximation* to $Lt(RG)$. It is still an approximation if $Lt(RG)$ is a *function* Thus as in Scott semantics for the "computable" functions each function is the limit of a series of approximations. But unlike the Scott semantics, each function is named by a term $Lt(RG)$, for which a series of approximations can be defined.

To help clarify the distinction between *defining* and *computing* and to prepare for later sections, the concept of an abstract computer intended to model the essential features of an actual computer is described in §8.2.

Even the simplest programming languages have recursive commands. How such commands can be defined in ITT is illustrated in §8.3 for two recursive commands of a very simple flowchart programming language adapted from an example of [139]. The definitions are used in the derivation of a sequent expressing a relationship between the two commands.

The requirement for a semantics of programming languages more complicated than the simple flowchart language requires the development of a theory of *domains*. In particular, *recursive* domains. As expressed in [72]

> It is the essential purpose of the theory of domains to study classes of spaces which may be used to give semantics for recursive definitions.

'Domain' has a very special meaning in what is now often called *Scott* semantics for programming languages: It is a complete partially ordered set satisfying additional conditions. The theory of domains developed within ITT in §8.4 has a narrower purpose and as a result the definition of domain is simpler. Only a narrower purpose need be served since the *logical* basis for recursions using recursion generators in ITT is simpler than the fixed point *algebraic* method used by Scott: A denotational semantics has already been provided in Chapter 4 for recursive predicates and in Chapter 5 for functions through the denotational semantics provided for ITT in Chapter 3. The narrower purpose remaining to be served is to find solutions for domain equations that arise naturally in the study of programming languages.

A *domain* in ITT is a predicate $B{:}[1]$ together with an *identity* predicate $\simeq_B{:}[1,1]$ on B; that is, $\simeq_B$ is reflexive, symmetric, and transitive. Six examples of domains are introduced in the Example of §4.7.2 that is taken from Milner and Tofte's paper [103]. The domains provide a basis for a semantics of the language described in the paper. Three *primitive* domains are needed, *Const*, *Var*, and *Exp* that are respectively the sets of constants, variables, and expressions of the language, the latter being recursively defined from the former two. For each of these domains intensional identity '=' serves as the identity. In addition there are three recursively *defined* domains *Val*, *Env*, and *Clos*. These are respectively the sets of values, environments, and closures. They are defined by the following *domain equations*:

1)
$$\begin{aligned} &Const \uplus Clos = Val \\ &Var \xrightarrow{\text{fin}} Val = Env \\ &Var \otimes Exp \otimes Env = Clos. \end{aligned}$$

Here the predicates $\uplus$, $\xrightarrow{\text{fin}}$, and $\otimes$ are respectively disjoint union, finite functions, and Cartesian product. Each is an example of a *binary* domain constructor since each defines a new domain in terms of two already defined domains. Domain constructors may be of any non-zero arity.

The domain equations (1) define *Val* in terms of *Clos*, *Clos* in terms of *Env*, and *Env* in terms of *Val*; they are described in §4.7.2 as the solutions of *simultaneous* Horn sequents.

The conditions that primitive domains and domain constructors must satisfy are described and definitions for some of them, including the finite function domain constructor $\xrightarrow{\text{fin}}$, are given in §8.4. No non-finite function domain constructor is defined although one is defined in both [72] and [3]. Whether this is an obstacle to defining a denotational semantics for programming languages within ITT remains to be seen. It is argued in §8.4.2 that it is not because a semantics for programming languages requires only function domain constructors that are defined on *potentially* infinite domains, as described in §4.9, and are therefore necessarily finite; recursive functions that are defined on *actually* infinite domains can be defined using recursion generators.

The chapter closes with a sketch in §8.5 of the role of specification languages in the development of large computer systems and of the possible advantages of ITT as a logical basis for such languages. Since for ITT to play such a role requires the development of a programmable meta-language for ITT, some essential and desirable features of such a language are described.

8.2. Definition, Derivation, and Computation

Chapter 4 emphasizes the *definition* of recursive predicates, and the *derivations* that *confirm* properties of them. A definition of a predicate is signaled by $\stackrel{\text{df}}{=}$. For example, a recursion generator $R\oplus$ can be defined:

$$\begin{aligned} R\oplus \stackrel{\text{df}}{=} (\lambda Z, x, y, z.[\quad &[\exists v{:}N].[x = 0 \wedge y = v \wedge z = v] \vee \\ &[\exists u, v, w{:}N].x = S(u) \wedge y = v \wedge z = S(w)] \quad]). \end{aligned}$$

This recursion generator can then be used in a definition

$$\oplus \overset{\text{df}}{=} Lt(R\oplus)$$

of the ternary predicate $\oplus$, the sum predicate of the natural numbers when they are represented by the members of N. It is from this definition that a derivation can be obtained for any property of the sum predicate that can be expressed in ITT. For example, in exercise 9 of §4.4, the derivable sequent ($\oplus$.2) expresses that $\oplus$ is functional; that the first two arguments of the predicate commute is expressed in the derivable sequent:

Com$\oplus$) $\qquad \vdash [\forall x, y, z.[\oplus(x, y, z) \rightarrow \oplus(y, x, z)]$.

A derivation *confirms* the property.

A *computation* can be understood to be the response to a request to find a term for which a particular formula is derivable. For example,

$$\oplus(S(0), S(0), ?)$$

can be understood to be a request to find a term t for which the sequent

$$\vdash \oplus(S(0), S(0), t)$$

is derivable. A derivation of

$$\vdash \oplus(S(0), S(0), S(S(0)))$$

confirms that $S(S(0))$ is such a term t.

The relationship between a logic like ITT and a programming language like the ones described in [148] is very distant. The logic programming languages Prolog [138] and Gödel [80] are more intimately connected, but as stressed by O'Donnell in [109] they have a *programming system* as a component that defines an output for given inputs.

Turing's description in [145] of what is now called *Turing machines* is the first precise definition of what constitutes a computer and its programs [117]. The basic features of a Turing machine are the essential features of any computing machine. At any give time a machine can be in any one of a finite but unbounded number of *states* each of which can be characterized by a finite string of characters; that is the states of a Turing machine are *potentially* but not *actually* infinite in number. The string of characters for a given state may include a pointer to a *command* in a finite list of commands called a *program*. A command is to be understood as an instruction to the machine to move from its current state to a successor state specified by the current state and the command to which it points. The machine may be started in a given state, called the *initial* state. Should the machine reach a state that does not include a pointer to a command, it is understood to halt at the current state and to terminate the computation described by the program and the initial state. The unification algorithm and the type assignment algorithms described for the logics TT and ITT can be understood in these terms.

A comparison of the results of the paper [108] by Nadathur and Miller with the results of Chapter 4 brings the distinction between definition and computation sharply into focus. The paper is primarily concerned with *computations* described by higher order logic programs while Chapter 4 is concerned with *definitions* of higher order recursions. In contrast to the paper, Chapter 4 treats non-well-founded predicates in the same simple manner as well-founded predicates. The same can be said for Jamie Andrews' paper [6] in which is described a higher-order logic programming system. It is based on the type-free higher order logic NaDSyL [59]. Slight changes in that logic permit Curry's fixed point combinator to be defined [37].

EXERCISES §8.2.

1. Provide a derivation for the sequent (Com$\oplus$). Use the rule LtInd with $R\oplus$.

8.3. Semantics for Recursive Commands

Here a flowchart language example from [139] is used to illustrate the point that a semantics for recursive commands of a programming language can be provided directly in ITT. An understanding of computation as described in §8.2 is assumed. The definitions of two recursive commands are used to derive a sequent expressing a relationship between them. Several recursion generators are defined in this section; each is assumed to be monotonic and it is left to the reader to establish this property. Here only $Lt(RG)$ is used; no use is made of $Gt(RG)$.

8.3.1. Primitive Syntax. A method was described in §4.9.1 for representing denumerably many characters by terms that are derivably distinct, and for constructing distinct representations of distinct strings of such characters. This method is exploited here to provide a syntax for a simple flowchart programming language. The primitive symbols of the language, printed boldface for readability, are defined:

$$
\begin{array}{ll}
\mathbf{false} & \stackrel{\text{df}}{=} \langle 0, Nil\rangle \\
\mathbf{true} & \stackrel{\text{df}}{=} \langle 1, Nil\rangle \\
\mathbf{CExp} & \stackrel{\text{df}}{=} \lambda x1, x2, x3.\langle 2, \langle x1, \langle x2, \langle x3, Nil\rangle\rangle\rangle\rangle \\
\mathbf{dummy} & \stackrel{\text{df}}{=} \langle 3, Nil\rangle \\
\mathbf{CCmd} & \stackrel{\text{df}}{=} \lambda x, z1, z2.\langle 4, \langle x, \langle z1, \langle z2, Nil\rangle\rangle\rangle\rangle \\
\mathbf{SeqCmd} & \stackrel{\text{df}}{=} \lambda z1, z2.\langle 5, \langle z1, \langle z2, Nil\rangle\rangle\rangle \\
\mathbf{WCmd} & \stackrel{\text{df}}{=} \lambda x, z.\langle 6, \langle x, \langle z, Nil\rangle\rangle\rangle \\
\mathbf{RWCmd} & \stackrel{\text{df}}{=} \lambda z, x.\langle 7, \langle z, \langle x, Nil\rangle\rangle\rangle
\end{array}
$$

Some justification for these definitions is necessary since the abstraction operators $\lambda x1$, $\lambda x2$, and $\lambda x3$ cannot be applied to type 1 terms, while at the same time it is assumed that, for example, $CExp(t_1, t_2, t_3)$:1 whenever t_1, t_2, t_3:1.

Assume that $x1, x2, x3$:1. Thus $\langle x1, \langle x2, \langle x3, Nil\rangle\rangle\rangle$:1 by secondary typing. Since the term 2 has secondary type 1, it follows that

$$(\lambda w.w(2, \langle x1, \langle x2, \langle x3, Nil\rangle\rangle\rangle)):[[1,1]]$$

so that the abstraction operators can be properly applied. Again by secondary typing therefore $CExp(t_1, t_2, t_3)$:1 whenever t_1, t_2, t_3:1.

These definitions have been chosen to ensure that the following theorem is provable:

THEOREM 56. *Let PS be any of the primitive symbols and let $\overline{x1}$ and $\overline{x2}$ be sequences of distinct variables of type* 1 *and of length appropriate for the symbol. Then*

$PS.1)$ $\quad \vdash \forall \overline{x1}, \overline{x2}.[PS(\overline{x1}) = PS(\overline{x2}) \rightarrow \overline{x1} = \overline{x2}]$.

Here $\overline{x1} = \overline{x2}$ abbreviates the conjunction of identities for the corresponding variables. Further, let $PS1$ and $PS2$ be distinct primitive symbols and let $\overline{x1}$ and $\overline{x2}$ be sequences of distinct variables of type 1 *and of length appropriate for the symbols. Then*

$PS.2)$ $\quad \vdash \forall \overline{x1}, \overline{x2}.\neg PS1(\overline{x1}) = PS2(\overline{x2})$.

A proof of the theorem is left to the reader.

8.3.2. Expressions. The rules for forming correct expressions in the language are implicitly stated in the following definition of the recursion generator $RExp$:

$$RExp \overset{\text{df}}{=} \lambda Z, x.[x = \mathbf{false} \vee x = \mathbf{true} \vee [\exists u, v, w{:}Z].x = \mathbf{CExp}(u, v, w)].$$

Define $Exp \overset{\text{df}}{=} Lt(RExp)$. The following sequent is derivable:

Exp) $\quad \vdash \forall x.[Exp(x) \leftrightarrow [x = \mathbf{false} \vee x = \mathbf{true} \vee [\exists x1, x2, x3{:}Exp].x = \mathbf{CExp}(x1, x2, x3)]$.

In addition each of the following sequents is derivable:

Exp.1) $\quad \vdash Exp(\mathbf{false})$

Exp.2) $\quad \vdash Exp(\mathbf{true})$

Exp.3) $\quad Exp(x1), Exp(x2), Exp(x3) \vdash Exp(\mathbf{CExp}(x1, x2, x3))$.

The rules for forming correct commands in the language are implicitly stated in the following definition of the recursion generator $RCmd$:

$$\begin{aligned} RCmd \overset{\text{df}}{=} \lambda Z, w.[&w = \mathbf{dummy} \\ &\vee \exists x, z1, z2.[Exp(x) \wedge Z(z1) \wedge Z(z2) \wedge w = \mathbf{CCmd}(x, z1, z2)] \\ &\vee \exists z1, z2.[Z(z1) \wedge Z(z2) \wedge w = \mathbf{SeqCmd}(z1, z2)] \\ &\vee \exists x, z.[Exp(x) \wedge Z(z) \wedge w = \mathbf{WCmd}(x, z)] \\ &\vee \exists x, z.[Exp(x) \wedge Z(z) \wedge w = \mathbf{RWCmd}(z, x)]].\end{aligned}$$

Define $Cmd \stackrel{\text{df}}{=} Lt(RCmd)$. The following sequent is derivable:

$$\begin{aligned}\text{Cmd)}\ \vdash \forall z.[Cmd(z) \leftrightarrow [\,&w = \mathbf{dummy}\\ &\vee[\exists x{:}Exp].[\exists z1, z2{:}Cmd].z = \mathbf{CCmd}(x, z1, z2)\\ &\vee[\exists z1, z2{:}Cmd].z = \mathbf{SeqCmd}(z1, z2)\\ &\vee[\exists x{:}Exp].[\exists z1{:}Cmd].z = \mathbf{WCmd}(x, z1)\\ &\vee[\exists x{:}Exp].[\exists z1{:}Cmd].z = \mathbf{RWCmd}(z1, x)\,].\end{aligned}$$

In addition each of the following sequents is derivable:

Cmd.1)		$\vdash Cmd(\mathbf{dummy})$
Cmd.2)	$Exp(x), Cmd(z1), Cmd(z2)$	$\vdash Cmd(\mathbf{CCmd}(x, z1, z2))$
Cmd.3)	$Cmd(z1), Cmd(z2)$	$\vdash Cmd(\mathbf{SeqCmd}(z1, z2))$
Cmd.4)	$Exp(x), Cmd(z)$	$\vdash Cmd(\mathbf{WCmd}(x, z))$
Cmd.5)	$Exp(x), Cmd(z)$	$\vdash Cmd(\mathbf{RWCmd}(z, x))$.

8.3.3. Expression semantics. The semantics for the language is defined in terms of a predicate $St{:}[1]$ that applies to the *states* of the computing system for which the language is a programming language. A predicate $Bv{:}[1]$, *Boolean values*, is defined:

$$Bv \stackrel{\text{df}}{=} \lambda v.[v = 0 \vee v = 1].$$

Let RES be the following recursion generator:

$$\begin{aligned}RES \stackrel{\text{df}}{=} \lambda Z, w1, w2, w3.[\\ &\exists u, v.[v = 0 \wedge St(u) \wedge w1 = \mathbf{false} \wedge w2 = u \wedge w3 = v)]\vee\\ &\exists u, v.[v = 1 \wedge St(u) \wedge w1 = \mathbf{true} \wedge w2 = u \wedge w3 = v)]\vee\\ &\exists x1, x2, x3, u, v.[\\ &\quad [[Z(x1, u, 1) \wedge Z(x2, u, v)] \vee [Z(x1, u, 0) \wedge Z(x3, u, v)]]\wedge\\ &\qquad w1 = \mathbf{CExp}(x1, x2, x3) \wedge w2 = u \wedge w3 = v)\,].\end{aligned}$$

Define $ExpSem \stackrel{\text{df}}{=} Lt(RES)$. The formula $ExpSem(x, u, v)$ is intended to express that the expression x in the state u has the Boolean value v; that is, the following sequent should be derivable:

ES) $\vdash [\forall x{:}Exp].[\forall u{:}St].[\exists! v{:}Bv].ExpSem(x, u, v).$

In addition each of the following sequents should be derivable:

ES.1) $\vdash [\forall u{:}St].[\forall v{:}Bv].[ExpSem(\mathbf{false}, u, v) \leftrightarrow v = 0]$

ES.2) $\vdash [\forall u{:}St].[\forall v{:}Bv].[ExpSem(\mathbf{true}, u, v) \leftrightarrow v = 1]$

ES.3) $$\begin{aligned}&[\forall x1, x2, x3{:}Exp].[\forall u{:}St].[\forall v{:}Bv].\\ &[[[ExpSem(x1, u, 1) \wedge ExpSem(x2, u, v)]\\ &\vee[ExpSem(x1, u, 0) \wedge ExpSem(x3, u, v)]] \leftrightarrow\\ &\qquad ExpSem(\mathbf{CExp}(x1, x2, x3), u, v)].\end{aligned}$$

Derivations of (ES.1), (ES.2), and (ES.3) are left to the reader. A derivation of (ES) follows in which ES abbreviates $ExpSem$ and

$$P \stackrel{\text{df}}{=} \lambda x.[\forall u{:}St].[\exists! v{:}Bv].ExpSem(x, u, v).$$

$$\begin{array}{ll}
-[\forall x{:}Exp].P(x) & \\
-\forall x.[RExp(P, x) \rightarrow P(x)] & \dashv \text{LtInd} \\
+RExp(P, x) & \text{(a)} \\
-P(x) & \\
-[\forall u{:}St].[\exists! v{:}Bv].ES(x, u, v) & \dashv \text{df } P \\
+St(u) & \\
-[\exists! v{:}Bv].ES(x, u, v) & \\
-[\exists v{:}Bv].[ES(x, u, v) \wedge [\forall v'{:}Bv].[ES(x, u, v') \rightarrow v = v']] & \text{(b)}
\end{array}$$

Because of the definition of $RExp$ and the node (a), the branch splits into three with each subbranch headed by one of the following nodes:

1. $+x = \textbf{false}$
2. $+x = \textbf{true}$
3. $+\exists x1, x2, x3.[P(x1) \wedge P(x2) \wedge P(x3) \wedge x = \textbf{CExp}(x1, x2, x3)]$

Consider branch (1) first; branch (2) is similar:

$$\begin{array}{ll}
+x = \textbf{false} & \\
-[\exists v{:}Bv].[ES(\textbf{false}, u, v) \wedge [\forall v'{:}Bv].[ES(\textbf{false}, u, v') \rightarrow v = v']] & \\
-[ES(\textbf{false}, u, 0) \wedge [\forall v'{:}Bv].[ES(\textbf{false}, u, v') \rightarrow 0 = v']] & \dashv [0/v] \\
-[\forall v'{:}Bv].[ES(\textbf{false}, u, v') \rightarrow 0 = v']] & \dashv \text{(ES.1)}, -\wedge \\
+Bv(v') & \\
+ES(\textbf{false}, u, v') & \\
-0 = v' & \\
+v' = 0 & \dashv \text{(ES.1)} \\
\hline
\end{array}$$

Consider now branch (3):

$$\begin{array}{ll}
+\exists x1, x2, x3.[P(x1) \wedge P(x2) \wedge P(x3) \wedge x = \textbf{CExp}(x1, x2, x3)] & \\
+P(x1) \quad +P(x2) \quad +P(x3) & \text{(c)} \\
+x = \textbf{CExp}(x1, x2, x3) & \text{(d)} \\
+[Bv(v1) \wedge ES(x1, u, v1) \wedge [\forall v'{:}Bv].[ES(x1, u, v') \rightarrow v1 = v'] & \\
+[Bv(v2) \wedge ES(x1, u, v2) \wedge [\forall v'{:}Bv].[ES(x1, u, v') \rightarrow v2 = v'] & \\
+[Bv(v3) \wedge ES(x1, u, v3) \wedge [\forall v'{:}Bv].[ES(x1, u, v') \rightarrow v3 = v'] & \\
 & \text{(e)} \dashv \text{(c) df } P \\
-[\exists v{:}Bv].[ES(\textbf{CExp}(x1, x2, x3), u, v) \wedge & \\
\quad [\forall v'{:}Bv].[ES(\textbf{CExp}(x1, x2, x3), u, v') \rightarrow v = v']] & \text{(f)} \dashv \text{(b),(d)}
\end{array}$$

$-[ES(\mathbf{CExp}(x1, x2, x3), u, 0) \wedge$
$\quad [\forall v' : Bv].[ES(\mathbf{CExp}(x1, x2, x3), u, v') \rightarrow 0 = v']]$ $\dashv [0/v](f)$
$-[ES(\mathbf{CExp}(x1, x2, x3), u, 1) \wedge$
$\quad [\forall v' : Bv].[ES(\mathbf{CExp}(x1, x2, x3), u, v') \rightarrow 1 = v']]$ $\dashv [1/v]$

L R
$-ES(\mathbf{CExp}(x1, x2, x3), u, 0)$

LL LR
$-ES(\mathbf{CExp}(x1, x2, x3), u, 1)$
$-[[ES(x1, u, 1) \wedge ES(x2, u, 0)]$
$\quad \vee[ES(x1, u, 0) \wedge ES(x3, u, 0)]]$ $\dashv$(ES.3)
$-[ES(x1, u, 1) \wedge ES(x2, u, 0)]$
$-[ES(x1, u, 0) \wedge ES(x3, u, 0)]$
$-[ES(x1, u, 1) \wedge ES(x2, u, 1)]$
$-[ES(x1, u, 0) \wedge ES(x3, u, 1)]$
$+[ES(x1, u, 0) \vee ES(x1, u, 1)]$
$+[ES(x2, u, 0) \vee ES(x2, u, 1)]$
$+[ES(x3, u, 0) \vee ES(x3, u, 1)]$

LR
$-[\forall v' : Bv].[ES(\mathbf{CExp}(x1, x2, x3), u, v') \rightarrow 1 = v']$
$+Bv(v')$
$+ES(\mathbf{CExp}(x1, x2, x3), u, v')$
$-1 = v'$
$+v' = 0$
$+ES(\mathbf{CExp}(x1, x2, x3), u, 0)$

R
$-[\forall v' : Bv].[ES(\mathbf{CExp}(x1, x2, x3), u, v') \rightarrow 0 = v']$

RL RR
similar to LR

RR
$-[\forall v' : Bv].[ES(\mathbf{CExp}(x1, x2, x3), u, v') \rightarrow 1 = v']$
$+Bv(v')$ $+ES(\mathbf{CExp}(x1, x2, x3), u, v')$
$-1 = v'$
$+0 = v'$
$+ES(\mathbf{CExp}(x1, x2, x3), u, 0)$
$+Bv(v'')$ $+ES(\mathbf{CExp}(x1, x2, x3), u, v'')$

$-0 = v''$
$+1 = v''$
$+ES(\mathbf{CExp}(x1, x2, x3), u, 1)$
$+[[ES(x1, u, 1) \wedge ES(x2, u, 0)]$
$\quad \vee[ES(x1, u, 0) \wedge ES(x3, u, 0)]]$ $\dashv$(ES.3)
$+[[ES(x1, u, 1) \wedge ES(x2, u, 1)]$
$\quad \vee[ES(x1, u, 0) \wedge ES(x3, u, 1)]]$ $\dashv$(ES.3)
$-[ES(x2, u, 0) \wedge ES(x2, u, 1)]$ $\dashv$(e)
$-[ES(x3, u, 0) \wedge ES(x3, u, 1)]$ $\dashv$(e)

ExpSem:[[1], [1], 1] is a function with arguments from *Exp* and *St* and values from *Bv*. An alternative way of presenting the same semantics is to *Curry* the function to turn it into a function with a single argument from *Exp* and values from a domain $[St \Rightarrow Bv]$ of the total functions with arguments from *St* and values from *Bv*. An advantage of doing so is to make the semantics more uniform in the sense that functions of only a single argument are needed. The domain $[St \Rightarrow Bv]$ can of course be defined as a predicate in ITT, with extensional identity as the identity. It is unnecessary to define it as a Scott domain as described in [72] or [2].

8.3.4. Command semantics. A particular command can be understood to be defined by a set of ordered pairs of members of *St*. A command *z* is defined for a particular state *ua* if there is a state *ub* for which the pair $\langle ua, ub\rangle$ is in the set for the command. If for a given *ua* there is at most a single such state *ub* then *z* is said to be a *deterministic* command; otherwise a *non-deterministic* command. Here it is assumed that all commands are deterministic. For non-deterministic commands universal quantification over the states *St* would have to be expressed in terms of universal quantification over a potentially infinite *St* as described in §4.9.3.

The command semantics is defined in terms of two recursive predicates *CmdSem* and *StT*. *CmdSem* has three arguments, a command and two states. The intended meaning of $CmdSem(z, ua, ub)$ is that the pair $\langle ua, ub\rangle$ is in the set of pairs for the command *z*. The *state transfer* predicate *StT* has four arguments, a member of *N*, a command and two states. The intended meaning of $StT(xn, z, ua, ub)$ is that the system is moved from state *ua* to state *ub* in *xn* steps by *z*, where in one step the system is moved from a state *ua'* to a state *ub'* if $CmdSem(z, ua', ub')$.

The command semantics defined may not assign a value to a command for a given state; it is possible for the system to be in an *infinite* loop so that no successor state can be defined. Nevertheless it is possible to prove several significant properties of the semantics.

A method of defining recursive predicates by simultaneous Horn sequents is described in §4.7.2. Here the method is applied in the definition of command semantics. Two simultaneous Horn sequents follow in which *CmdSem* and *StT* are the defined constants and *RCS* and *RST* are the simultaneous recursion generators.

HCS) $RCS(CmdSem, z, ua, ub, StT) \quad \vdash CmdSem(z, ua, ub)$
HST) $RST(CmdSem, StT, xn, z, ua, ub) \quad \vdash StT(xn, z, ua, ub)$

Here are the definitions of *RCS*, *RST*, and *RCST*, the latter being the recursion generator defined by *RCS* and *RST*; see §4.7.2. In the fourth and fifth disjuncts of the definition of *RCS* the predicate *Least*, defined in exercise (6) of §4.4, is used with first argument $M(Z2, x, z1, ua)$ where

$$M \stackrel{\text{df}}{=} (\lambda Z2, x, z1, ua, xn.[\exists u1{:}St].[Z2(xn, z1, ua, u1) \wedge ExpSem(x, u1, 0)]).$$

$$\begin{aligned}
RCS \stackrel{\text{df}}{=} \lambda Z1, z, ua, ub, Z2.[St(ua) \wedge St(ub) \wedge [\\
&[ua = ub \wedge z = \textbf{dummy}] \vee \\
&[\exists z1, z2{:}Cmd].[\\
&\quad [\exists u{:}St].[Z1(z1, ua, u) \wedge Z1(z2, u, ub)] \wedge \\
&\qquad z = \textbf{SeqCmd}(z1, z2)] \vee \\
&[\exists x{:}Exp].[\exists z1, z2{:}Cmd].[\\
&\quad [\ [ExpSem(x, ua, 1) \wedge Z1(z1, ua, ub)] \vee \\
&\qquad [ExpSem(x, ua, 0) \wedge Z1(z2, ua, ub)]\ \ ] \wedge \\
&\qquad z = \textbf{CCmd}(x, z1, z2)] \vee \\
&[\exists x{:}Exp].[\exists z1{:}Cmd].[\\
&\quad \exists yn.[Least(M(Z2, x, z1, ua), yn) \wedge Z2(yn, z1, ua, ub)] \wedge \\
&\qquad z = \textbf{WCmd}(x, z1)] \vee \\
&[\exists x{:}Exp].[\exists z1{:}Cmd].[\\
&\quad \exists yn.[Least(M(Z2, x, z1, ua), yn) \wedge Z2(S(yn), z1, ua, ub)] \wedge \\
&\qquad z = \textbf{RWCmd}(z1, x)].
\end{aligned}$$

$$\begin{aligned}
RST \stackrel{\text{df}}{=} \lambda Z1, Z2, xn, z, ua, ub.[St(ua) \wedge St(ub) \wedge Cmd(z) \wedge \\
&[\ [xn = 0 \wedge ua = ub] \vee \\
&\quad [\exists yn{:}N].[\exists u{:}St].[xn = S(yn) \wedge Z2(yn, z, ua, u) \wedge \\
&\qquad Z1(z, u, ub)]\ ].
\end{aligned}$$

The single recursion generator defined from *RCS* and *RST* as described in §4.7.2 is

$$\begin{aligned}
RCST \stackrel{\text{df}}{=} \lambda Z, \overline{x_1}, \overline{x_2}.[\\
&RCS((\lambda\overline{x_1}.\exists\overline{x_2}.Z(\overline{x_1}, \overline{x_2})), \overline{x_1}, (\lambda\overline{x_2}.\exists\overline{x_1}.Z(\overline{x_1}, \overline{x_2}))) \wedge \\
&RST((\lambda\overline{x_1}.\exists\overline{x_2}.Z(\overline{x_1}, \overline{x_2})), (\lambda\overline{x_2}.\exists\overline{x_1}.Z(\overline{x_1}, \overline{x_2})), \overline{x_2})\ \ ].
\end{aligned}$$

Here $\overline{x_1}$ is $z1, ua1, ub1$ and $\overline{x_2}$ is $xn, z2, ua2, ub2$.

The predicates *CmdSem* and *StT* are defined:

$$CmdSem \stackrel{\text{df}}{=} \lambda z1, ua1, ub1.\exists xn, z2, ua2, ub2. Lt(RCST)(z1, ua1, ub1, xn, z2, ua2, ub2).$$

$$StT \stackrel{\text{df}}{=} \lambda xn, z2, ua2, ub2.\exists z1, ua1, ub1.$$
$$Lt(RCST)(z1, ua1, ub1, xn, z2, ua2, ub2).$$

Then by Theorem 44 of §4.7.2, each of the sequents (HCS) and (HST) and its converse is derivable provided both the following sequents are derivable:

a1) $\vdash \exists z, ua, ub.CmdSem(z, ua, ub)$

a2) $\vdash \exists xn, z, ua, ub.StT(xn, z, ua, ub)$.

The derivations of these two sequents, under the assumption that the sequent $\vdash \exists u.St(u)$ is derivable, are left as exercises.

Let the predicate *Fnc* be defined:

$$Fnc \stackrel{\text{df}}{=} (\lambda Z.\forall v, v'.[Z(v) \wedge Z(v') \rightarrow v{=}v']).$$

A predicate is said to tbe *functional* if it satisfies *Fnc*. Given that only deterministic commands are admitted and given the intended meaning of the formula $CmdSem(z, ua, ub)$, $CmdSem(z, ua)$ should be functional; that is, the following sequent should be derivable:

CS) $\vdash [\forall z{:}Cmd].[\forall ua{:}St].Fnc(CmdSem(z, ua))$.

A derivation is given below that makes use of the following derivable sequent:

ST) $\vdash [\forall z{:}Cmd].[\,[\forall ua{:}St].Fnc(CmdSem(z, ua)) \rightarrow$
$[\forall xn{:}N].[\forall ua{:}St].Fnc(StT(xn, z, ua))]$.

A derivation of (ST) follows:

$-[\forall z{:}Cmd].[\,[\forall ua{:}St].Fnc(CmdSem(z, ua)) \rightarrow$
$\quad [\forall xn{:}N].[\forall ua{:}St].Fnc(StT(xn, z, ua))]$.
$+Cmd(z)$
$+[\forall ua{:}St].Fnc(CmdSem(z, ua))$ (a)
$-[\forall xn{:}N].[\forall ua{:}St].Fnc(StT(xn, z, ua))]$

$\dashv$NInd

L R

$-[\forall ua{:}St].Fnc(StT(0, z, ua))]$
$+St(ua)$
$-Fnc(StT(0, z, ua))$
$-\forall u, u'.[StT(0, z, ua, u) \wedge StT(0, z, ua, u') \rightarrow u{=}u'])$
$+StT(0, z, ua, u)$
$+StT(0, z, ua, u')$
$-u = u'$
$+RST(CmdSem, StT, 0, z, ua, u)$ $\dashv$(Lt.3)
$+[St(ua) \wedge St(u) \wedge [\,[0{=}0 \wedge ua{=}u] \vee$
$\quad [\exists yn{:}N].[\exists u'{:}St].[0{=}S(yn) \wedge StT(yn, z, ua, u') \wedge$
$\quad\quad CmdSem(z, u', u)]\,]$ $\dashv$df *RST*
$+St(ua)$ $St(u)$
$+[\,[0{=}0 \wedge ua{=}u] \vee$

$[\exists yn{:}N].[\exists u'{:}St].[0{=}S(yn) \wedge StT(yn,z,ua,u') \wedge$
$CmdSem(z,u',u)]\,]$

$\dashv +\vee$

LL | LR

$+[0{=}0 \wedge ua{=}u]$
$+ua = u$
$+ua = u'$ similar derivation
$+u = u'$

LR

$+[\exists yn{:}N].[\exists u'{:}St].[0{=}S(yn) \wedge Z2(yn,z,ua,u') \wedge Z1(z,u',u)]\,]$
$+N(yn)$ $0{=}S(yn)$

$\dashv$(S.2)

R

$-\forall xn.[\,[\forall ua{:}St].Fnc(StT(xn,z,ua)) \rightarrow$
$[\forall ua{:}St].Fnc(StT(S(xn),z,ua))]$
$+[\forall ua{:}St].Fnc(StT(xn,z,ua))$ (b)
$-[\forall ua{:}St].Fnc(StT(S(xn),z,ua))$
$+St(ua)$
$-Fnc(StT(S(xn),z,ua))$
$-\forall u,u'.[StT(S(xn),z,ua,u) \wedge StT(S(xn),z,ua,u') \rightarrow u{=}u']$)
$+StT(S(xn),z,ua,u)$ (c)
$+StT(S(xn),z,ua,u')$ (c')
$-u = u'$ (d)
$+St(ua)$ $+St(u)$ $\dashv$(c), df RST, $+\wedge$
$+[\,[S(xn){=}0 \wedge ua{=}u] \vee$
$[\exists yn{:}N].[\exists u'{:}St].[S(xn){=}S(yn) \wedge StT(yn,z,ua,u') \wedge$
$CmdSem(z,u',u)]\,]$ $\dashv$(c), df RST, $+\wedge$

RL | RR

$+[S(xn){=}0 \wedge ua{=}u]$
$+S(xn){=}0$

$\dashv$(S.2)

RR

$+[\exists yn{:}N].[\exists u1{:}St].[S(xn) = S(yn) \wedge$
$StT(yn,z,ua,u1) \wedge CmdSem(z,u1,u)]\,]$
$+N(yn)$ $+St(u1)$ $+S(xn) = S(yn)$
$+StT(yn,z,ua,u1)$ $+CmdSem(z,u1,u)$
$+xn{=}yn$ $\dashv$(S.1)
$+StT(xn,z,ua,u1)$
$+StT(xn,z,ua,u1')$ $+CmdSem(z,u1',u')$ similar from $\dashv$(c')

$+u1{=}u1'$ $\dashv$(b)
$+CmdSem(z, u1, u)$
$+CmdSem(z, u1, u')$
$+u = u'$ $\dashv$(a)

$\dashv$(d)

A derivation of (CS) follows in which

$$P \stackrel{\text{df}}{=} \lambda z.\forall ua.Fnc(CmdSem(z, ua)).$$

$-\forall z.[Cmd(z) \rightarrow P(z)]$
$-\forall z.[Lt(RCmd)(z) \rightarrow P(z)]$ $\dashv$df Cmd
$-\forall z.[RCmd(P, z) \rightarrow P(z)]$ $\dashv$LtInd
$+RCmd(P, z)$ (a)
$-P(z)$
$-\forall ua, ub, ub'.[CmdSem(z, ua, ub) \wedge$
$CmdSem(z, ua, ub') \rightarrow ub = ub']$
$+CmdSem(z, ua, ub)$ (b)
$+CmdSem(z, ua, ub')$ (b')
$-ub = ub'$ (c)

There are five subbranches, one for each of the disjunctions of $RCmd$ in §8.3.2 arising from (a). The first and the last will be completed; the remaining three cases are left to the reader.

1)
$+z = \mathbf{dummy}$
$+CmdSem(\mathbf{dummy}, ua, ub)$
$+CmdSem(\mathbf{dummy}, ua, ub')$
$+ua = ub \quad +ua = ub'$
$+ub = ub'$

$\dashv$(c)

5)
$+\exists x, z1.[Exp(x) \wedge P(z1) \wedge z = \mathbf{RWCmd}(z1, x)]$
$+Exp(x) \quad +P(z1)$
$+z = \mathbf{RWCmd}(z1, x)$ (d)
$+\forall ua.Fnc(CmdSem, z1, ua)$ $\dashv$df P
$+\forall ua, ub, ub'.[CmdSem(z1, ua, ub) \wedge CmdSem(z1, ua, ub') \rightarrow ub = ub']$
$+CmdSem(\mathbf{RWCmd}(z1, x), ua, ub)$ (e)$\dashv$(b),(d)
$+CmdSem(\mathbf{RWCmd}(z1, x), ua, ub')$ (e') $\dashv$(b'),(d)
$+RCS(CmdSem, \mathbf{RWCmd}(z1, x), ua, ub, StT)$ $\dashv$(Lt.3)

There are five disjuncts for RCS. Only the last needs to be considered; each branch corresponding to one of the others can be closed because of Theorem 56 of §8.3.1; this theorem will be referred to as T.56. The last disjunct is

$+[St(ua) \wedge St(ub) \wedge$
$\quad [\exists x':Exp].[\exists z1':Cmd].[$
$\quad\quad \exists yn.[Least(M(StT, x', z1', ua), yn) \wedge$
$\quad\quad\quad StT(S(yn), z1', ua, ub)] \wedge$
$\quad\quad\quad\quad \mathbf{RWCmd}(z1, x) = \mathbf{RWCmd}(z1', x')].$
$+St(ua) \quad St(ub)$
$+Exp(x') \quad +Cmd(z1')$
$+\exists yn.[Least(M(StT, x', z1', ua), yn) \wedge$
$\quad StT(S(yn), z1', ua, ub)]$
$+\mathbf{RWCmd}(z1, x) = \mathbf{RWCmd}(z1', x')]$
$+z1{=}z1' \quad x{=}x'$ $\dashv$T.56
$+Exp(x) \quad +Cmd(z1)$
$+\exists yn.[Least(M(StT, x, z1, ua), yn) \wedge$
$\quad StT(S(yn), z1, ua, ub)]$
$+Least(M(StT, x, z1, ua), yn)$
$+StT(S(yn), z1, ua, ub)]$

The last two conclusions have been derived from (e). The next two nodes can be justified by repeating the derivation using (e') instead of (e).

$+Least(M(StT, x, z1, ua), yn')$
$+StT(S(yn'), z1, ua, ub')]$
$+yn{=}yn'$ $\dashv$Least.1
$+StT(S(yn), z1, ua, ub')]$
$+ub{=}ub'$ $\dashv$(ST)
$\dashv$(c)

8.3.5. Example theorem. The following derivable sequent is an illustration of the kind of result that can be proved about a programming language when its semantics is defined in ITT:

RWW) $\vdash [\forall x:Exp].[\forall z:Cmd].[\forall ua, ub:St].[CmdSem(\mathbf{RWCmd}(z, x), ua, ub)$
$\leftrightarrow CmdSem(\mathbf{SeqCmd}(\mathbf{WCmd}(x, z), z), ua, ub)].$

A derivation follows:

$-[\forall x:Exp].[\forall z:Cmd].[\forall ua, ub:St].[CmdSem(\mathbf{RWCmd}(z, x), ua, ub)$
$\quad \leftrightarrow CmdSem(\mathbf{SeqCmd}(\mathbf{WCmd}(x, z), z), ua, ub)]$
$+Exp(x) \quad +Cmd(z) \quad +St(ua) \quad +St(ub)$
$-[CmdSem(\mathbf{RWCmd}(z, x), ua, ub)$
$\quad \leftrightarrow CmdSem(\mathbf{SeqCmd}(\mathbf{WCmd}(x, z), z), ua, ub)]$

L R

$+CmdSem(\mathbf{RWCmd}(z, x), ua, ub)$
$-CmdSem(\mathbf{SeqCmd}(\mathbf{WCmd}(x, z), z), ua, ub)$
$+RCS(CmdSem, \mathbf{RWCmd}(z, x), ua, ub, StT)$ $\dashv$Lt.3

$-RCS(CmdSem, \mathbf{SeqCmd}(\mathbf{WCmd}(x, z), z), ua, ub, StT)$ (a)$\dashv$Lt.2
$+[St(ua) \wedge St(ub) \wedge$ $\dashv$df RCS
$\quad [\exists x'{:}Exp].[\exists z1'{:}Cmd].[\exists yn.[Least(M(StT, x', z1', ua), yn) \wedge$
$\qquad StT(S(yn), z1', ua, ub)] \wedge \mathbf{RWCmd}(z, x) = \mathbf{RWCmd}(z1', x')]$
$+Exp(x') \quad +Cmd(z1')$
$+\exists yn.[Least(M(StT, x', z1', ua), yn) \wedge$
$\quad StT(S(yn), z1', ua, ub)]$
$+\mathbf{RWCmd}(z, x) = \mathbf{RWCmd}(z1', x')]$
$+z{=}z1' \quad x{=}x'$ $\dashv$T.56
$+\exists yn.[Least(M(StT, x, z, ua), yn) \wedge$
$\quad StT(S(yn), z, ua, ub)]$
$+Least(M(StT, x, z, ua), yn)$ (b)
$+StT(S(yn), z, ua, ub)]$
$+RST(StT, S(yn), z, ua, ub)]$ $\dashv$Lt.3
$+[St(ua) \wedge St(ub) \wedge Cmd(z) \wedge$
$\quad [[S(yn) = 0 \wedge ua = ub] \vee$
$\qquad [\exists yn'{:}N].[\exists u{:}St].[S(yn) = S(yn') \wedge StT(yn', z, ua, u) \wedge$
$\qquad\quad CmdSem(z, u, ub)]]]$ $\dashv$df ST
$+[[S(yn) = 0 \wedge ua = ub] \vee$
$\quad [\exists yn'{:}N].[\exists u{:}St].[S(yn) = S(yn') \wedge StT(yn', z, ua, u) \wedge$
$\qquad CmdSem(z, u, ub)]]$
$+[\exists yn'{:}N].[\exists u{:}St].[S(yn) = S(yn') \wedge StT(yn', z, ua, u) \wedge$ $\dashv$(S.2)
$\quad CmdSem(z, u, ub)]$
$+[\exists u{:}St].[S(yn) = S(yn') \wedge StT(yn', z, ua, u) \wedge$
$\quad CmdSem(z, u, ub)]$
$+St(u) S(yn) = S(yn') \quad +yn{=}yn'$ $\dashv$(S.1)
$+StT(yn, z, ua, u)$ (c)
$+CmdSem(z, u, ub)$ (d)

From the second disjunct of the definition of RCS follows from (a).

$-[St(ua) \wedge St(ub) \wedge$
$\quad [\exists z1, z2{:}Cmd].[$
$\qquad [\exists u'{:}St].[CmdSem(z1, ua, u') \wedge CmdSem(z2, u', ub)] \wedge$
$\qquad\quad \mathbf{SeqCmd}(\mathbf{WCmd}(x, z), z) = \mathbf{SeqCmd}(z1, z2)]$
$-[\exists z1, z2{:}Cmd].[$
$\quad [\exists u'{:}St].[CmdSem(z1, ua, u') \wedge CmdSem(z2, u', ub)] \wedge$
$\qquad \mathbf{SeqCmd}(\mathbf{WCmd}(x, z), z) = \mathbf{SeqCmd}(z1, z2)]$
$-Cmd(\mathbf{WCmd}(x, z)) \wedge Cmd(z)$
$\quad [\exists u'{:}St].[CmdSem(\mathbf{WCmd}(x, z), ua, u') \wedge CmdSem(z, u', ub)] \wedge$
$\qquad \mathbf{SeqCmd}(\mathbf{WCmd}(x, z), z) = \mathbf{SeqCmd}(\mathbf{WCmd}(x, z), z)]$
$-[\exists u'{:}St].[CmdSem(\mathbf{WCmd}(x, z), ua, u') \wedge CmdSem(z, u', ub)] \wedge$
$\quad \mathbf{SeqCmd}(\mathbf{WCmd}(x, z), z) = \mathbf{SeqCmd}(\mathbf{WCmd}(x, z), z)]$

$-[\exists u'{:}St].[CmdSem(\mathbf{WCmd}(x,z),ua,u') \wedge CmdSem(z,u',ub)]$
$-[CmdSem(\mathbf{WCmd}(x,z),ua,u) \wedge CmdSem(z,u,ub)]$
$-CmdSem(\mathbf{WCmd}(x,z),ua,u)$ $\dashv$(d)
$-RCS(CmdSem,\mathbf{WCmd}(x,z),ua,u)$ $\dashv$Lt.2
$-[St(ua) \wedge St(u)]\wedge$
$[\exists x'{:}Exp].[\exists z1{:}Cmd].[$
$\exists yn'.[Least(M(StT,x',z1,ua),yn')\wedge$
$StT(yn',z1,ua,u)]\wedge$
$\mathbf{WCmd}(x,z) = \mathbf{WCmd}(x',z1)]$ $\dashv$df RCS
$-[\exists x'{:}Exp].[\exists z1{:}Cmd].[$
$\exists yn'.[Least(M(StT,x',z1,ua),yn')\wedge$
$StT(yn',z1,ua,u)]\wedge$
$\mathbf{WCmd}(x,z) = \mathbf{WCmd}(x',z1)]$
$-\exists yn'.[Least(M(StT,x,z,ua),yn')\wedge$
$StT(yn',z,ua,u)]\wedge$
$\mathbf{WCmd}(x,z) = \mathbf{WCmd}(x,z)]$
$-[Least(M(StT,x,z,ua),yn)\wedge$
$StT(yn,z,ua,u)\wedge$
$\mathbf{WCmd}(x,z) = \mathbf{WCmd}(x,z)]$
$-[StT(yn,z,ua,u) \wedge \mathbf{WCmd}(x,z) = \mathbf{WCmd}(x,z)]$ $\dashv$(b)
$-\mathbf{WCmd}(x,z) = \mathbf{WCmd}(x,z)$ $\dashv$(c)

R
Similar to L.

EXERCISES §8.3.

1. Provide derivations for (Exp) and (Cmd).
2. Provide derivations for the sequents (a1) and (a2) of §8.3.4.
3. Provide derivations for the sequents

$$\vdash [\forall z{:}Cmd].\forall ua,ub.[CmdSem(z,ua,ub) \rightarrow [St(ua) \wedge St(ub)]]$$

$$\vdash [\forall z{:}Cmd].[\forall ua,ub{:}St].[\forall xn,xn'{:}N].$$
$$[StT(xn,z,ua,ub) \wedge StT(xn',z,ua,ub) \rightarrow xn = xn']$$

4. Complete the R branch of the derivation in §8.3.5.

8.4. Recursive Domains

By a *domain* is meant an ordered pair $\langle B,\simeq\rangle$ where $\simeq$ is an identity on B. The predicate B is referred to as the *base* of the domain and $\simeq$ as the *identity*. The predicate DOM applies to domains:

$$DOM \stackrel{\text{df}}{=} \lambda W.\exists X,Y.[W = \langle X,Y\rangle \wedge [\forall u,v,w{:}X].$$
$$[Y(u,u)\wedge[Y(u,v) \rightarrow Y(v,u)]\wedge[Y(u,v)\wedge Y(v,w) \rightarrow Y(u,w)]].$$

There are *primitive* domains defined directly in ITT and *recursive* domains defined recursively by *domain equations* using primitive domains and *domain constructors*. Primitive domains are discussed and several examples offered in §8.4.1. The reasons for allowing only finite functions as domain constructors are offered in §8.4.2; several examples illustrate the point. The domain constructors needed for the domain equations (1) are defined in §8.4.3. A method of solving domain equations is illustrated in §8.4.4 with the Example of §4.7.2 that is taken from Milner and Tofte's paper [103] and has already been discussed in the introduction §8.1 to this chapter.

8.4.1. Primitive domains. Consider again the example domain equations of §8.1:

1)
$$
\begin{aligned}
&Const \uplus Clos = Val\\
&Var \xrightarrow{\text{fin}} Val = Env\\
&Var \otimes Exp \otimes Env = Clos.
\end{aligned}
$$

The bases of the primitive domains $Const, Var$, and Exp appearing in the equations are respectively the sets of constants, variables, and expressions of the language; the expressions being defined in terms of variables and constants by the grammar of the language. The identity for each primitive domain in this case is $=$. Thus each of the sequents

$$\vdash DOM(\langle Const, =\rangle) \quad \vdash DOM(\langle Var, =\rangle) \quad \vdash DOM(\langle Exp, =\rangle)$$

can be assumed to be derivable.

In addition to the three primitive domains appearing in the equations (1), examples from [71] include *Ide* the *identifiers*, *Num* the *numerals*, *Bv* the *Boolean values*, and *Com* the *commands*. As these examples illustrate the members of the base of a primitive domain are *syntactic* objects. The base of a primitive domain may contain any number of provably distinct members. This means that if Bs is the base of a primitive domain and $\simeq_{Bs}$ its identity, then Bs:[1] and there are any number of terms s_i for which

$$\vdash D(s_i) \text{ and } s_i \simeq_{Bs} s_j \vdash$$

are derivable whenever $1 \leq i < j$. This can be accomplished by exploiting the properties of the numerals N and ordered pair $\langle\,\rangle$ as was done in §8.3.

8.4.2. Finite functions and domain constructors. A typical use of functions in defining a semantics for a programming language is the second of the domain equations (1) where an *environment* is defined to be an assignment of values to finitely many variables. Assuming that a domain Var of variables and a domain Val of values is known, the definition is comprehensible and precise. The notation

$$[Var \xrightarrow{\text{fin}} Val]$$

is understood to name the set of all functions with source a finite subset of Var and target members of Val. Gordon in [71] on the other hand defines environments as

$$[Ide \longrightarrow Dv + \{unbound\}]$$

where *Ide* are identifiers, *Dv* denotable values, and 'unbound' is a constant. This notation names the set of all *acceptable* functions with source *Ide* and values either a member of *Dv* or 'unbound'. Here 'acceptable function' is a special subclass of all the functions of the same source and target that is defined by the Scott semantics. The point of introducing 'unbound' as a possible value appears to be to account for the fact that only finitely many members of *Ide* can be assigned a value at any given time, although the number is potentially infinite in the sense discussed in §4.9. But this definition of environment makes it possible for an environment to assign values to infinitely many members of *Ide*. To avoid this, and incidentally simplify the definition, Gordon's definition could be

$$[Ide \xrightarrow{\text{fin}} Dv].$$

Apart from the simplification, whether this change has any practical effects is unknown.

Consider an example of Stoy from Chapter 12 of [139]. He defines the domain S of *states* to be

$$[\,[L \longrightarrow V] \times [L \longrightarrow T]\,]$$

where L is a domain of *locations*, V a domain of *values*, T a domain of *truth values*, and $\times$ is Cartesian product. Again '$\longrightarrow$' is an acceptable function from source to target. But as Scott's quote in §4.9 notes, L is a *potentially* infinite domain so that Stoy's definition of S could be taken to be

$$[[L \xrightarrow{\text{fin}} V] \times [L \xrightarrow{\text{fin}} T]\,]$$

again simplifying the definition and properly narrowing the meaning.

It is these examples and others that suggest that replacing 'acceptable function', that is '$\longrightarrow$', by 'finite function', that is '$\xrightarrow{\text{fin}}$' properly narrows the definitions of domains.

8.4.3. Domain constructors. The three defined domains *Val*, *Env*, and *Clos* of (1) have as their bases respectively the sets of values, environments, and closures. The domain constructors used in the equations are $\uplus$, $\xrightarrow{\text{fin}}$, and $\otimes$. They are respectively disjoint union, finite functions, and Cartesian product and must be defined as ternary predicates to ensure that the following sequents are derivable:

DOM)

$$\vdash [\forall W1, W2{:}DOM].[\uplus(W1, W2, W) \rightarrow DOM(W)]$$
$$\vdash [\forall W1, W2{:}DOM].[\xrightarrow{\text{fin}} (W1, W2, W) \rightarrow DOM(W)]$$
$$\vdash [\forall W1, W2{:}DOM].[\otimes(W1, W2, W) \rightarrow DOM(W)].$$

Consider for example the first of the domain constructors used in (1). It is defined

$$\uplus \stackrel{\text{df}}{=} \lambda W1, W2, W.\exists X1, Y1, X2, Y2.[W1 = \langle X1, Y1\rangle \wedge W2 = \langle X2, Y2\rangle \wedge$$
$$W = \langle Bs_{\uplus}(X1, X2), \simeq_{\uplus} (Y1, Y2)\rangle$$

where $Bs_{\uplus}$ and $\simeq_{\uplus}$ are defined:

$$Bs_{\uplus} \stackrel{\text{df}}{=} \lambda X1, X2, u.[[\exists x1{:}X1].u = \langle x1, 0\rangle \vee [\exists x2{:}X2].u = \langle x2, 1\rangle]$$

$$\simeq_{\uplus} \stackrel{\text{df}}{=} \lambda Y1, Y2, u, v.\exists x1, x2.[[Y1(x1, x2) \wedge u = \langle x1, 0\rangle \wedge v = \langle x2, 0\rangle] \vee$$
$$[Y2(x1, x2) \wedge u = \langle x1, 0\rangle \wedge v = \langle x2, 0\rangle]].$$

That the first of the sequents (DOM) is derivable is left as an exercise. The domain constructor $\otimes$ can be similarly defined:

$$\otimes \stackrel{\text{df}}{=} \lambda W1, W2, W.\exists X1, Y1, X2, Y2.[W1 = \langle X1, Y1\rangle \wedge W2 = \langle X2, Y2\rangle \wedge$$
$$W = \langle Bs_{\otimes}(X1, X2), \simeq_{\otimes} (Y1, Y2)\rangle$$

where $Bs_{\otimes}$ and $\simeq_{\otimes}$ are defined:

$$Bs_{\otimes} \stackrel{\text{df}}{=} \lambda X1, X2, u.[\exists x1{:}X1].[\exists x2{:}X2].u = \langle x1, x2\rangle$$
$$\simeq_{\otimes} \stackrel{\text{df}}{=} \lambda Y1, Y2, u, v.\exists x1, x2, x3, x4.[Y1(x1, x2) \wedge Y2(x3, x4) \wedge$$
$$u = \langle x1, x3\rangle \wedge v = \langle x2, x4\rangle].$$

A derivation of the last of the sequences (DOM) is also left as an exercise.

Consider now the *finite function* constructor $\stackrel{\text{fin}}{\longrightarrow}$. Its definition takes a slightly different form from the first two domain constructors:

$$\stackrel{\text{fin}}{\longrightarrow} \stackrel{\text{df}}{=} \lambda W1, W2, W.\exists X1, Y1, X2, Y2.[W1 = \langle X1, Y1\rangle \wedge W2 = \langle X2, Y2\rangle \wedge$$
$$W = \langle Bs_{fin}(X1, Y1, X2), \simeq_{fin} (X1, Y1, X2, Y2)\rangle].$$

The definitions of Bs_{fin} and $\simeq_{fin}$, require a prior understanding of finite functions.

A finite function with argument from a domain with base $B1$ and value from a domain with base $B2$ can be understood to be a list of ordered pairs $\langle b1, b2\rangle$, $b1 \in B1$ and $b2 \in B2$, satisfying a functionality constraint: If both $\langle b1, b2\rangle$ and $\langle b1', b2'\rangle$ are on the list, then $b1 \simeq_1 b1'$ does not hold, where $\simeq_1$ is the identity of the first domain. Two such functions are identical if they are lists of the same ordered pairs, where 'same' is defined in terms of $\simeq_1$ and $\simeq_2$, the latter being the identity for the second domain.

These remarks are the motivation for the following definition:

$$LF \stackrel{\text{df}}{=} \lambda X1, Y, X2.Lt(RF(X1, Y, X2))$$

where $RF(X1, Y, X2)$ is the recursion generator, with parameters $X1, X2{:}[1]$ and $Y{:}[1, 1]$ defined below.

In the contexts in which LF is used $DOM(\langle X1, Y\rangle)$ may be assumed. The intended meaning of $LF(X1, Y, X2)(Z, z)$ is that z:1 is a list of pairs $\langle x1, x2\rangle$, with $x1 \in X1$ and $x2 \in X2$, for which for no other pair $\langle x1', x2'\rangle$ in the list does $Y(x1, x1')$ hold; the members of Z:[1] are the pairs in the list z. Thus

$$B_{fin} \stackrel{\text{df}}{=} \lambda X1, Y, X2, z.\exists Z.LF(X1, Y, X2)(Z, z).$$

$$\simeq_{fin} \stackrel{\text{df}}{=} \lambda X1, Y1, X2, Y2, z1, z2.\exists Z1, Z2.[\; LF(X1, Y1, X2)(Z1, z1) \wedge LF(X1, Y1, X2)(Z2, z2) \wedge \equiv(Z1, Z2)].$$

where $\equiv(Z1, Z2)$ abbreviates:

$$[\forall x1{:}X1].[\forall x2{:}X2].[\; [Z1(\langle x1, x2\rangle) \rightarrow \exists u, v.[Y1(u, x1) \wedge Y2(v, x2) \wedge Z2(\langle u.v\rangle)]]\wedge [Z2(\langle x1, x2\rangle) \rightarrow \exists u, v.[Y1(u, x1) \wedge Y2(v, x2) \wedge Z1(\langle u.v\rangle)]] \;].$$

The recursion generator RF is defined:

$$RF \stackrel{\text{df}}{=} \lambda X1, Y, X2, W, Z, z.[\exists Za, za, Zb, zb, x1, x2.\; [\,[Z1 =_e \oslash \wedge z2 = Nil] \vee [W(X1, Y, X2)(Za, za) \wedge X1(x1) \wedge \forall u.[\exists v.Za(\langle u, v\rangle) \rightarrow \neg Y(u, x1)] \wedge X2(x2) \wedge Zb =_e \lambda u.[Za(u) \vee u = \langle x1, x2\rangle] \wedge zb = \langle\langle x1, x2\rangle, za\rangle] \wedge Z =_e Zb, z = zb]].$$

With LF so defined, the following sequents are derivable:

LF.1. $\vdash \forall X1, Y, X2, Z1, z.$
$[LF(X1, Y, X2)(Z1, z) \rightarrow \forall Z2.[LF(X1, Y, X2)(Z2, z) \rightarrow Z1 =_e Z2]].$

LF.2. $\vdash \forall X1, Y, X2, Z1, z.$
$[LF(X1, Y, X2)(Z1, z) \rightarrow \forall Z2.[Z1 =_e Z2 \rightarrow LF(X1, Y, X2)(Z2, z)]].$

A derivation of (LF.2) is left as an exercise. A derivation of (LF.1) follows in which

$$P \stackrel{\text{df}}{=} \lambda X1, Y, X2, Z1, z.\forall Z2.[LF(X1, Y, X2)(Z2, z) \rightarrow Z1 =_e Z2].$$

$-\forall X1, Y, X2, Z1, z.[LF(X1, Y, X2)(Z1, z) \rightarrow P(X1, Y, X2)(Z2, z)]$
$-\forall Z1, z.[LF(X1, Y, X2)(Z1, z) \rightarrow P(X1, Y, X2)(Z2, z)]$
$-\forall Z1, z.[RF(X1, Y, X2)(P(X1, Y, X2), Z1, z)$
$\quad \rightarrow P(X1, Y, X2)(Z2, z)]$ $\quad\dashv$LtInd
$+RF(X1, Y, X2)(P(X1, Y, X2), Z1, z)$
$-P(X1, Y, X2)(Z2, z)$
$-\forall Z2.[LF(X1, Y, X2)(Z2, z) \rightarrow Z1 =_e Z2]$ $\quad\dashv$df P
$+LF(X1, Y, X2)(Z2, z)$ (a)
$-Z1 =_e Z2$ (b)
$+\exists Za, za, Zb, zb, x1, x2.[[Z1 =_e \oslash \wedge z = Nil] \vee$
$\quad [P(X1, Y, X2)(Za, za) \wedge X1(x1) \wedge \forall u.[\exists v.Za(\langle u, v\rangle) \rightarrow \neg Y(u, x1)]$
$\quad \wedge X2(x2) \wedge Z1 =_e \lambda u.[Za(u) \vee u = \langle x1, x2\rangle]$
$\quad \wedge z = \langle\langle x1, x2\rangle, za\rangle]]$ $\quad\dashv$df RF

$+[[Z1 =_e \oslash \wedge z = Nil] \vee$
$[P(X1, Y, X2)(Za, za) \wedge X1(x1) \wedge \forall u.[\exists v.Za(\langle u, v\rangle) \to \neg Y(u, x1)]$
$\wedge X2(x2) \wedge Z1 =_e \lambda u.[Za(u) \vee u = \langle x1, x2\rangle] \wedge z = \langle\langle x1, x2\rangle, za\rangle]]$

L R

$+[Z1 =_e \oslash \wedge z = Nil]$
$+Z1 =_e \oslash$ (c)
$+z = Nil]$ (d)
$+[[Z2 =_e \oslash \wedge z = Nil] \vee$
$[LF(X1, Y, X2)(Za, za) \wedge X1(x1) \wedge \forall u.[\exists v.Za(\langle u, v\rangle) \to \neg Y(u, x1)]$
$\wedge X2(x2) \wedge Z2 =_e \lambda u.[Za(u) \vee u = \langle x1, x2\rangle]$
$\wedge z = \langle\langle x1, x2\rangle, za\rangle]]$ $\dashv$ Lt.3, (a)

LL LR

$+[Z2 =_e \oslash \wedge z = Nil]$
$+[Z2 =_e \oslash$
$+Z1 =_e Z2$ $\dashv$ (c)

$\dashv$ (b)

LR

$+[LF(X1, Y, X2)(Za, za) \wedge X1(x1) \wedge \forall u.[\exists v.Za(\langle u, v\rangle) \to \neg Y(u, x1)]$
$\wedge X2(x2) \wedge Z2 =_e \lambda u.[Za(u) \vee u = \langle x1, x2\rangle] \wedge z = \langle\langle x1, x2\rangle, za\rangle]]$
$+z = \langle\langle x1, x2\rangle, za\rangle$

$\dashv$ (d)

R

$+[P(X1, Y, X2)(Za, za) \wedge X1(x1) \wedge \forall u.[\exists v.Za(\langle u, v\rangle) \to \neg Y(u, x1)]$
$\wedge X2(x2) \wedge Z1 =_e \lambda u.[Za(u) \vee u = \langle x1, x2\rangle] \wedge z = \langle\langle x1, x2\rangle, za\rangle]$
$+Z1 =_e \lambda u.[Za(u) \vee u = \langle x1, x2\rangle]$ (e)
$+[[Z2 =_e \oslash \wedge z = Nil] \vee$
$[LF(X1, Y, X2)(Za, za) \wedge X1(x1) \wedge \forall u.[\exists v.Za(\langle u, v\rangle) \to \neg Y(u, x1)]$
$\wedge X2(x2) \wedge Z2 =_e \lambda u.[Za(u) \vee u = \langle x1, x2\rangle]$
$\wedge z = \langle\langle x1, x2\rangle, za\rangle]]$ $\dashv$ Lt.3, (a)

RL RR

similar to LR

RR

$+[LF(X1, Y, X2)(Za, za) \wedge X1(x1) \wedge \forall u.[\exists v.Za(\langle u, v\rangle) \to \neg Y(u, x1)]$
$\wedge X2(x2) \wedge Z2 =_e \lambda u.[Za(u) \vee u = \langle x1, x2\rangle] \wedge z = \langle\langle x1, x2\rangle, za\rangle]]$
$+Z2 =_e \lambda u.[Za(u) \vee u = \langle x1, x2\rangle]$
$+Z1 =_e Z2$ $\dashv$ (e)

$\dashv$ (b)

To test the correctness of the definition of $\xrightarrow{\text{fin}}$, a derivation of the second of the sequences of (DOM) follows.

$-[\forall W1, W2{:}DOM].[\xrightarrow{\text{fin}} (W1, W2, W) \rightarrow DOM(W)]$
$+DOM(W1) \quad +DOM(W2)$
$+ \xrightarrow{\text{fin}} (W1, W2, W)$ (a)
$-DOM(W)$ (b)
$+\exists X, Y.[W1 = \langle X, Y\rangle \wedge [\forall u, v, w{:}X].[Y(u, u)\wedge$
$\quad [Y(u, v) \rightarrow Y(v, u)] \wedge [Y(u, v) \wedge Y(v, w) \rightarrow Y(u, w)]]$
$+[W1 = \langle X1, Y1\rangle \wedge [\forall u, v, w{:}X1].[Y1(u, u)\wedge$
$\quad [Y1(u, v) \rightarrow Y1(v, u)] \wedge [Y1(u, v) \wedge Y1(v, w) \rightarrow Y1(u, w)]]$
$+[W2 = \langle X2, Y2\rangle \wedge [\forall u, v, w{:}X2].[Y2(u, u)\wedge$
$\quad [Y2(u, v) \rightarrow Y2(v, u)] \wedge [Y2(u, v) \wedge Y2(v, w) \rightarrow Y2(u, w)]]$
$+W1 = \langle X1, Y1\rangle$ (c)
$+[\forall u, v, w{:}X1].[Y1(u, u)\wedge$
$\quad [Y1(u, v) \rightarrow Y1(v, u)] \wedge [Y1(u, v) \wedge Y1(v, w) \rightarrow Y1(u, w)]]$
$+W2 = \langle X2, Y2\rangle$ (c')
$+[\forall u, v, w{:}X2].[Y2(u, u)\wedge$
$\quad [Y2(u, v) \rightarrow Y2(v, u)] \wedge [Y2(u, v) \wedge Y2(v, w) \rightarrow Y2(u, w)]]$
$+ \xrightarrow{\text{fin}} (\langle X1, Y1\rangle, \langle X2, Y2\rangle, W)$ $\dashv$(a),(c),(c')
$+\exists X1', Y1', X2', Y2'.[\langle X1, Y1\rangle = \langle X1', Y1'\rangle$
$\quad \wedge\langle X2, Y2\rangle = \langle X2', Y2'\rangle\wedge$
$\quad\quad W = \langle B_{fin}(X1', Y1', X2'), \simeq_{fin} (X1', Y1', X2', Y2')\rangle]$ $\dashv$df $\xrightarrow{\text{fin}}$
$+[\langle X1, Y1\rangle = \langle X1', Y1'\rangle \wedge \langle X2, Y2\rangle = \langle X2', Y2'\rangle\wedge$
$\quad\quad W = \langle B_{fin}(X1', Y1', X2'), \simeq_{fin} (X1', Y1', X2', Y2')\rangle]$
$+X1 = X1' \quad +Y1 = Y1' \quad +X2 = X2' \quad +Y2 = Y2'$
$+W = \langle B_{fin}(X1, Y1, X2), \simeq_{fin} (X1, Y1, X2, Y2)\rangle$
$-DOM(\langle B_{fin}(X1, Y1, X2), \simeq_{fin} (X1, Y1, X2, Y2)\rangle)$ $\dashv$(b)
$-\exists X, Y.[\langle B_{fin}(X1, Y1, X2), \simeq_{fin} (X1, Y1, X2, Y2)\rangle = \langle X, Y\rangle\wedge$
$\quad [\forall u, v, w{:}X].[Y(u, u) \wedge [Y(u, v) \rightarrow Y(v, u)]$
$\quad\quad \wedge[Y(u, v) \wedge Y(v, w) \rightarrow Y(u, w)]]$ $\dashv$df DOM

Substitute $B_{fin}(X1, Y1, X2)$ for X and $\simeq_{fin}(X1, Y1, X2, Y2)$ for Y, and draw the following conclusion by $-\wedge$ from $-t{=}t$, where t is equal to $\langle B_{fin}(X1, Y1, X2), \simeq_{fin} (X1, Y1, X2, Y2)\rangle$

$-[[\forall u, v, w{:}B_{fin}(X1, Y1, X2)].[$
$\quad \simeq_{fin} (X1, Y1, X2, Y2)(u, u)\wedge$
$\quad [\simeq_{fin} (X1, Y1, X2, Y2)(u, v) \rightarrow\simeq_{fin} (X1, Y1, X2, Y2)(v, u)]\wedge$
$\quad [\simeq_{fin} (X1, Y1, X2, Y2)(u, v)\wedge \simeq_{fin} (X1, Y1, X2, Y2)(v, w) \rightarrow$
$\quad\quad\quad \simeq_{fin} (X1, Y1, X2, Y2)(u, w)]\ ]$
$+B_{fin}(X1, Y1, X2)(u) \quad +B_{fin}(X1, Y1, X2)(v) \quad +B_{fin}(X1, Y1, X2)(w)$

$$
\begin{aligned}
&-[\simeq_{fin}(X1, Y1, X2, Y2)(u, u) \wedge \\
&\quad [\simeq_{fin}(X1, Y1, X2, Y2)(u, v) \rightarrow \simeq_{fin}(X1, Y1, X2, Y2)(v, u)] \wedge \\
&\quad [\simeq_{fin}(X1, Y1, X2, Y2)(u, v) \wedge \simeq_{fin}(X1, Y1, X2, Y2)(v, w) \rightarrow \\
&\qquad\qquad \simeq_{fin}(X1, Y1, X2, Y2)(u, w)]\ ]
\end{aligned}
$$

Three branches result, one from each of the three conjuncts. The first and the last will be left to the reader. The second branch follows:

$$
\begin{array}{ll}
-[\simeq_{fin}(X1, Y1, X2, Y2)(u, v) \rightarrow \simeq_{fin}(X1, Y1, X2, Y2)(v, u)] & \\
+\simeq_{fin}(X1, Y1, X2, Y2)(u, v) & \text{(d)} \\
-\simeq_{fin}(X1, Y1, X2, Y2)(v, u) & \text{(d')} \\
+\exists Z1, Z2.[LF(X1, Y1, X2)(Z1, u) \wedge & \\
\quad LF(X1, Y1, X2)(Z2, v) \wedge \equiv (Z1, Z2)] & \dashv \text{df} \simeq_{fin}, \text{(d)} \\
-\exists Z1, Z2.[LF(X1, Y1, X2)(Z1, v) \wedge & \\
\quad LF(X1, Y1, X2)(Z2, u) \wedge \equiv (Z2, Z1)] & \dashv \text{df} \simeq_{fin}, \text{(d')} \\
+[LF(X1, Y1, X2)(Z1', u) \wedge LF(X1, Y1, X2)(Z2', v) \wedge \equiv (Z1', Z2')] & \\
-[LF(X1, Y1, X2)(Z1', v) \wedge LF(X1, Y1, X2)(Z2', u) \wedge \equiv (Z2', Z1')] & \\
\hline
& \dashv \text{df} \equiv, \text{(LF.2)}
\end{array}
$$

The constructors $\uplus$, $\otimes$, and $\xrightarrow{\text{fin}}$ are taken from [103]. An additional constructor used in [139] and in [71] is the finite powerdomain constructor fpd. It is a constructor with a single input domain and with base output the finite sets of members of the input base with an appropriate output identity. It can be defined in ITT using lists to represent finite sets.

8.4.4. Solving domain equations: an example. A method for solving domain equations is illustrated here using the Example of §8.1 which has been taken from [103] and previously discussed in §8.1; the equations are repeated here for convenience.

$$
\begin{aligned}
1)\qquad & Const \uplus Clos = Val \\
& Var \xrightarrow{\text{fin}} Val = Env \\
& Var \otimes Exp \otimes Env = Clos.
\end{aligned}
$$

Const, *Var*, and *Exp* are the bases of the primitive domains of (1) defined in §8.4.1; in each case $=$ is the identity. The domain constructors $\uplus$, $\xrightarrow{\text{fin}}$, and $\otimes$ are defined as ternary predicates in §8.4.3 for which the sequences (DOM) are derivable. The domain with base $Var \otimes Exp$ is not primitive although it will be treated as such in the solution of the equations (1); it is defined $\langle B_{\otimes}(Var, Exp), \simeq_{\otimes}(Var, Exp)\rangle$. The domains with bases *Val*, *Env*, and *Clos* are the domains defined by (1); they are the 'unknowns' of the equations. Let them be $\langle Var, \simeq_{var}\rangle$, $\langle Env, \simeq_{env}\rangle$, and $\langle Clos, \simeq_{clos}\rangle$. The equations (1) can then be understood to be the following extensional identities:

2)
$$\vdash B_{\uplus}(Const, Clos) =_e Val$$
$$\vdash \simeq_{\uplus} (=, \simeq_{clos}) =_e \simeq_{val}$$
$$\vdash B_{fin}(Var, =, Val) =_e Env$$
$$\vdash \simeq_{fin} (Var, =, Val, \simeq_{val}) =_e \simeq_{env}$$
$$\vdash B_{\otimes}(B_{\otimes}(Var, Exp), Env) =_e Clos$$
$$\vdash \simeq_{\otimes} (\simeq_{\otimes} (=, =), \simeq_{env}) =_e \simeq_{clos}.$$

These suggest the following sets of simultaneous Horn sequents:

3a)
$$B_{\uplus}(Const, Clos)(x) \vdash Val(x)$$
$$B_{fin}(Var, =, Val)(x) \vdash Env(x)$$
$$B_{\otimes}(B_{\otimes}(Var, Exp), Env)(x) \vdash Clos(x)$$

3b)
$$\simeq_{\uplus} (=, \simeq_{clos})(u, v) \vdash \simeq_{val} (u, v)$$
$$\simeq_{fin} (Var, =, Val, \simeq_{val})(u, v) \vdash \simeq_{env} (u, v)$$
$$\simeq_{\otimes} (\simeq_{\otimes} (=, =), \simeq_{env} (u, v)) \vdash \simeq_{clos} (u, v).$$

Simultaneous recursion generators $RG1$, $RG2$, and $RG3$ of type [[1],1] and $RGI1$, $RGI2$, and $RGI3$ of type [[1,1],1,1] are defined as follows:

$$RG1 \stackrel{\text{df}}{=} \lambda Z, z.B_{\uplus}(Const, Z)(z)$$
$$RG2 \stackrel{\text{df}}{=} \lambda Z, z.B_{fin}(Var, =, Z)(z)$$
$$RG3 \stackrel{\text{df}}{=} \lambda Z, z.B_{\otimes}(B_{\otimes}(Var, Exp), Z)(z)$$

$$RGI1 \stackrel{\text{df}}{=} \lambda ZZ, u, v. \simeq_{\uplus} (=, ZZ)(u, v)$$
$$RGI2 \stackrel{\text{df}}{=} \lambda Z, ZZ, u, v. \simeq_{fin} (Var, =, Z, ZZ)(u, v)$$
$$RGI3 \stackrel{\text{df}}{=} \lambda ZZ, u, v. \simeq_{\otimes} (\simeq_{\otimes} (=, =), ZZ(u, v)).$$

With these definitions the set (3a) is the set (a) of the example in §4.7. A recursion generator RGS was defined there as follows:

$$RGS \stackrel{\text{df}}{=} \lambda Z, x, y, z.[\ RG1(\lambda z.\exists x, y.Z(x, y, z))(x) \wedge RG2(\lambda x.\exists y, z.Z(x, y, z))(y) \wedge RG3(\lambda y.\exists x, z.Z(x, y, z))(z)\ ].$$

It defines a Horn sequent

HSS)
$$RGS(C, x, y, z) \vdash C(x, y, z).$$

Let C and the predicates Val, Env, and $Clos$ be defined:

$$C \stackrel{\text{df}}{=} Lt(RGS)$$
$$Val \stackrel{\text{df}}{=} \lambda x.\exists y, z.C(x, y, z)$$
$$Env \stackrel{\text{df}}{=} \lambda y.\exists x, z.C(x, y, z)$$
$$Clos \stackrel{\text{df}}{=} \lambda z.\exists x, y.C(x, y, z).$$

Then the sequents (3a) and their converses are derivable provided the sequent

$C\exists$)
$$\vdash \exists x, y, z.C(x, y, z)$$

is derivable. $(C\exists)$ is derivable under the assumption that each of the primitive domains *Const*, *Var*, and *Exp* is not empty; that is, when each of the following sequents is derivable:

PD$\exists$) $\qquad \vdash \exists x.Const(x) \quad \vdash \exists x.Var(x) \quad \vdash \exists x.Exp(x).$

Since *RGS* is monotonic, it follows from (ItLt.1) of §4.8.1 that the sequent

$$\vdash \forall x, y, z.[\ [\exists u{:}N].It(RGS, \oslash)(u, x, y, z) \rightarrow C(x, y, z)]$$

is derivable. Hence to derive $(C\exists)$ it is sufficient to derive

$$\vdash \exists x, y, z.[\exists u{:}N].It(RGS, \oslash)(u, x, y, z)$$

that is, it is sufficient to find a natural number n for which

b) $$\vdash \exists x, y, z.It(RGS, \oslash)(S^n(0), x, y, z)$$

is derivable.

To this end define the following predicates,

$$Val_n \stackrel{\text{df}}{=} \lambda x.\exists y, z.It(RGS, \oslash)(S^n(0), x, y, z)$$
$$Env_n \stackrel{\text{df}}{=} \lambda y.\exists x, z.It(RGS, \oslash)(S^n(0), x, y, z)$$
$$Clos_n \stackrel{\text{df}}{=} \lambda z.\exists x, y.It(RGS, \oslash)(S^n(0), x, y, z)$$

and consider their extensions for $n = 0$, $n = 1$, $n = 2$, and $n = 3$:

$$\begin{array}{lll} Val_0 =_e \oslash & Val_1 =_e Const & Val_2 =_e Const \\ Env_0 =_e \oslash & Env_1 =_e \oslash & Env_2 =_e [Var \xrightarrow{\text{fin}} Const] \\ Clos_0 =_e \oslash & Clos_1 =_e \oslash & Clos_2 =_e \oslash \end{array}$$

$$Val_3 =_e Const$$
$$Env_3 =_e [Var \xrightarrow{\text{fin}} Const]$$
$$Clos_3 =_e [Var \otimes Exp \otimes [Var \xrightarrow{\text{fin}} Const]\ ]$$

where now $\oslash{:}[1]$. Thus (b) is derivable when $n = 3$.

The Horn sequents (3b) can be treated in a similar fashion since *Val* has now been defined.

Clearly the method employed for solving the domain equations (1) can be generalized for any set of domain equations under appropriate assumptions.

EXERCISES §8.4.

1. Complete the two cases omitted from the derivation of the second of the sequents (DOM).
2. Provide a definition for the finite power domain constructor fpd and a derivation for the sequent
$$\vdash [\forall W1, W2{:}DOM].[fpd(W1, W2, W) \rightarrow DOM(W)].$$
3. Provide a derivation for (LF.2) of §8.4.3 using the result of Exercise 2 of §2.5.

4. Domain equations adapted from [71] are:

$$State = (Ide \overset{\text{fin}}{\rightarrow} Value)$$

$$Value = (Bv \uplus Proc)$$

$$Proc = (State \overset{\text{fin}}{\rightarrow} State)$$

Express them as Horn sequents and prove that, under appropriate assumptions, each Horn sequent and its converse is derivable.

8.5. Logical Support for Specification Languages

The motivation for specification languages is aptly summarized by Spivey in the introduction to the reference manual for the Z specification language [136]:

> Formal specifications use mathematical notation to describe in a precise way the properties which an information system must have, without unduly constraining the way in which these properties are achieved. They describe what a system must do without saying *how* it is to be done. This *abstraction* makes formal specifications useful in the process of developing a computer system, because they allow questions about what the system does to be answered confidently, without the need to disentangle the information from a mass of detailed program code, or to speculate about the meaning of phrases in an imprecisely-worded prose description.
>
> A formal specification can serve as a single, reliable reference point for those who investigate the customer's needs, those who implement programs to satisfy those needs, those who test the results, and those who write instruction manuals for the system. Because it is independent of the program code, a formal specification of a system can be completed early in its development. Although it might need to be changed as the design team gains in understanding and the perceived needs of the customer evolve, it can be a valuable means of promoting a common understanding among all those concerned with the system.

Naturally a formal specification that is to "serve as a single, reliable reference point" must be consistent. For example, since the semantic networks as described by Sowa in [135] have been shown in [21] to be inconsistent, their use in specifications must be suspect. But there are many formal logics, including ITT, that have been proved consistent.

A logic that supports a particular specification language is one in which the language can be defined and logical conclusions drawn from sentences of the language. For example, a first order logic as described in §1.6 extended with identity supports the language Z. But because a specification language is intended to assist humans, the *presentation* of its sentences, that is the graphical presentation of the sentences as they appear on the screen of monitors or printed on paper, is one of its most important features. The *internal* representation of the sentences is irrelevant to a user, although the representation can affect the efficiency of the processing of the language.

The logic ITT as described in Chapter 3, like any higher-order logic, can clearly support many different specification languages. The simplicity of its treatment of recursion and of partial functions may provide an advantage for it over other higher-order logics. But this possibility can only be tested when a computer system has been implemented that can assist a user in the construction of derivations. In its simplest form such a system would be a "bookkeeper" that checks for the syntactic correctness of formulas and of semantic trees, and maintains a list of open branches of semantic trees being processed. In a more sophisticated form the system might provide elementary automatic theorem proving and assistance with the creative task of finding eigen terms for applications of the $-\exists$ and $+\forall$ rules.

Even the "bookkeeper" form of the computer system requires providing a programmable meta-language for ITT. But this task has been accomplished many times for other logics; all of the following systems, for example, have found wide application: Isabelle [114], Lambda Prolog [107], LF Logical Framework [116], and the Coq Proof Assistant [41]. But given the treatment of Horn sequent definitions in Chapter 4, a logic programming language would form a "natural" programmable meta-language for ITT or its intuitionist/constructive variant HITT. The discussion by Hill and Gallagher in [81] as to how a logic programming language such as Gödel [80] might be adapted to serve as a programmable meta-language suggests a point of departure.

REFERENCES

[1] S. Abramsky, Dov M. Gabbay, and T.S.E. Maibaum (editors), ***Handbook of logic in computer science, background: Mathematical structures***, vol. 1, Clarendon Press, Oxford, 1992.

[2] S. Abramsky, Dov M. Gabbay, and T.S.E. Maibaum (editors), ***Handbook of logic in computer science, Semantic structures***, vol. 3, Clarendon Press, Oxford, 1994.

[3] SAMSON ABRAMSKY and ACHIM JUNG, *Domain theory*, In Abramsky et al. [2], pp. 1–168.

[4] HARVEY ABRAMSON and VERONICA DAHL, ***Logic grammars***, Springer-Verlag, 1989.

[5] PETER ACZEL, ***Non–well–founded sets***, CSLI Lecture Notes, no. 14, CSLI, 1988.

[6] JAMES H. ANDREWS, *A weakly-typed higher order logic with general lambda terms and y combinator*, ***Proceedings, works in progress track, 15th international conference on theorem proving in higher order logics***, no. CP-2002-211736, NASA Conference Publication, 2002, pp. 1–11.

[7] ———, *Cut elimination for a weakly-typed higher order logic*, Technical Report 611, Department of Computer Science, University of Western Ontario, December 2003.

[8] PETER B. ANDREWS, *Resolution in type theory*, ***The Journal of Symbolic Logic***, vol. 36 (1971), pp. 414–432.

[9] ———, ***An introduction to mathematical logic and type theory: To truth through proof***, Academic Press, 1986.

[10] K.R. APT and M.H. VAN EMDEN, *Contributions to the theory of logic programming*, ***Journal of the ACM***, vol. 29 (1982), no. 3, pp. 841–862.

[11] ALFRED JULES AYER, ***Language, truth and logic***, second ed., Victor Gollancz Ltd, 1950.

[12] MAURICE JAY BACH, ***A specification of data structures with application to data base systems***, Ph.D. thesis, Faculty of Pure Science, Columbia University, 1978.

[13] H.P. BARENDREGT, ***The lambda calculus, its syntax and semantics***, revised ed., North-Holland, 1985.

[14] Jon Barwise (editor), ***Handbook of mathematical logic***, North Holland, 1977.

[15] Jon Barwise and John Etchmendy, ***The liar: An exercise in truth and circularity***, Oxford University Press, 1987.

[16] Jon Barwise and Lawrence Moss, ***Vicious circles***, CSLI, 1996.

[17] Michael J. Beeson, ***Foundations of constructive mathematics***, Springer-Verlag, 1980.

[18] Paul Benacerraf and Hilary Putnam, ***Philosophy of mathematics, selected readings***, Cambridge University Press, 1983, 2nd edition, 1st edition published by Prentice-Hall Inc., 1964.

[19] E.W. Beth, *Semantic entailment and formal derivability*, ***Mededelingen de Koninklijke Nederlandse Akademie der Wetenschappen, Afdeeling Letterkunde, Nieuwe Reeks***, vol. 18 (1955), no. 13, pp. 309–342.

[20] ———, *Semantic construction of intuitionistic logic*, ***Mededelingen de Koninklijke Nederlandse Akademie der Wetenschappen, Afdeeling Letterkunde, Nieuwe Reeks***, vol. 19 (1956), no. 11, pp. 357–388.

[21] Michael Julian Black, *Naive semantic networks*, Final Paper for Directed Studies in Computer Science, Dept of Computer Science, University of B.C., January 22,1985.

[22] K.A. Bowen and R.A. Kowalski (editors), ***Fifth international logic programming conference***, MIT Press, 1988.

[23] Rudolf Carnap, ***The logical syntax of language***, Kegan Paul, Trench, Trubner and Co. Ltd, London, 1937, English translation by Amethe Smeaton of Logische Syntax der Sprache, 1934.

[24] Rudolph Carnap, ***Meaning and necessity, a study in semantics and modal logic***, University of Chicago Press, 1947.

[25] Noam Chomsky, ***Language and problems of knowledge***, MIT Press, 1988.

[26] A. Church and J.B. Rosser, *Some properties of conversion*, ***Transactions of the American Mathematical Society***, vol. 39 (1936), pp. 11–21.

[27] Alonzo Church, *Schröder's anticipation of the simple theory of types*, ***The Journal of Unified Science (Erkenntnis)***, vol. IX (1939), pp. 149–152.

[28] ———, *A formulation of the simple theory of types*, ***The Journal of Symbolic Logic***, vol. 5 (1940), pp. 56–68.

[29] ———, ***The calculi of lambda conversion***, Princeton University Press, 1941.

[30] ———, *A formulation of the logic of sense and denotation*, ***Structure, method and meaning, essays in honor of Henry M. Sheffer*** (Horace M. Kallen, Paul Henle and Susanne K. Langer, editors), The Liberal Arts Press, New York, 1951.

[31] ———, ***Introduction to mathematical logic volume I***, Princeton University Press, 1956.

[32] NINO B. COCCHIARELLA, ***Logical investigations of predication and the problem of universals***, Bibliopolis Press, Naples, 1986.

[33] ———, *Conceptual realism versus Quine on classes and higher-order logic*, ***Synthese***, vol. 90 (1992), pp. 379–436.

[34] A. COLMERAUER, H. KANOUI, P. ROUSSEL, and R. PASERO, *Un systeme de communication homme-machine en francais*, Technical report, Groupe de Recherche en Intelligence Artificielle, Université d'Aix-Marseille, 1973.

[35] HASKELL B. CURRY, ***Outline of a formalist philosophy of mathematics***, North-Holland, 1951.

[36] ———, ***A theory of formal deducibility***, Notre Dame Mathematical Lectures, No. 6, 1957.

[37] HASKELL B. CURRY and ROBERT FEYS, ***Combinatory logic***, North-Holland, 1958.

[38] Martin Davis (editor), ***The undecidable, basic papers on undecidable propositions, unsolvable problems and computable functions***, Raven Press, Hewlett New York, 1965.

[39] KEITH DEVLIN, ***The maths gene: How mathematical thinking evolved and why numbers are like gossip***, Basic Books, 2000.

[40] ———, ***The maths gene: Why everyone has it, but most people don't use it***, Weidenfeld and Nicolson, 2000.

[41] GILLES DOWAK, AMY FELTY, HUGO HERBELIN, GERARD HUET, CHET MURTHY, CATHERINE PARENT, CHRISTINE PAULIN-MOHRING, and BENJAMIN WERNER, *The coq proof assistant user's guide*, Rapport Techniques 154, INRIA, Rocquencourt, France, 1993.

[42] WILLIAM M. FARMER, *A partial functions version of Church's simple theory of types*, ***The Journal of Symbolic Logic***, vol. 55 (1990), pp. 1269–1290.

[43] S. FEFERMAN, *Categorical foundations and foundations of category theory*, ***Logic, foundations of mathematics and computability theory*** (R.E. Butts and J. Hintikka, editors), D. Reidel, Dordrecht–Holland, 1977, pp. 149–169.

[44] ———, *Towards useful type-free theories*, ***The Journal of Symbolic Logic***, vol. 49 (1984), pp. 75–111.

[45] ———, ***In the light of logic***, Oxford University Press, 1998.

[46] M.C. FITTING and R. MENDELSOHN, ***First-order modal logic***, Synthese Library, vol. 277, Kluwer Academic, 1998.

[47] MELVIN FITTING, ***First-order logic and automated theorem proving***, 2 ed., Springer-Verlag, 1996.

[48] ———, ***Types, tableaus, and Gödel's god***, Kluwer Academic, 2000.

[49] GOTTLOB FREGE, *Begriffsschrift, a formula language, modeled upon that of arithmetic, for pure thought*, In van Heijenoort [147], A translation of a booklet of 1879.

[50] Dov M. Gabbay, C.J. Hogger, and J.A. Robinson (editors), ***Handbook of logic in artificial intelligence and logic programming, logic programming***, vol. 5, Clarendon Press, Oxford, 1998.

[51] Peter Geach and Max Black (editors), ***Translations from the philosophical writings of Gottlob Frege***, 2nd ed., Blackwell, Oxford, 1960.

[52] GERHARD GENTZEN, *Investigations of logical deductions*, In Szabo [140], translation of a paper of 1935.

[53] P. C. GILMORE, *Attributes, sets, partial sets and identity*, ***Logic and foundations of mathematics*** (D. van Dalen, J.G. Dijkman, S.C. Kleene, and A.S. Troelstra, editors), Wolters-Noordhoff Publishing, Groningen, 1968, pp. 53–69.

[54] PAUL C. GILMORE, *An abstract computer with a Lisp-like machine language without a label operator*, ***Computer programming and formal systems*** (P. Braffort and D. Hirschberg, editors), North-Holland, Amsterdam, 1963, pp. 71–86.

[55] ———, *A consistent naive set theory: Foundations for a formal theory of computation*, IBM Research Report RC 3413, IBM Thomas J. Watson Research Laboratory, June 1971.

[56] ———, *Purely functional programs as recursions*, IBM Research Report RC 4088, IBM Thomas J. Watson Research Laboratory, October 1972.

[57] ———, *Defining and computing many–valued functions*, ***Parallel computers — parallel mathematics, Proceedings of the IMACS (AICA)-GI symposium, Technical University of Munich*** (M. Feilmeier, editor), 1977, pp. 17–23.

[58] ———, *Combining unrestricted abstraction with universal quantification*, ***To h.b. curry: Essays on combinatorial logic, lambda calculus and formalism*** (J.P. Seldin and J.R. Hindley, editors), Academic Press, 1980, This is a revised version of [55], pp. 99–123.

[59] ———, *Natural deduction based set theories: A new resolution of the old paradoxes*, ***The Journal of Symbolic Logic***, vol. 51 (1986), pp. 393–411.

[60] ———, *A foundation for the entity relationship approach: How and why*, ***Proceedings of the 6th entity relationship conference*** (S.T. March, editor), North-Holland, 1988, pp. 95–113.

[61] ———, *NaDSyL and some applications*, ***Computational logic and proof theory, The Kurt Gödel colloquium*** (Georg Gottlob, Alexander Leitsch, and Daniele Mundici, editors), Lecture Notes in Computer Science, vol. 1289, Springer-Verlag, 1997, pp. 153–166.

[62] ———, *An impredicative simple theory of types*, Presented at the Fourteenth Workshop on Mathematical Foundations for Programming Systems, Queen Mary College, London, May 1998.

[63] ———, *An intensional type theory: Motivation and cut-elimination*, ***The Journal of Symbolic Logic***, vol. 66 (2001), pp. 383–400.

[64] ———, *A nominalist motivated intuitionist type theory: A foundation for recursive function theory*, In preparation, 2005.

[65] PAUL C. GILMORE and GEORGE K. TSIKNIS, *A formalization of category theory in NaDSet*, ***Theoretical Computer Science***, vol. 111 (1993), pp. 211–253.

[66] ———, *Logical foundations for programming semantics*, ***Theoretical Computer Science***, vol. 111 (1993), pp. 253–290.

[67] J.Y. GIRARD,, 1994, Letter to author, March 16, 1994.

[68] KURT GÖDEL, *The consistency of the continuum hypothesis*, 1940.

[69] ———, *On formally undecidable propositions of Principia Mathematica and related systems I*, In van Heijenoort [147], translation of 1931 paper, pp. 596–616.

[70] NELSON GOODMAN and W.V. QUINE, *Steps toward a constructive nominalism*, ***The Journal of Symbolic Logic***, vol. 12 (1947), pp. 105–122.

[71] MICHAEL J.C. GORDON, ***The denotational description of programming languages, An introduction***, Springer-Verlag, New York, Heidelberg, Berlin, 1979.

[72] C.A. GUNTER and D.S. SCOTT, *Semantic domains*, ***Handbook of theoretical computer science, Volume B: Formal models and semantics*** (Jan van Leeuwen, editor), vol. B, MIT Press/Elsevier, 1990, pp. 633–674.

[73] LEON HENKIN, *The completeness in the first order functional calculus*, ***The Journal of Symbolic Logic***, vol. 14 (1949), pp. 150–166.

[74] ———, *Completeness in the theory of types*, ***The Journal of Symbolic Logic***, vol. 15 (1950), pp. 81–91.

[75] ———, *Some notes on nominalism*, ***The Journal of Symbolic Logic***, vol. 18 (1951), pp. 19–29.

[76] JACQUES HERBRAND, *Investigations in proof theory: The properties of true propositions*, In van Heijenoort [147], Translation of portion of PhD thesis, pp. 525–581.

[77] AREND HEYTING, *Die formalen Regeln der intuitionistischen Logik*, ***Sitzungberichte der Preussischen Akademie der Wissenschafter, Physikalisch-mathematische Klasse***, (1930), pp. 42–56.

[78] D. HILBERT and P. BERNAYS, ***Grundlagen der Mathematik I***, second ed., vol. 1, Springer-Verlag, 1968, The first edition was published in 1934.

[79] ———, ***Grundlagen der Mathematik II***, second ed., vol. 2, Springer-Verlag, 1970, The first edition was published in 1939.

[80] PATRICIA HILL and JOHN LLOYD, ***The Gödel programming language***, MIT Press, 1994.

[81] P.M. HILL and J. GALLAGHER, *Meta-programming in logic programming*, In Gabbay et al. [50], pp. 421–497.

[82] J.R. HINDLEY, B. LERCHER, and J.P. SELDIN, ***Introduction to combinatory logic***, Cambridge University Press, 1972.

[83] R. HINDLEY, *The principal type-scheme of an object in cobinatory logic*, ***Transactions of the American Mathematical Society***, vol. 146 (1969), pp. 29–65.

[84] K.J.J. HINTIKKA, *Form and content in quantification theory*, ***Acta Philosophica Fennica***, vol. 8 (1955), pp. 7–55.

[85] WILFRED HODGES, ***Logic, an introduction to elementary logic***, Penguin Books Ltd, 1991, reprinting of book published by Pelican Books Ltd, 1977.

[86] ALFRED HORN, *On sentences which are true of direct unions of algebras*, ***The Journal of Symbolic Logic***, vol. 16 (1951), pp. 14–21.

[87] GREGOR KICZALES, JOHN LAMPING, ANURAG MENDHEKAR, CHRIS MAEDA, CRISTINA VIDEIRA LOPES, JEAN-MARC LOINGTIER, and JOHN IRWIN, *Aspect-oriented programming*, ***Proceedings of the European conference on object-oriented programming***, vol. LNCS 1241, Springer-Verlag, 1997.

[88] STEPHEN COLE KLEENE, ***Introduction to metamathematics***, North-Holland, 1952.

[89] R.A. KOWALSKI, *Predicate logic as a programming language*, ***Information Processing***, vol. 74 (1974), pp. 569–574.

[90] SAUL A. KRIPKE, *Semantical analysis of intuitionistic logic I*, ***Formal systems and recursive functions***, North-Holland, 1965, pp. 92–130.

[91] J. LAMBEK, *Are the traditional philosophies of mathematics really incompatible?*, ***The Mathematical Intelligencer***, vol. 16 (1994), pp. 56–62.

[92] ———, *Categorical logic*, ***Encyclopedia of mathematics***, Kluwer Academic, 1998, pp. 107–111.

[93] J. LAMBEK and P.J. SCOTT, ***Introduction to higher order categorical logic***, Cambridge University Press, 1994.

[94] J.W. LLOYD, ***Foundations of logic programming***, 2nd ed., Springer-Verlag, 1987.

[95] S. MACLANE, *Categorical algebra and set-theoretical foundations*, ***Axiomatic set theory, proceedings of a symposium in pure mathematics*** (Providence, R.I.), vol. XIII, American Mathematical Society, 1967, pp. 231–240.

[96] P. MARTIN-LÖF, ***Notes on constructive mathematics***, Almqvist & Wiksell, Stockholm, 1970.

[97] ———, ***Intuitionistic type theory***, Bibliopolis, Napoli, 1984.

[98] J. MCCARTHY, *Recursive functions of symbolic expressions and their computation by machine*, ***Communications of the Association for Computing Machinery***, vol. 3 (1960), pp. 184–195.

[99] Bernard Meltzer and Donald Michie (editors), ***Machine intelligence 5***, American Elsevier, 1970.

[100] R. MILNER, *A theory of type polymorphism in programming languages*, ***Journal Computing System Science***, vol. 17 (1978), pp. 348–375.

[101] ———, ***A calculus of communicating systems***, Lecture Notes in Computer Science, no. 92, Springer-Verlag, 1980.

[102] ———, *Calculi for synchrony and asynchrony*, ***Theoretical Computer Science***, vol. 25 (1983), pp. 267–310.

[103] ROBIN MILNER and MADS TOFTE, *Co-induction in relational semantics*, ***Theoretical Computer Science***, vol. 87 (1991), pp. 209–220.

[104] RICHARD MONTAGUE, *English as a formal language*, In Thomason [143], pp. 188–221.

[105] ———, *Syntactical treatment of modality, with corollaries on reflexion principles and finite axiomatization*, In Thomason [143], pp. 286–302.

[106] P.D. MOSSES, *Denotational semantics*, In Gabbay et al. [50], pp. 499–590.

[107] G. NADATHUR and D. MILLER, *An overview of lambda prolog*, In Bowen and Kowalski [22], pp. 810–827.

[108] GOPALAN NADATHUR and DALE MILLER, *Higher-order logic programming*, In Gabbay et al. [50], pp. 577–632.

[109] MICHAEL J. O'DONNELL, *Introduction: Logic and logic programming languages*, In Gabbay et al. [50], pp. 1–67.

[110] S. OWRE, N. SHANKAR, and J. M. RUSHBY, *The PVS specification language (beta release)*, Technical report, Computer Science Laboratory, SRI International, Menlo Park CA 94025, June 1993.

[111] D. PARK, *Fixpoint, induction and proofs of program properties*, In Meltzer and Michie [99], pp. 59–78.

[112] BARBARA H. PARTEE and HERMANN L.W. HENDRIKS, *Montague grammar*, In van Benthem and ter Meulen [146].

[113] L.C. PAULSON, ***Logic and computation: Interactive proof with Cambridge LCF***, Cambridge Tracts in Theoretical Computer Science, vol. 2, Cambridge University Press, 1990.

[114] ———, ***Isabelle: A generic theorem prover***, Lecture Notes in Computer Science, vol. 828, Springer Verlag, 1994.

[115] GIUSEPPE PEANO, *The principles of arithmetic, presented by a new method*, In van Heijenoort [147], A reprinting of a 1889 paper, pp. 83–97.

[116] F. PFENNING, *Logic programming in the LF logical framework*, ***Logical frameworks***, Cambridge University Press, 1991, pp. 149–181.

[117] I.C.C. PHILLIPS, *Recursion theory*, In Abramsky et al. [1], pp. 79–187.

[118] AXEL POIGNÉ, *Basic category theory*, In Abramsky et al. [1], pp. 413–634.

[119] DAG PRAWITZ, ***Natural deduction, a proof-theoretical study***, Stockholm Studies in Philosphy, vol. 3, Almquist and Wiksell, Stockholm, 1965.

[120] WILLARD VAN ORMAN QUINE, ***Mathematical logic***, revised ed., Harvard University Press, 1951.

[121] JOHN C. REYNOLDS, ***Theories of programming languages***, Cambridge University Press, 1998.

[122] ABRAHAM ROBINSON, ***Non-standard analysis***, North-Holland, 1966.

[123] J.A. ROBINSON, *A machine-oriented logic based on resolution*, ***Journal of the Association for Computing Machinery***, vol. 12 (1965), pp. 23–41.

[124] JOHN RUSHBY, *Formal methods and the certification of critical systems*, Technical Report SRI-CSL-93-07, Computer Science Laboratory,Stanford Research Institute, November 1993.

[125] BERTRAND RUSSELL, ***Introduction to mathematical philosophy***, George Allen and Unwin Ltd, London, 1919, Reprinted many times.

[126] ———, ***The principles of mathematics***, second ed., George Allen and Unwin Ltd, London, 1937.

[127] JEAN E. SAMMET, ***Programming languages: History and fundamentals***, Prentic-Hall, Inc., 1969.

[128] K. SCHÜTTE, *Syntactical and semantical properties of simple type theory*, ***The Journal of Symbolic Logic***, vol. 25 (1960), pp. 305–326.

[129] DANA SCOTT, *Mathematical concepts in programming language semantics*, ***AFIPS conference proceedings***, vol. 40, AFIPS press, 1972.

[130] WILFRED SELLARS, *Abstract entities*, ***Review of Metaphysics***, vol. 16 (1963), pp. 625–671.

[131] ———, *Classes as abstract entities and the Russell paradox*, ***Review of Metaphysics***, vol. 17 (1963), pp. 67–90.

[132] JOSEPH R. SHOENFIELD, *Axioms of set theory*, In Barwise [14], pp. 321–344.

[133] THORALF SKOLEM, *Some remarks on axiomatized set theory*, In van Heijenoort [147], An English translation of a 1922 paper, pp. 290–301.

[134] R.M. SMULLYAN, ***First-order logic***, Dover Press, New York, 1994, Revised Edition, first published by Springer-Verlag, Berlin, 1968.

[135] J.F. SOWA, ***Conceptual structures: Information processing in mind and machine***, Addison-Wesley, Reading, Mass., 1984.

[136] J.M. SPIVEY, ***The Z notation: A reference manual***, Prentice Hall, New York, 1992, Second Edition.

[137] S.W.P. STEEN, ***Mathematical logic***, Cambridge University Press, 1972.

[138] LEON STERLING and EHUD SHAPIRO, ***The art of Prolog***, The M.I.T. Press, 1986.

[139] JOSEPH E. STOY, ***Denotational semantics: The Scott-Strachey approach to programming language theory***, The MIT Press, 1977.

[140] M.E. Szabo (editor), ***The collected papers of Gerhard Gentzen***, North-Holland, 1969.

[141] ALFRED TARSKI, *A lattice-theoretical fixpoint theorem and its applications*, ***Pacific Journal of Mathematics***, vol. 5 (1955), pp. 285–309.

[142] ———, *The concept of truth in formalized languages*, ***Logic, semantics, metamathematics, papers from 1923 to 1938***, Oxford University Press, 1956, An English translation by J.H. Woodger of a paper of 1936., pp. 152–268.

[143] Richmond H. Thomason (editor), ***Formal philosophy, selected papers of Richard Montague***, Yale University Press, New Haven and London, 1974.

[144] A.S. TROELSTRA, *Aspects of constructive mathematics*, In Barwise [14], pp. 973–1052.

[145] A.M. TURING, *On computable numbers, with an application to the Entsheidungsproblem*, In Davis [38], reprinting of paper of 1936-37, pp. 115–153.

[146] Johann van Benthem and Alice ter Meulen (editors), ***Handbook of logic and language***, Elsevier, 1997.

[147] Jean van Heijenoort (editor), ***From Frege to Gödel, A source book in mathematical logic, 1879-1931***, Harvard University Press, 1967.

[148] Richard L. Wexelblat (editor), ***A history of programming languages***, Academic Press, 1981.

[149] ALFRED NORTH WHITEHEAD and BERTRAND RUSSELL, ***Principia mathematica***, vol. 1, Cambridge University Press, 1925, second edition.

[150] NIKLAUS WIRTH, ***Algorithms + data structures = programs***, Prentice-Hall, 1976.

[151] ERNST ZERMELO, *Investigations in the foundations of set theory*, In van Heijenoort [147], translation of paper of 1908, pp. 199–215.

INDEX

Lecture Notes in Logic

General Remarks

This series is intended to serve researchers, teachers, and students in the field of symbolic logic, broadly interpreted. The aim of the series is to bring publications to the logic community with the least possible delay and to provide rapid dissemination of the latest research. Scientific quality is the overriding criterion by which submissions are evaluated.

Books in the Lecture Notes in Logic series are printed by photo-offset from master copy prepared using LaTeX and the ASL style files. For this purpose the Association for Symbolic Logic provides technical instructions to authors. Careful preparation of manuscripts will help keep production time short, reduce costs, and ensure quality of appearance of the finished book. Authors receive 50 free copies of their book. No royalty is paid on LNL volumes.

Commitment to publish may be made by letter of intent rather than by signing a formal contract, at the discretion of the ASL Publisher. The Association for Symbolic Logic secures the copyright for each volume.

The editors prefer email contact and encourage electronic submissions.

Editorial Board

Editorial Policy

1. Submissions are invited in the following categories:

i) Research monographs
ii) Lecture and seminar notes
iii) Reports of meetings
iv) Texts which are out of print

Those considering a project which might be suitable for the series are strongly advised to contact the publisher or the series editors at an early stage.

2. Categories i) and ii). These categories will be emphasized by Lecture Notes in Logic and are normally reserved for works written by one or two authors. The goal is to report new developments quickly, informally, and in a way that will make them accessible to non-specialists. Books in these categories should include

– at least 100 pages of text;
– a table of contents and a subject index;
– an informative introduction, perhaps with some historical remarks, which should be accessible to readers unfamiliar with the topic treated;

In the evaluation of submissions, timeliness of the work is an important criterion. Texts should be well-rounded and reasonably self-contained. In most cases the work will contain results of others as well as those of the authors. In each case, the author(s) should provide sufficient motivation, examples, and applications. Ph.D. theses will be suitable for this series only when they are of exceptional interest and of high expository quality.

Proposals in these categories should be submitted (preferably in duplicate) to one of the series editors, and will be refereed. A provisional judgment on the acceptability of a project can be based on partial information about the work: a first draft, or a detailed outline describing the contents of each chapter, the estimated length, a bibliography, and one or two sample chapters. A final decision whether to accept will rest on an evaluation of the completed work.

3. Category iii). Reports of meetings will be considered for publication provided that they are of lasting interest. In exceptional cases, other multi-authored volumes may be considered in this category. One or more expert participant(s) will act as the scientific editor(s) of the volume. They select the papers which are suitable for inclusion and have them individually refereed as for a journal. Organizers should contact the Managing Editor of Lecture Notes in Logic in the early planning stages.

4. Category iv). This category provides an avenue to provide out-of-print books that are still in demand to a new generation of logicians.

5. Format. Works in English are preferred. After the manuscript is accepted in its final form, an electronic copy in LaTeX format will be appreciated and will advance considerably the publication date of the book. Authors are strongly urged to seek typesetting instructions from the Association for Symbolic Logic at an early stage of manuscript preparation.

Lecture Notes in Logic

From 1993 to 1999 this series was published under an agreement between the Association for Symbolic Logic and Springer-Verlag. Since 1999 the ASL is Publisher and A K Peters, Ltd. is Co-publisher. The ASL is committed to keeping all books in the series in print.

Current information may be found at `http://www.aslonline.org`, the ASL Web site. Editorial and submission policies and the list of Editors may also be found above.

Previously published books in the *Lecture Notes in Logic* are:

1. *Recursion Theory.* J. R. Shoenfield. (1993, reprinted 2001; 84 pp.)
2. *Logic Colloquium '90; Proceedings of the Annual European Summer Meeting of the Association for Symbolic Logic, held in Helsinki, Finland, July 15–22, 1990.* Eds. J. Oikkonen and J. Väänänen. (1993, reprinted 2001; 305 pp.)
3. *Fine Structure and Iteration Trees.* W. Mitchell and J. Steel. (1994; 130 pp.)
4. *Descriptive Set Theory and Forcing: How to Prove Theorems about Borel Sets the Hard Way.* A. W. Miller. (1995; 130 pp.)
5. *Model Theory of Fields.* D. Marker, M. Messmer, and A. Pillay. (1996; 154 pp.)
6. *Gödel '96; Logical Foundations of Mathematics, Computer Science and Physics; Kurt Gödel's Legacy. Brno, Czech Republic, August 1996, Proceedings.* Ed. P. Hajek. (1996, reprinted 2001; 322 pp.)
7. *A General Algebraic Semantics for Sentential Objects.* J. M. Font and R. Jansana. (1996; 135 pp.)
8. *The Core Model Iterability Problem.* J. Steel. (1997; 112 pp.)
9. *Bounded Variable Logics and Counting.* M. Otto. (1997; 183 pp.)
10. *Aspects of Incompleteness.* P. Lindstrom. (1997, 2nd ed. 2003; 163 pp.)
11. *Logic Colloquium '95; Proceedings of the Annual European Summer Meeting of the Association for Symbolic Logic, held in Haifa, Israel, August 9–18, 1995.* Eds. J. A. Makowsky and E. V. Ravve. (1998; 364 pp.)
12. *Logic Colloquium '96; Proceedings of the Colloquium held in San Sebastian, Spain, July 9–15, 1996.* Eds. J. M. Larrazabal, D. Lascar, and G. Mints. (1998; 268 pp.)
13. *Logic Colloquium '98; Proceedings of the Annual European Summer Meeting of the Association for Symbolic Logic, held in Prague, Czech Republic, August 9–15, 1998.* Eds. S. R. Buss, P. Hájek, and P. Pudlák. (2000; 541 pp.)
14. *Model Theory of Stochastic Processes.* S. Fajardo and H. J. Keisler. (2002; 136 pp.)
15. *Reflections on the Foundations of Mathematics; Essays in Honor of Solomon Feferman.* Eds. W. Seig, R. Sommer, and C. Talcott. (2002; 444 pp.)

16. *Inexhaustibility; A Non-exhaustive Treatment.* T. Franzén. (2004; 255 pp.)

17. *Logic Colloquium '99; Proceedings of the Annual European Summer Meeting of the Association for Symbolic Logic, held in Utrecht, Netherlands, August 1–6, 1999.* Eds. J. van Eijck, V. van Oostrom, and A. Visser. (2004; 208 pp.)

18. *The Notre Dame Lectures.* Ed. P. Cholak. (2005, 185 pp.)

19. *Logic Colloquium 2000; Proceedings of the Annual European Summer Meeting of the Association for Symbolic Logic, held in Paris, France, July 23–31, 2000.* Eds. R. Cori, A. Razborov, S. Todorčević, and C. Wood. (2005; 408 pp.)

20. *Logic Colloquium '01; Proceedings of the Annual European Summer Meeting of the Association for Symbolic Logic, held in Vienna, Austria, August 1–6, 2001.* Eds. M. Baaz, S. Friedman, and J. Krajíček. (2005, 486 pp.)

21. *Reverse Mathematics 2001.* Ed. S. Simpson. (2005, 401 pp.)

22. *Intensionality.* Ed. R. Kahle. (2005, 265 pp.)

23. *Logicism Renewed: Logical Foundations for Mathematics and Computer Science.* P. Gilmore. (2005, 234 pp.)